Fluid Dynamics

Volume I

Fluid Dynamics
Volume I

Edited by **Bradley Foster**

NY RESEARCH PRESS
New York

Published by NY Research Press,
23 West, 55th Street, Suite 816,
New York, NY 10019, USA
www.nyresearchpress.com

Fluid Dynamics: Volume I
Edited by Bradley Foster

International Standard Book Number: 978-1-63238-199-6 (Hardback)

Printed in China.

Contents

Preface

Fluid Dynamics deals with the natural science of fluids in motion. A sub-discipline of fluid mechanics, fluid dynamics offers a systematic structure to solve practical problems. Under the mechanics of fluid flow, liquid and gases, both are considered fluids. Various empirical and semi-empirical laws are derived from flow measurement. Properties of fluids such as velocity, density, pressure, temperature are calculated using different relations and equations through experiments. These relations are used in petroleum industries, water supply, sewage flow, industrial flow etc. It is even used in traffic engineering, where traffic is assumed to flow like fluid.

Fluid dynamics has a wide range of applications such as determining the mass flow rate of fluid, calculating forces and moments on an object in a fluid flow, understanding nebulae in interstellar space, predicting weather patterns among numerous others. The concept of pressure is also an important focus area in fluid dynamics. Bernoulli's equation is significant in the study of all fluid flows.

I'd like to thank all the contributors who've made this book an enlightening one. Their expertise and knowledge shall prove to be beneficial for all readers. I would like to thank the publishing house and my family for their support.

Editor

Numerical Calculation of Viscous-Elastic Fluid Flooding Residual Oil Film in the Complex Pore

Lili Liu[1], Chao Yu[1], Lihui Wang[2], Lu Liu[1]
[1]Department of Petroleum Engineering, Northeast Petroleum University, Daqing, China
[2]Daqing Petroleum Equipment Group, Daqing, China

ABSTRACT

In order to analyze the stress and deformation of different permeability of residual oil film in the complex pore, which are affected by the viscous-elasticity of the fluid, the hydrodynamic displacement mechanism is explored from the standpoint of hydrodynamics, that is, the residual oil film displaced by alternating injection of different concentrations of the polymer solution, viscous-elastic fluid flow equation is established in the complex pore by choosing continuity equation, motion equation and the upper convected Maxwell constitutive equation. The flow field is computed by using the method of numerical analysis. Not only the stress and deformation of residual oil film on the different permeability of micro pores, but also the analysis of the flooding mechanism of alternating injection of different concentrations of the polymer solution is got. The results show that the larger the viscous-elasticity of polymer solution is, the bigger the normal deviatoric stress acting on the residual oil film is; the distribution of normal deviatoric stress has the abrupt change. The stronger the viscous-elasticity of the polymer solution is, the bigger the horizontal stress difference acting on the residual oil film is and the more obvious the deformation is; the high-concentration polymer solution is suitable for high-permeability micro pores. Low-concentration polymer solution is suitable for medium and low-permeability micro pores. Alternating injection of polymer solution can improve Volumetric Sweep Efficiency and increase the deformation of residual oil film, which is conducive to enhancing oil recovery.

Keywords: Complex Pore; Residual Oil Film; Deformation; Viscous-Elasticity; Permeability

1. Introduction

After water flooding, polymer flooding has a favourable effect on flooding effects, which largely enhance the oil recovery [1]. Because of stratigraphic shearing and dilution of the formation water, simple injection of high-concentration polymer makes the plugging effect decrease with time. Because high-permeability layer is very easy to break, simple injection of low-concentration polymer solution makes low concentration of polymer cross-flow in high-permeability pore, hardly get into medium and low-permeability layer. In order to solve the drawbacks of general injection method, alternating injection is taken to control mobility well. Each permeable layer takes what he needs. Swept volume is also expanded. Zhang Jihong [2] and others got results from experiment that alternating injection of gel and surfactant flooding after polymer flooding was possible to further enhance oil recovery. Zhang Zhilong [3] and others explored flooding situation of alternating injection of foam and AS after polymer flooding. In this paper, the force and deformation of residual oil film acted by polymer solution are analyzed theoretically and the result that alternating injection of different concentrations of the polymer solution can also enhance oil recovery.

2. Calculation Model

In order to analyze the force and deformation of the residual oil film in different permeability micro pores, Two-dimensional parallel channel model is shown in **Figure 1**. Low, medium and the high micro pores width, respectively, are taken as 10 μm, 20 μm, 30 μm, taking micro pores width proportion H_1:H_2:H_3 as 1:2:3. **Figure 2** is the corresponding calculation model.

In **Figure 2**, both model building and grid designing are taken account into inlet effect and boundary effect.

3. Basic Equation

Polymer solution with complicated rheological proper

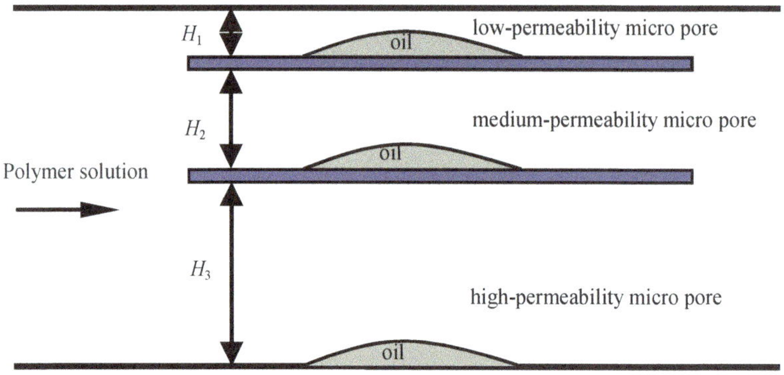

Figure 1. Parallel channel model.

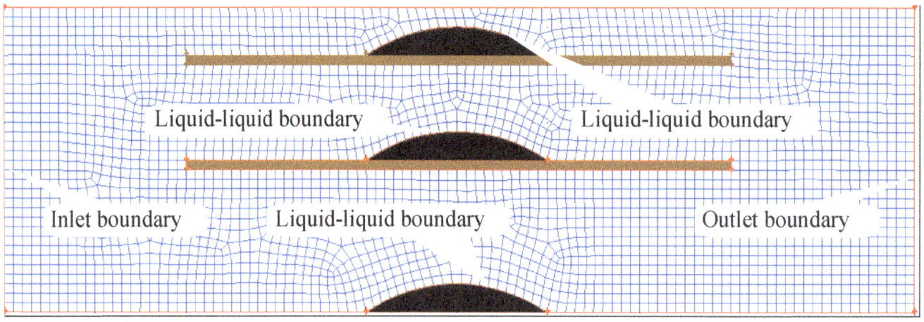

Figure 2. Computational grid and boundary condition

belongs to viscous-elastic fluid. Based on experiment research, the upper convected Maxwell constitutive equation is suitable to describe the viscous-elastic polymer solution. Continuity equation, motion equation and constitutive equation [4] are used as follows:

Continuity equation:

$$\nabla \cdot \boldsymbol{u} = 0 \tag{1}$$

where: $\boldsymbol{u}$ is velocity vector.

Momentum equation:

$$\mathrm{d}\boldsymbol{u}/\mathrm{d}t = \boldsymbol{F} + \mathrm{div}\boldsymbol{P}/\rho \tag{2}$$

where: $\boldsymbol{F}$ is mass force; ρ is density; $\boldsymbol{P}$ is stress tensor.

$$\boldsymbol{P} = -p\boldsymbol{I} + \boldsymbol{T} \tag{3}$$

where: $\boldsymbol{I}$ is unit tensor; $-p$ is static pressure; $\boldsymbol{T}$ is deviatoric stress tensor.

Constitutive equation:

$$\boldsymbol{T} + \lambda \overset{\nabla}{\boldsymbol{T}} = \eta \boldsymbol{A}_1 \tag{4}$$

where: λ is relaxation time, which is the characterization of fluid elasticity; η is shearing viscosity; $\boldsymbol{A}_1$ is one order Rivlin-Ericksen tensor.

4. Calculation Result

Weissenberger number is a dimensionless quantity which is used to describe elasticity of fluid. The bigger its value is, the stronger the elasticity of the fluid is. Calculation formula of Weissenberger number is:

$$We = \lambda v / l$$

where: We is Weissenberger number; v is characteristic velocity; l is characteristic length.

In order to analyze the force acting on the oil film from fluid of different viscous-elastic flowing through different permeability micro pores, the stress field is calculated when viscous-elastic fluid flows residual oil film under the different conditions that We is 0.1, 0.2, 0.3 respectively. In addition, normal deviatoric stress τ_{nn} and the difference of horizontal stress p_{nh} on the residual oil film surface are calculated.

4.1. The Change of the Normal Deviatoric Stress [5] Acting on the Residual Oil Film in Different Permeability Cores

As shown in **Figure 3**, in the different permeability micro pores, the higher of We of viscous-elastic fluid is, the larger normal deviatoric stress is; with the increase of We, the higher compression on upstream of residual oil film is, the lower tension on downstream of residual oil film is. When We = 0.3, in the downstream of oil film, only a small area of close to oil film center is in tension, and others are all in larger compression. The result of two ends compressed and center tensioned makes the small area of oil center extrude from oil film. Under the continual action of polymer solution, this small oil film will be pulled, it won't rupture until it achieves stress limit, finally, separate from ontology. Compared with the low-

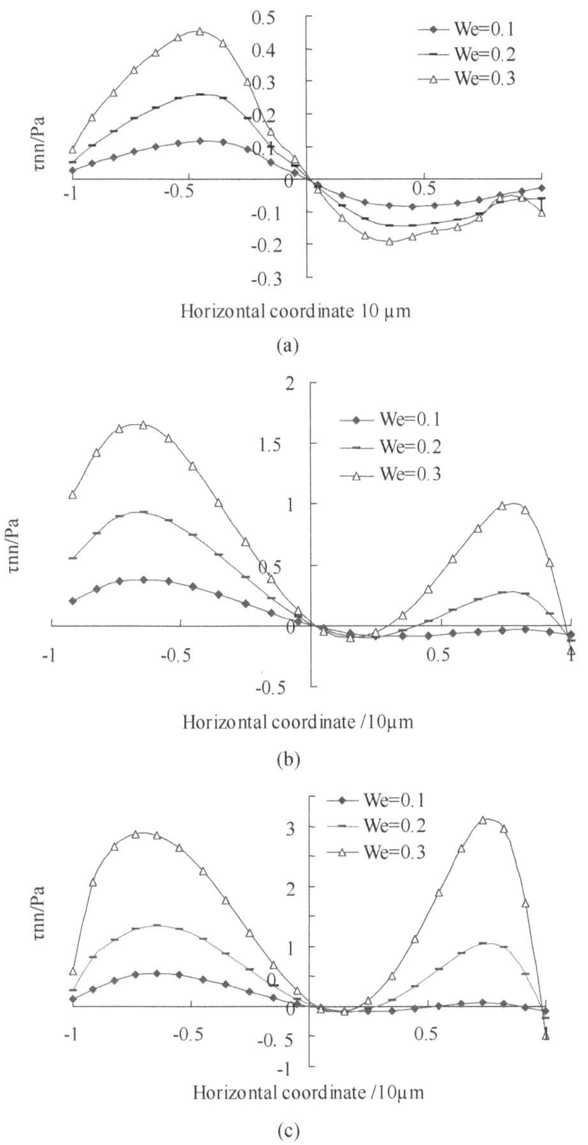

Figure 3. Normal deviatoric stress acting on residual oil film in the different permeability micro pores. (a) low-permeability micro pore; (b) medium-permeability micro pore; (c) high-permeability micro pore. Note: compression is positive, tension is negative.

permeability micro pore, the normal deviatoric stress of high-permeability micro pore is 6 times than that of it with the same oil film.

When the polymer solution of the *We* = 0.3 is injected, the normal deviatoric stress acting on residual oil film differ greatly from low and medium-permeability micro pores from high-permeability micro pores. High viscous-elasticity of the high-concentration of polymer significantly increase the force and deformation of the residual oil film, but high concentration makes polymer molecules overlap and molecular clew size large, which makes them difficultly enter the medium and low-permeability layer, especially in low-permeability layer.

Therefore, the high concentration of high molecular weight polymers mainly enriched in the high-permeability layer, generating large normal deviatoric stress to the residual oil film in high-permeability layer. This shows that the force and deformation of residual oil film in the micro pore are affected by viscous-elasticity of driving agent. High-concentration of polymer is available for high-permeability micro pore.

When the polymer solution of the *We* = 0.3 is injected, the normal deviatoric stress acting on residual oil film is less the same between low and medium-permeability micro pore and high-permeability micro pore. This is due to the low-concentration polymer solution can be relatively easy to enter the low-permeability layer. Compared to the high-concentration polymer solution, the total amount of the low-concentration polymer solution entered is large. Due to the increase in the amount of the polymer entered low and medium-permeability micro pore, the degree of deformation increases. So the low-concentration polymer is suitable for low-permeability micro pore.

4.2. Changes of Horizontal Stress Difference Acting on the Residual Oil Film in Different Permeability Cores

Both the microscopic visualization flooding result [6] and the theoretical result [7] showed that viscous-elastic polymer solution generated a greater level of thrust on residual oil film. The main reason for the deformation of oil film is stress difference which is produced by asymmetry of normal deviatoric stress of the film surface. Therefore, it is necessary to further analyze horizontal thrust of flooding by different driving fluids, as it is the size of the horizontal stress difference. The greater the value is, the more obvious the film deformation is. As seen from **Figure 4**, the horizontal stress difference of the viscous-elastic fluid increases with the increasing of We. Horizontal stress difference which high concentration of the polymer solution acting on the high-permeability oil film is significantly higher than that of low-permeability of oil film, the high-concentration polymer is suitable for high-permeability micro pore. Horizontal stress difference which low-concentration polymer acted on oil film in the medium-permeability micro pore has little difference. So the low-concentration polymer is suitable for low-permeability micro pore.

4.3. Deformation of the Residual Oil Film in the Different Permeability Cores [8]

As can be seen from **Figure 5**, high concentration polymer solution is injected; the deformation of high-permeability of the residual oil film is more evident than that of the low and medium-permeability. The deformation of residual oil film of high, medium and low-permeability

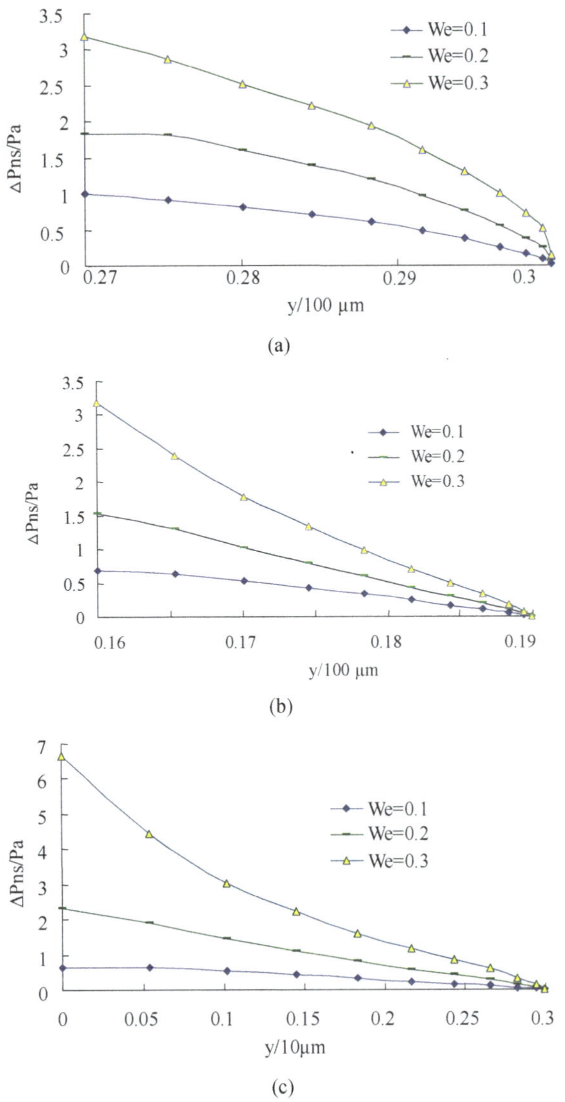

Figure 4. Horizontal stress difference acting on the residual oil film in different permeability cores. (a) low-permeability micro pore; (b) medium-permeability micro pore; (c) high-permeability micro pore.

have little difference when low concentration of polymer solution is injected.

Through the analysis of the force and deformation of residual oil film, we find that alternating injection different viscous-elastic driving fluid can enhance oil recovery. General injection is taken, force and deformation of the residual oil is very small in low and medium-permeability micro pore, displacing effect is not good, the degree of use is low; force and deformation of the residual oil is large in high-permeability micro pore. Therefore, alternating injection takes advantage of principle that different driving agent flooding will make different force and deformation of the residual oil, which can improve excessive absorption of fluid in high-permeability micro pore and insufficient absorption of fluid in low and me-

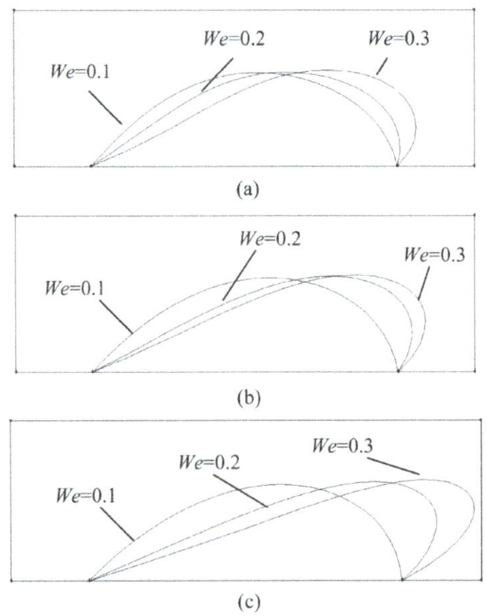

Figure 5. Deformation of the residual oil film in the different permeability micro pore. (a) low-permeability micro pore; (b) medium-permeability micro pore; (c) high-permeability micro pore.

dium-permeability micro pore, make more low-concentration polymer into low and medium-permeability micro pore, Each permeability layer takes what he needs; fluidity is controlled well, the residual oil of force and deformation is more, and the swept volume is expanded, at the same time, it will play a better role in flooding.

5. Conclusions

- The normal deviatoric stress of viscous-elastic polymer solution acting on oil film increases with the increasing of *We*. The distribution of normal deviatoric stress occurs corresponding to mutations, which are more conducive to the deformation of the oil film.
- The horizontal stress difference of driving fluid acting on oil film increases with the elasticity of driving fluid increased.
- For high-permeability micro pore, high-concentration polymer solution is injected which can acquire the largest normal deviatoric stress and the most obvious deformation. For low and medium-permeability micro pore, the normal deviatoric stress of injecting low-concentration polymer solution is more or less the same with that in the high-permeability micro pore.
- Alternating injection different viscous-elasticity polymer solution can enhance oil recovery, which is consistent with the results of laboratory experiments.

REFERENCES

[1] X. G, Yue, Y. F. Wang, K. L. Wang, *et al.*, "Enhance Oil Recovery Foundation," Publishing House of Oil Industry,

Beijing, 2002.

[2] J. H. Zhang, X. Dong, Y. Z. Ye, *et al.*, "Flooding Effects of Alternating Injection of Gel and Surfactant after Polymer Flooding," *Journal of Daqing Petroleum Institute Contents*, Vol. 34, No. 2, 2010, pp. 85-88.

[3] Z. L. Zhang and M. S. Wang, "Flooding Physical Simulation of Alternating Injection of Foam with the AS System after polymer Flooding," *Oilfield Chemistry*, Vol. 23, No. 2, 2006, pp. 166-172.

[4] S. F. Ha, "Constitutive Equation and Computational Analytical Theory of Non-Newtonian Fluids," Science Press, Beijing, 2002.

[5] L. L. Liu, L. H. Wang, Y. Zhang, *et al.*, "Mechanical Analysis of Stress Acted on Residual Oil Film by Polymer Solution," *Xinjiang Petroleum Geology*, Vol. 31, No. 5, 2010, pp. 533-535.

[6] D. M. Wang, J. C. Cheng, Q. Y. Yan, *et al.*, "Viscous-Elastic Polymer Can Increase Micro Scale Displacement in Cores," SPE 63227, 2000, pp. 2-8.

[7] H. J. Yin, D. M. Wang, H. Y. Zhong, *et al.*, "Study on Flow Behaviors of Viscoelastic Polymer Solution in Micro Pore with Dead End," SPE 101950, 2001, pp. 4-15.

[8] X. H. Liu, "A Biomechanical Model for Simulating the Deformation of a Leukocyte Adhered to the Surface of a Blood Vessel under Steady Shear Flow," *Journal of Biomedical Engineering*, Vol. 20, No. 1, 2003, pp. 30-34.

2

FEM-BEM Coupling for the MHD Pipe Flow in an Exterior Region

S. Han Aydin
Department of Mathematics, Karadeniz Technical University, Trabzon, Turkey

ABSTRACT

In this study, the magnetohydrodynamic (MHD) flow through a circular pipe under the influence of a transverse magnetic field when the outside medium is also electrically conducting is solved numerically by using FEM-BEM coupling approach. The coupled partial differential equations defined for the interior medium are transformed into homogenous modified Helmholtz equations. For the exterior medium on an infinite region, the Laplace equation is considered for the exterior magnetic field. Unknowns in the equations are also related with the corresponding Dirichlet and Neumann type coupled boundary conditions. Unknown values of the magnetic field on the boundary and for the exterior region are obtained by using BEM, and the unknown velocity and magnetic field inside the pipe are obtained by using SUPG type stabilized FEM. Computations are carried for very high values of magnetic Reynolds numbers Rm_1, Reynolds number Re and magnetic pressure Rh of the fluid. The results show that using stabilized method enables us to get stable and accurate numerical approximations consistent with the physical configuration of the problem over rough mesh which also results a cheap computational cost.

Keywords: FEM-BEM Coupling; SUPG; MHD Pipe Flow

1. Introduction

The objective of this work is to present MHD pipe flow of circular cross-section under the influence of a transverse magnetic field numerically by using FEM-BEM coupling approach when the outside medium is also electrically conducting. It is already known that there are many applications of MHD pipe flow such as the design of the cooling systems with liquid metals for nuclear reactors, electromagnetic pumps, MHD generators, and flowmeters measuring blood pressure, etc. The exact solution of the problem can be obtained only for some special cases [1,2]. Therefore, there are many numerical methods applied to the solution of the MHD pipe flow (see [3-9] and references there in).

The boundary element method is a technique which offers a great advantage to analyze stationary problems especially with infinite domains [10]. BEM has the advantage of discretizing only the pipe wall and gives the solution both in the exterior and interior regions at any point. Due to the regularity conditions, it is possible to limit the discretization to a finite boundary.

The finite element method is one of the most preferable numerical method applied to the solution of many physical and engineering problems. Applications of the finite element method to MHD problems are already performed by many researchers [11-14]. But, in standard Galerkin formulations, there exist numerical instabilities for the high Hartmann number value cases. It is possible to eliminate these, either using finer mesh which increases the computational cost, or considering stabilized finite element methods [15-18]. One of the most popular of such numerical methods is referred as the streamline upwind Petrov-Galerkin (SUPG) method [19]. The finite element method of the SUPG type reduces the oscillations in the standard Galerkin method and achieves stability by adding mesh-dependent perturbation terms to the formulation. These terms enhance the coercivity of the formulation by acting like artificial diffusion in the direction of the flow.

The present paper deals with the solution of the general MHD problem inside of a pipe when the surrounding medium is also electrically conducting, and has small magnetization compared to the fluid inside the pipe. The flow is assumed as incompressible and viscous, and the fluid inside the pipe is electrically conducting. The outside medium has also electrical conductivity which is small compared to fluid. However, the continuity of both induced magnetic fields inside and outside of the pipe is maintained through the conducting wall of the pipe. Therefore, this problem is a generalization of the special cases of MHD flow through a pipe. Thus, we consider in this paper, the most general case of MHD flow in a pipe which has electrically conducting wall with the same conductivity of outside infinite medium. This way, partial differential equations defined inside the pipe and outside infinite region are solved simultaneously with the coupled induced current conditions on the pipe wall. Equations defined inside the pipe are decoupled and transformed to homogenous modified Helmholtz equations for which the BEM formulations are given with fundamental solution $K_0(x)$. BEM formulation in the exterior region is carried with the logarithmic fundamental solution of Laplace equation. Using this formulation, unknown magnetic field values are obtained on the boundary of the pipe.

The paper is organized as follows: In Section 2, we describe the mathematical modelling of the equations. BEM and FEM coupling numerical solution procedure is presented in Section 3. Application of the prescribed algorithm is discussed in Section 4 by presenting some numerical experiments obtained from different values of magnetic Reynolds numbers Rm_1, Reynolds number Re and magnetic pressure Rh of the fluid. Conclusions are drawn in Section 5.

2. Mathematical Model

The MHD flow is governed by a set of coupled PDE's obtained from Navier-Stokes equations for conducting fluids, and Maxwell's equations for electromagnetic field through Ohm's law. We consider MHD flow in a straight pipe (channel, duct) of sufficient length, and of cross-section a circle in the xy-plane. The fluid is flowing through the pipe due to an applied constant pressure gradient $\frac{\partial p}{\partial z}$, and is viscous, incompressible, electrically conducting. The electrical permitivity and magnetic permeability of the fluid are assumed to be close to those of the external space. The axis of the pipe is coincident with the z-axis, and the y-axis is parallel with the magnetic induction at infinity. Thus, externally applied magnetic field with an intensity B_0 is assumed to be in y-direction. It determines the appearance within the fluid of an induced magnetic field (induced current). We assume that the wall of the pipe and the outside medium are also electrically conducting having the same electrical conductivity and magnetic permeability since the thickness of pipe wall is assumed to be very small (**Figure 1**).

Thus, the non-dimensional form of the equations to be solved inside the pipe and its exterior are [20,21]

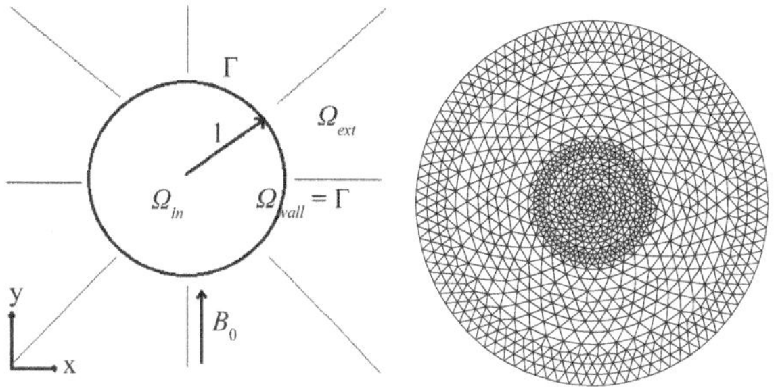

Figure 1. Problem definition (left) and discretization of the domain (pipe and external region) (left).

$$\left.\begin{aligned}\nabla^2 V(x,y) + ReRh\frac{\partial B}{\partial y}(x,y) &= -1\\ \nabla^2 B(x,y) + Rm_1\frac{\partial V}{\partial y}(x,y) &= 0\end{aligned}\right\} \text{ in } \Omega_{in} \tag{1}$$

$$\nabla^2 B_{ext}(x,y) = 0 \quad \text{in } \Omega_{ext} \tag{2}$$

with the no-slip condition on the pipe wall

$$V = 0 \quad \text{on } \partial\Omega_{in} = \Gamma \tag{3}$$

and continuity conditions for the induced magnetic fields

$$B(x,y) = B_{ext}(x,y), \quad \text{on } \Gamma \tag{4}$$

$$\frac{1}{Rm_1}\frac{\partial B(x,y)}{\partial n} = \frac{1}{Rm_2}\frac{\partial B_{ext}(x,y)}{\partial n'}, \quad \text{on } \Gamma \tag{5}$$

$$\lim_{x^2+y^2\to\infty} B_{ext}(x,y) = 0 \tag{6}$$

where n and n' are unit outward normals on Γ for the regions Ω_{in} and Ω_{ext}, respectively. Rm_1 and Rm_2 are the magnetic Reynolds numbers inside the pipe and in external medium.

MHD Equation (1) are decoupled first by denoting $V_1 = V$ and $B_1 = \frac{ReRh}{M}B$, where $M = \sqrt{ReRhRm_1}$ is the Hartmann number of the fluid, the system (1) becomes [1,20]

$$\left.\begin{aligned}\nabla^2 V_1 + M\frac{\partial B_1}{\partial y} &= -1\\ \nabla^2 B_1 + M\frac{\partial V_1}{\partial y} &= 0.\end{aligned}\right\} \text{ in } \Omega_{in} \tag{7}$$

As a further transformation, define new variables $U_1(x,y)$ and $U_2(x,y)$ as

$$\begin{aligned}U_1 &= V_1 + B_1\\ U_2 &= V_1 - B_1\end{aligned} \tag{8}$$

in order to decouple the Equations (7) as

$$\begin{aligned}\nabla^2 U_1 + M\frac{\partial U_1}{\partial y} &= -1\\ \nabla^2 U_2 - M\frac{\partial U_2}{\partial y} &= -1.\end{aligned} \tag{9}$$

For obtaining homogeneous equations, let's define $W_1(x,y)$ and $W_2(x,y)$

$$W_1 = U_1 + \frac{1}{M}y \qquad W_2 = U_2 - \frac{1}{M}y \tag{10}$$

which result in

$$\nabla^2 W_1 + M\frac{\partial W_1}{\partial y} = 0 \qquad \nabla^2 W_2 - M\frac{\partial W_2}{\partial y} = 0. \tag{11}$$

Finally, the transformation with

$$u_1 = W_1 e^{ky} \qquad u_2 = W_2 e^{-ky} \tag{12}$$

where $k = M/2$, reduces the equations to two homogeneous modified Helmholtz equations

$$\nabla^2 u_1 - k^2 u_1 = 0 \qquad \nabla^2 u_2 - k^2 u_2 = 0. \tag{13}$$

Now, the system of Equations (1) and (2) are transformed to

$$\left.\begin{array}{l}\nabla^2 u_1 - k^2 u_1 = 0\\ \nabla^2 u_2 - k^2 u_2 = 0\end{array}\right. \text{ in } \Omega_{in} \tag{14}$$

$$\nabla^2 B_{ext} = 0 \quad \text{in } \Omega_{ext} \tag{15}$$

with the boundary conditions (4)-(6) on Γ which are also transformed in terms of new variables [21]

$$e^{-ky}u_1 + e^{ky}u_2 = 0 \tag{16}$$

$$e^{-ky}u_1 - e^{ky}u_2 = \frac{2ReRh}{M}B_{ext} + \frac{2}{M}y \tag{17}$$

$$e^{-ky}\frac{\partial u_1}{\partial n} - e^{ky}\frac{\partial u_2}{\partial n} = \frac{2M}{Rm_2}\frac{\partial B_{ext}}{\partial n'} + \frac{2}{M}\frac{\partial y}{\partial n}. \tag{18}$$

3. FEM-BEM Coupling

Using the Equations (14)-(15) with the boundary conditions (16)-(18), it is possible to obtain $u_1, \frac{\partial u_1}{\partial n}, u_2, \frac{\partial u_2}{\partial n}, B_{ext}, \frac{\partial B_{ext}}{\partial n'}$ on the boundary of the pipe wall using boundary element method only discretizing the boundary. After calculating these unknown values on the boundary, one can also able to obtain exterior values of B_{ext} again using BEM. It is also possible to calculate the unknown values V and B at any point inside the pipe as done in [21].

However, for the large values of Rm_1, Re or Rh which increase the value of the Hartmann number M, BEM solutions obtained using less number of discretization points brings some numerical instabilities because of the convection dominated behavior of the Equation (1). Therefore, these equations should be solved over coarse mesh using stabilization technique.

One of the well-known stabilization method is streamline upwind Petrov-Galerkin (SUPG) finite element method [19] which is applicable for the convection dominated case of the convection diffusion type equations.

In order to apply SUPG stabilization, transformed form of the equations

$$\nabla^2 U_1 + M\frac{\partial U_1}{\partial y} = -1$$

$$\nabla^2 U_2 - M\frac{\partial U_2}{\partial y} = -1.$$

with the boundary conditions

$$U_1 = B_1 \text{ and } U_2 = -B_1 \quad \text{on } \Gamma$$

are used. The weak formulation of the problem (9) can be stated as [22]: Find $U_1, U_2 \in V = H_0^1(\Omega)^2$ such that

$$\left.\begin{array}{l}(\nabla U_1, \nabla v) - M\left(\frac{\partial U_1}{\partial y}, v\right) = (1, v)\\ (\nabla U_2, \nabla v) + M\left(\frac{\partial U_2}{\partial y}, v\right) = (1, v)\end{array}\right\}, \quad \forall v \in V. \tag{19}$$

To introduce a finite element method, we begin by partitioning the domain into triangular elements in a standard way (e.g. no overlapping, no vertex on the edge of a neighboring elements, etc.) and let Ω_h be such a partition of Ω and a finite dimensional space V_h which is related to the choice of partition and satisfies $V_h \subset V$. Now the Galerkin finite element formulation of the problems reads; find $U_{1h}, U_{2h} \in V_h$ such that

$$\left.\begin{array}{l}(\nabla U_{1h}, \nabla v_h) - M\left(\frac{\partial U_{2h}}{\partial y}, v_h\right) = (1, v_h)\\ (\nabla U_{2h}, \nabla v_h) + M\left(\frac{\partial U_{2h}}{\partial y}, v_h\right) = (1, v_h)\end{array}\right\}, \quad \forall v_h \in V_h. \tag{20}$$

Finally, if we insert the stabilization terms to the formulation, the SUPG type stabilized FEM variational formulation is written as [18,19]: find $U_{1h}, U_{2h} \in V_h$ such that

$$\begin{bmatrix}(\nabla U_{1h}, \nabla v_h) - M\left(\frac{\partial U_{2h}}{\partial y}, v_h\right) - (1, v_h)\\ (\nabla U_{2h}, \nabla v_h) + M\left(\frac{\partial U_{2h}}{\partial y}, v_h\right) - (1, v_h)\end{bmatrix} + \tau_K \begin{bmatrix}\left(-M\frac{\partial U_{1h}}{\partial y} - 1\right)\left(-M\frac{\partial v_h}{\partial y}\right)\\ \left(M\frac{\partial U_{2h}}{\partial y} - 1\right)\left(M\frac{\partial v_h}{\partial y}\right)\end{bmatrix} = 0 \tag{21}$$

$\forall v_h \in V_h$ with the stabilization parameter

$$\tau_K = \begin{cases} \dfrac{h_K}{2M} & \text{if } Pe_k \geq 1 \\ \dfrac{h_K^2}{12} & \text{if } Pe_k < 1 \end{cases} \quad (22)$$

where h_K is the diameter of the element K, $Pe_K = \dfrac{Mh_K}{6}$ is the Peclet number. It is seen from t formulation that the SUPG method is computationally cheap and practically easy to implement. It is also very effective way of obtaining numerically stable solutions for convection dominated problems.

4. Numerical Results

We consider a long pipe of circle cross-section defined by $\{(x,y): x^2+y^2 \leq 1\}$. The boundary of the circular region in IR^2 is discretized by using $N = 64$ constant boundary elements on which the node is at the center. Also, in order to calculate unknown on internal and external region, 316 points (corresponding to 694 triangular elements) inside the pipe and 720 points (corresponding 1408 triangular elements) are used (**Figure 1**). Discretized system of linear equations obtained from BEM formulation in order to obtain unknown values at the boundary are solved by using the solver DGESV and equations obtained from FEM formulation for the unknown values inside the pipe region are solved by using the sparse solver UMFPACK which are LAPACK driver routines from Netlib library. The behaviors of the velocity of the fluid and inside and outside induced currents (induced magnetic fields) are visualized in terms of contour plots for very high values of magnetic Reynolds numbers Rm_1, Reynolds number Re and magnetic pressure number Rh of the fluid.

Before we present numerical results, let's define the notations; Labels "F-BEM", "BEM" and "Ref" refers to solutions FEM and BEM Coupling, BEM only and Reference solutions, respectively. As a Reference solution, we take the solution of the same problem over very fine mesh.

For the small values of the Hartmann number, both solution obtained either FEM-BEM Coupling or BEM only are almost equal. Therefore, only the high values of the Rm_1, Re and Rh = 10 are presented. From **Figure 2**, for the values $Rm_1 = 1000$, $Rm_2 = 1$, Re = 1 and Rh = 10, it is seen that, although induced magnetic field (induced current) solution are almost same and smooth, for the velocity values, there are some numerical instabilities in the solutions obtained from BEM only method which are source from the convection dominated behavior of the Equations (21). However, FEM-BEM coupling solution technique eliminates these disturbances and shows the well-known characteristic of the MHD flow that is the existence of the boundary layer formation as Hartmann number getting large.

Figure 3 shows equal velocity and induced current lines, respectively, for very large value of Reynolds number Re = 100 when Rm_1 = 100, Rm_2 = 1 and Rh = 10. It seen from that, the effect of the stabilization in FEM-BEM Coupling is also seen from the induced current solution additional to the velocity solutions. The similar behaviour is also seen for the high value of the magnetic pressure number Rh = 100 from **Figure 4**. The accuracy of the FEM-BEM Coupling approach to BEM only is also displayed in terms of the maximum values of the velocity and induced currents by comparing with the Reference solution (**Table 1**). It is known that, for large values of the parameters, the problem has unstable numerical results especially in maximum and minimum values. This is also seen from the table that BEM solutions give higher maximum values compared to reference solution.

5. Conclusion

We consider a FEM-BEM coupling approach for the approximate solution of the MHD flow through a circular pipe under the influence of a transverse magnetic field when the outside medium is also electrically conducting. Coupled equations with coupled boundary conditions are solved first on the boundary of the pipe by transforming the modified Helmholtz equations and Laplace equation. Then, velocity and induced magnetic field inside the pipe are calculated by considering SUPG typed stabilized finite element method, and induced magnetic on the external region is calculated with constant boundary ele-

Table 1. Comparison maximum values of the velocity and induced current.

Parameters			Velocity			Induced Current		
Rm_1	Re	Rh	F-BEM	BEM	Ref	F-BEM	BEM	Ref
1000	1	10	0.0089	0.0096	0.0091	0.078	0.091	0.085
100	100	10	0.0022	0.0028	0.0016	0.0008	0.001	0.0009
100	1	100	0.0052	0.0058	0.0052	0.0081	0.0091	0.0087

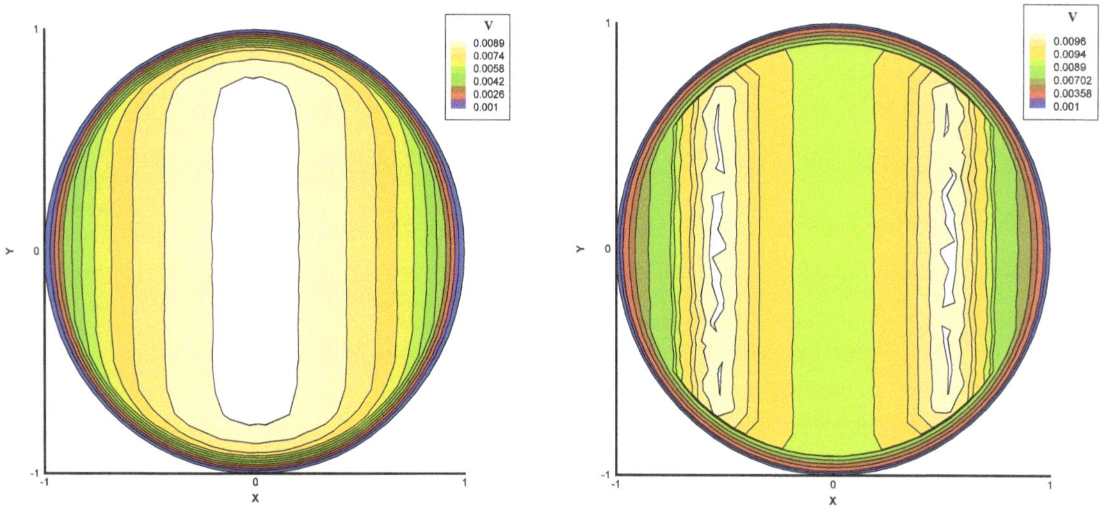

[Velocity]

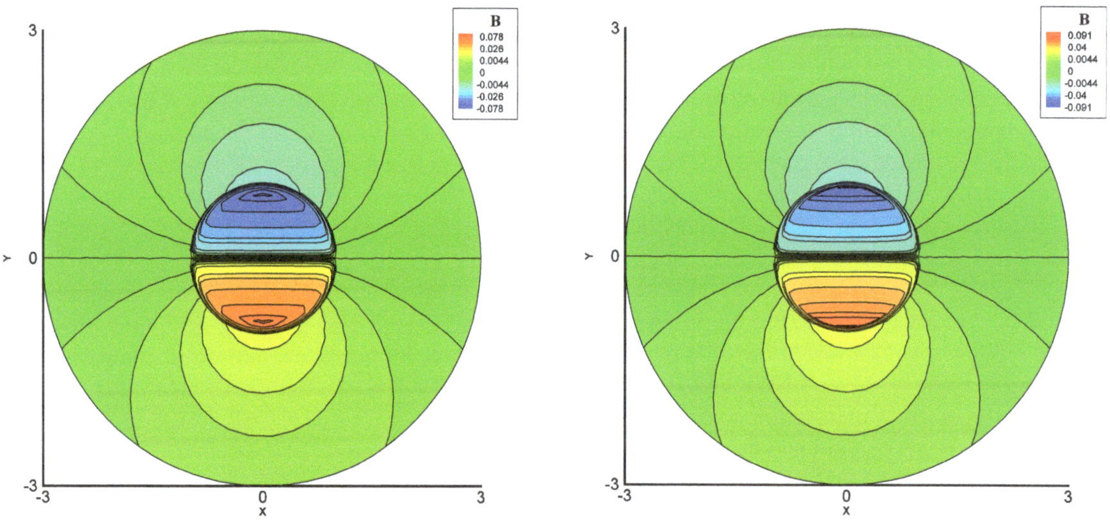

[Induced currents]

Figure 2. F-BEM (left) and BEM (right) contours for $Rm_1 = 1000$, $Rm_2 = 1$, $Re = 1$, $Rh = 10$.

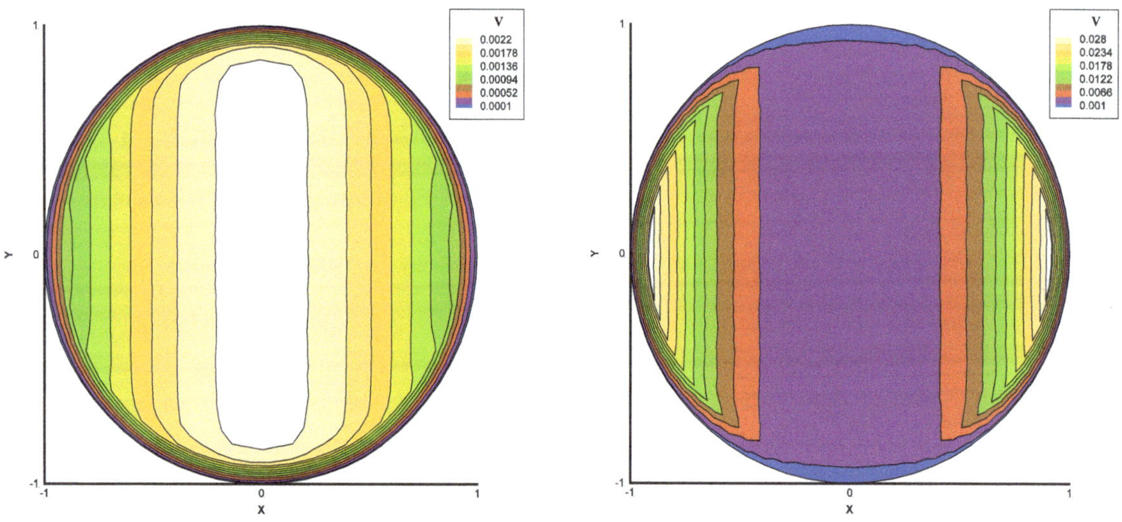

[Velocity]

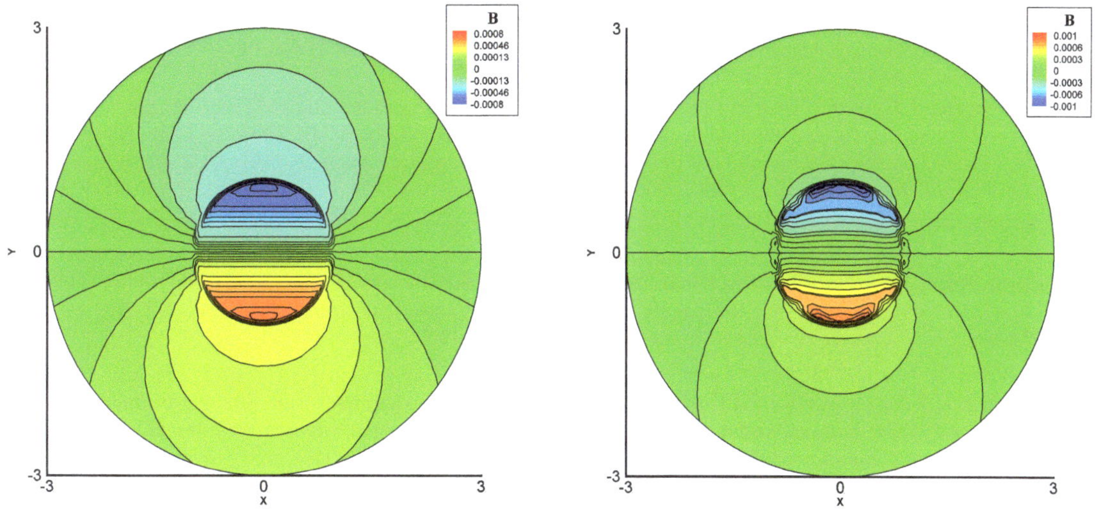

[Induced currents]

Figure 3. F-BEM (left) and BEM (right) contours for $Rm_1 = 100$, $Rm_2 = 1$, $Re = 100$, $Rh = 10$.

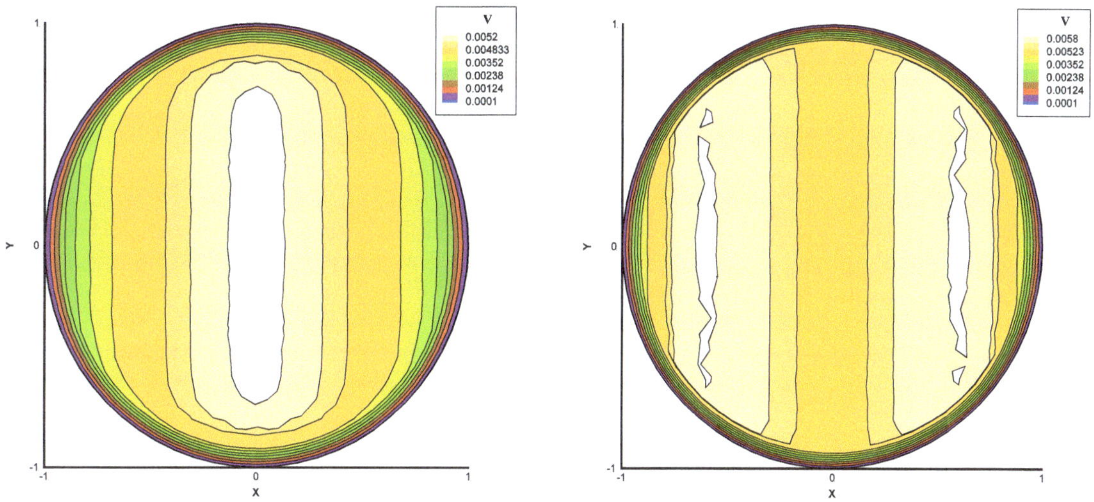

Figure 4. F-BEM (left) and BEM (right) velocity contours for $Rm_1 = 100$, $Rm_2 = 1$, $Re = 1$, $Rh = 100$.

ment method due to the regularity conditions at infinity. Obtained solutions show the accuracy and efficiency of the proposed numerical scheme for the high values of the Hartmann number. Stabilized FEM-BEM Coupling approach is very effective not only in removing disturbances in numerical solutions but also in the maximum and minimum values of the unknowns.

6. Acknowledgements

This work is supported by Karadeniz Technical University by Scientific Research Project (BAP) under Grant Number #10464.

REFERENCES

[1] L. Dragos, "Magnetofluid Dynamics," Abacus Press, Preston, 1975, pp. 92-99.

[2] J. A. Shercliff, "Steady Motion of Conducting Fluid in a Pipe under Transverse Magnetic Fields," *Journal of Fluid Mechanics*, Vol. 1, No. 6, 1956, pp. 644-666.

[3] A. Carabineanu, A. Dinu and I. Oprea, "The Application of the Boundary Element Method to the Magnetohydrodynamic Duct Flow," *Zeitschrift für Angewandte Mathematik und Physik*, Vol. 46, No. 6, 1995, pp. 971-981.

[4] T. W. H. Sheu and R. K. Lin, "Development of a Convection-Diffusion-Reaction Magnetohydrodynamic Solver on Nonstaggared Grids," *International Journal for Numerical Methods in Fluids*, Vol. 45, No. 11, 2004, pp. 1209-1233.

[5] K. E. Barrett, "Duct Flow with a Transverse Magnetic Field at High Hartmann Numbers," *International Journal for Numerical Methods in Engineering*, Vol. 50, 2001, pp. 1893-1906.

[6] H. W. Liu and S. P. Zhu, "The Dual Reciprocity Boun-

dary Element Method for Magnetohydrodynamic Channel Flows," *ANZIAM Journal*, Vol. 44, 2002, pp. 305-322.

[7] M. Tezer-Sezgin and S. H. Aydn, "Dual Reciprocity Boundary Element Method for Magnetohydrodynamic Flow Using Radial Basis Functions," *International Journal of Computational Fluid Dynamics*, Vol. 16, No. 1, 2002, pp. 49-63.

[8] M. Tezer-Sezgin, "Boundary Element Method Solution of MHD Flow in Rectangular Duct," *International Journal for Numerical Methods in Fluids*, Vol. 18, No. 10, 1994, pp. 937-952.

[9] M. Dehghan and D. Mirzai, "Meshless Local Boundary Integral Equation (LBIE) Method for the Unsteady Magnetohydrodynamic (MHD) Flow in Rectangular and Circular Pipes," *Computer Physics Communications*, Vol. 180, No. 9, 2009, pp. 1458-1466.

[10] F. Paris and J. Canas, "Boundary Element Method, Fundamentals and Applications," Oxford University Press Inc., New York, 1997, pp. 64-78.

[11] A. J. Meir, "Finite Element Analysis of Magnetohydrodynamic Pipe Flow," *Applied Mathematics and Computation*, Vol. 57, No. 2-3, 1993, pp. 177-196.

[12] M. Tezer-Sezgin and S. Koksal, "FEM for Solving MHD Flow in a Rectangular Duct," *International Journal for Numerical Methods in Engineering*, Vol. 28, No. 2, 1989, pp. 445-459.

[13] N. B. Salah, A. Soulaimani, W. G. Habash and M. Fortin, "A Conservative Stabilized Finite Element Method for the Magneto-Hydrodynamic Equations," *International Journal for Numerical Methods in Fluids*, Vol. 29, No. 5, 1999, pp. 535-554.

[14] N. B. Salah, A. Soulaimani and W. G. Habashi, "A Finite Element Method for Magnetohydrodynamics," *Computer Methods in Applied Mechanics and Engineering*, Vol. 190, No. 43-44, 2001, pp. 5867-5892.

[15] A. I. Nesliturk and M. Tezer-Sezgin, "The Finite Element Method for MHD Flow at High Hartmann Numbers", *Computer Methods in Applied Mechanics and Engineering*, Vol. 194, No. 9-11, 2005, pp. 1201-1224.

[16] J. F. Gerbeau, "A Stabilized Finite Element Method for the Incompressible Magnetohydrodynamic Equations," *Numerische Mathematik*, Vol. 87, No. 1, 2000, pp. 83-111.

[17] R. Codina and N. H. Silva, "Stabilized Finite Element Approximation of the Stationary Magnetohydrodynamics Equations," *Computational Mechanics*, Vol. 38, No. 4-5, 2006, pp. 344-355.

[18] S. H. Aydin, A. I. Nesliturk and M. Tezer-Sezgin, "Two-Level Finite Element Method with a Stabilizing Subgrid for the Incompressible MHD Equations," *International Journal for Numerical Methods in Fluids*, Vol. 62, No. 2, 2010, pp. 188-210.

[19] A. N. Brook and T. J. R. Hughes, "Streamline Upwind/ Petrov-Galerkin Formulations for Convection Dominated Flows with Particular Emphasis on the İncompressible Navier-Stokes Equations," *Computer Methods in Applied Mechanics and Engineering*, Vol. 32, No. 1-3, 1982, pp. 199-259.

[20] A. Carabineanu and E. Lungu, "Pseudospectral Method for MHD Pipe Flow," *International Journal for Numerical Methods in Engineering*, Vol. 68, No. 2, 2006, pp. 173-191.

[21] M. Tezer-Sezgin and S. H. Aydn, "BEM Solution of MHD Flow in a Pipe Coupled with Magnetic İnduction of Exterior Region," *Computing*, Vol. 95, No. 1, 2013, pp. 751-770.

[22] J. N. Reddy, "An Introduction to the Finite Element Method," McGraw-Hill, New York, 1993.

A Bio-Hydraulic Modelling Approach to Control the Hemodialysis Patient Status—Calculus and Validation

Mohamed Habib Sellami[1,2,3]

[1]Laboratory of Thermal Radiation, Department of Physics, Faculty of Science Tunis, University Tunis El Manar, Tunis, Tunisia
[2]High School for Rural Engineering, University of Jendouba, Jendouba, Tunisia
[3]Tunisian Association of Science, Technology and Development, Tunisia

ABSTRACT

We have developed a model permitting the control of the hemodialysis patient status. We have represented the functioning of each body compartment by a set of differential equations that express the mass balance of the carbonate and bicarbonate concentrations in each one. Experimental data from bibliography were analyzed and compared to our calculus. Results were well satisfying.

Keywords: Systolic Pressure; Sympathetic Activity; Baroreceptor; Biomarker; Glomarular Filtration Rate

1. Introduction

With advances in technology, dialysis has now become cheaper and less onerous. As there is a shortage of cadaveric donors, only one-third of those on the list will receive a kidney transplant. This means that the majority of such patients will spend the rest of their lives on dialysis [1]. The yearly mortality among patients being treated with dialysis is mainly due to cardiovascular diseases and respiratory infection. Hypertension continues to be a major risk factor for cardiovascular disease. Good evidence suggests that improvements in dialysis efficiency reduce morbidity and mortality of hemodialysis patients. Dialysis efficiency has also been related to better control of carbon dioxide arterial blood pressures, systolic and diastolic blood pressures, cardiac output, brain blood flow, pulmonary blood flow, renal blood flow, urea/creatinine levels and sympathetic nerve activity [1]. That's why a physiologic model based on an integration over all human body compartments can help to control those parameters. In fact, the new emphasis of research in dialysis modelling is on developing models that more closely resemble the physiology of the body [2-4]. In this direction, by using the mass balance equations principles and the notions of control theory, this paper focuses on modelling the interaction between the different physiological control systems in human (sympathetic nerve, cardiopulmonary and renal) and their link with an important health problem in the worldwide: hemodialysis. The model proposed can help to detect the eventual dysfunction at each physiologic level by controlling the systolic/diastolic pressures, carbon dioxide arterial pressure and the sympathetic nerve activity. For this goal, I will begin with introducing relationships among blood pressures (systolic/diastolic pressures), respiration, peripheral resistance, and sympathetic nerve activity [2-4]. Then, I will propose a quantitative description of the respiratory processes responsible for the regulation of acid-base balance in man [5-8]. This will be done by formulating the carbon dioxide concentration and its arterial blood pressure for the brain compartment, lung compartment and body tissue compartment. The buffer content of whole blood, the bicarbonate content of plasma and the blood flows which are very useful for the fixed purpose, will also be treated [9-11]. I will incorporate the glomerular filtration rate and the kidney/dialyser clearance [12-17] in order to test the effect of their change on the health indicators (biomarkers) of hemodialysis patient status (acid-base buffer systems, the pH of human plasma, concentration of bicarbonate content of plasma, the partial pressure of carbon dioxide, the sympathetic nerve activity and systolic/diastolic blood pressures [18]).

2. Cardiac Compartment: Relationships between Systolic/Diastolic Pressures and Sympathetic Activity

Sympathetic activity can be thought of as a complex output of the central nervous system, providing subtle control over end organ function. This control is exerted in a number of frequency bands, including rhythms related to the cardiac cycle, respiratory cycle and kidney cycle. Determining the blood pressures (diastolic and systolic pressures, arterial pressure, partial pressure of carbon dioxide in arterial blood) at different level of the body can help to deduce the sympathetic activity and after detect the eventual dysfunction if the frequency is not in the normal band. At any end-diastolic volume, an increase in sympathetic activity results in an increase in stroke volume due both to an increase in venous return and in contractility. The diastolic blood pressure (respectively the systolic blood pressure) defined as the force against the artery wall when the heart is at rest (respectively when the heart is beating) can be deduced from the following equations [2-4]:

$$\frac{dP_{di}}{dt} = -\frac{P_{di}}{\delta_{vr}} \tag{1}$$

$$P_s = P_{d_{i-1}} + \tau_{cc\text{-}i}\frac{t-t_i}{0.125}\exp\left(1-\frac{t-t_i}{0.125}\right) \tag{2}$$

$$\delta_{vr} = 2.2 - 1.2\left[C_{v\text{-}Ne} + \left(10 - C_{v\text{-}Ne}\right)\frac{\left(C_{v\text{-}Ne}\right)^{1.5}}{31.62+\left(C_{v\text{-}Ne}\right)^{1.5}}\right] \tag{3}$$

$$\tau_{cc\text{-}i} = \tau'_{cc\text{-}i} + \left(70-\tau'_{cc\text{-}i}\right)\frac{\left(\tau'_{cc\text{-}i}\right)^{2.5}}{\left(\tau'_{cc\text{-}i}\right)^{2.5}+40996} \tag{4}$$

$$\tau'_{cc\text{-}i} = 25 + 40C_{c\text{-}Ne} + 10t_{i-1} \tag{5}$$

$C_{v\text{-}Ne}$ is the vascular concentration of norepinephrine (Ne)

$C_{c\text{-}Ne}$ is the cardiac concentration of norepinephrine (Ne)

P_{di} is the diastolic blood pressure at the beat i

$P_{d_{i-1}}$ is the diastolic blood pressure of the previous beat (i – 1)

P_S is the systolic blood pressure

t is the duration of heart period

t_i is the duration of contraction (i) onset

δ_{vr} is the time-varying relaxation of the Windkessel arteries

$\tau_{cc\text{-}i}$ is the cardiac contractility at the beat i

The increase in end organs function (heart, kidney, pulmonary) is mediated by sympathetic neurons which release norepinephrine in the vicinity of the authorythmic cells. The role of norepinephrine is to increase the rate of depolarisation of the pacemaker pre-potential by rising sodium and calcium conductance, while simultaneously decreasing potassium conductance [4]. If not measured, the vascular concentration of norepinephrine and the cardiac concentration of norepinephrine can be deduced from those relations [2,3]:

$$\frac{dC_{v\text{-}Ne}}{dt} = -\frac{C_{v\text{-}Ne}}{2} + 1.2\varphi_{sa}\left(t-4.2\right) \tag{6}$$

$$\frac{dC_{c\text{-}Ne}}{dt} = -\frac{C_{c\text{-}Ne}}{2} + 1.2\varphi_{sa}\left(t-1.65\right) \tag{7}$$

$$\varphi_{sa} = \max\left[0, 0.8-0.7\varphi_{ba}+0.035\left(1-\varphi_{ri}\right)\right] \tag{8}$$

$$\varphi_{ba} = 0.02\left(P-50\right) + 0.00125\frac{dP}{dt} \tag{9}$$

$$\varphi_{ri} = \cos\left(2\pi\theta\right) \tag{10}$$

$$\frac{d\theta}{dt} = \frac{1}{T_{resp}} - 1.3\left(\varphi_{ba}-1.3\right) \tag{11}$$

$C_{v\text{-}Ne}$ is the vascular concentration of norepinephrine (Ne)

$C_{c\text{-}Ne}$ is the cardiac concentration of norepinephrine (Ne)

P is the blood pressure

T_{resp} is a constant respiratory period.

φ_{sa} is the efferent sympathetic activity

φ_{ba} is the baroreceptor activity

φ_{ri} is the respiratory influences

θ is an instantaneous phase of respiration, ($0 < \theta < 0.5$ correspond to expiratory period and $0.5 < \theta < 1$, correspond to inspiratory period).

3. Modelling Partial Pressure of Carbon Dioxide in the Arterial Blood

It has been observed that an increase in carbon dioxide concentration in the arterial blood greatly depresses neuronal activity in particular the sympathetic activity [2-4]. In fact any dysfunction of the kidney/dialyser provokes a deviation of the carbon dioxide concentration or its pressure from their normal values that will probably influences the metabolic processes and after the systolic/diastolic pressures and the sympathetic activity. To determine the carbon dioxide arterial blood pressures and concentrations at all body's compartments, I will apply the mass balance equation for the brain, lung and body tissue.

3.1. Formulation of Carbon Dioxide Exchange in the Brain Compartment

For the brain, the rate of change of carbon dioxide volume is determined by a balance between the net rate of carbon dioxide transported to the brain tissue by the blood and the net rate at which carbon dioxide is produced in the brain tissue. Also the carbon dioxide concentration in brain tissue is, generally, equilibrated with the carbon dioxide concentration in the venous blood leaving the brain. This can be formulated by [9-11]:

$$V_{\text{B-CO}_2}\frac{\mathrm{d}C_{\text{B-CO}_2}(t)}{\mathrm{d}t} = \Gamma_{\text{mrB-CO}_2} + \Phi_{\text{B}}(t)\left[C_{\text{a-CO}_2}(t) - C_{\text{B-CO}_2}(t)\right] \quad (12)$$

$$\Phi_{\text{B}}(t) = \Phi_{\text{B0}}\left(1 + 0.03\left(P_{\text{a-CO}_2}(t) - 40\right)\right) \quad (13)$$

$$C_{\text{B-CO}_2} = 0.0065 P_{\text{B-CO}_2} + 0.0244 \quad (14)$$

$C_{\text{B-CO}_2}(t)$ is the total CO_2 concentration in the brain tissue at instant t

$C_{\text{a-CO}_2}(t)$ is the total CO_2 concentration in arterial blood at instant t

$P_{\text{a-CO}_2}$ is the partial pressure of CO_2 in the arterial blood

$P_{\text{B-CO}_2}$ is the partial pressure of CO_2 in brain tissue

$V_{\text{B-CO}_2}$ is the effective brain tissue volume for CO_2 generally equal to 0.9

$\Gamma_{\text{mrB-CO}_2}$ is the metabolic rate of carbon dioxide production in the brain (generally equal to 0.042)

$\Phi_B(t)$ is the brain blood flow at instant t

Φ_{B0} is a parameter indicates the normal brain blood flow for $P_{\text{a-CO}_2} = 40$ mmHg

3.2. Formulation of Carbon Dioxide Exchange in the Lung Compartment

Carbon dioxide can exist in the blood in many different forms: as free carbon dioxide, and in chemical combinations with water (bicarbonate), haemoglobin, and plasma protein. Carbon dioxide elimination is accomplished by bringing fresh air into the lungs, there is no other way to excrete the body's carbon dioxide production. In fact, haemoglobin is the key chemical compound that combines with oxygen from the lungs and carries the oxygen from the lungs to cells throughout the body. The blood also transports carbon dioxide, which is the waste product of this energy production process, back to the lungs from which it is exhaled into the air. At every instant t the change of carbon dioxide volume is determined by a balance between the net rate of carbon dioxide diffusing into the lungs from the blood and that expired from the lungs. The formulations that express those phenomena are [9-11]:

$$V_{\text{A-CO}_2}\frac{\mathrm{d}P_{\text{a-CO}_2}(t)}{\mathrm{d}t} = \Phi_{\text{Av}}\left[P_{\text{I-CO}_2} - P_{\text{a-CO}_2}(t)\right] + 863\Phi_{\text{p}}(t)\left[C_{\text{v-CO}_2}(t) - C_{\text{a-CO}_2}(t)\right] \quad (15)$$

$$\Phi_{\text{Av}} = H_{\text{p}}\mathrm{e}^{-0.05P_{\text{a-O}_2}}\left(P_{\text{a-CO}_2} - I_{\text{p}}\right) + H_{\text{c}}\left(P_{\text{a-CO}_2} - I_{\text{c}}\right) - 2.4 \quad (16)$$

$$\Phi_{\text{Av}} = H_{\text{p}}\mathrm{e}^{-0.05P_{\text{a-O}_2}}\left(P_{\text{a-CO}_2} - I_{\text{p}}\right) + H_{\text{c}}\left(P_{\text{a-CO}_2} - I_{\text{c}}\right) - 2.4 \quad (17)$$

$$\Phi_p(t) = \frac{P_{\text{a-p}}(t) - P_{\text{v-p}}(t)}{\tau_{\text{pr-p}}} \quad (18)$$

$$\tau_{\text{pr-p}} = \frac{P_a}{\Gamma_{\text{hr}}V_s} \quad (19)$$

$C_{\text{a-CO}_2}$ concentration of bound and dissolved CO_2 in arterial blood

$C_{\text{v-CO}_2}$ concentration of bound and dissolved CO_2 in the mixed venous blood

H_c the central controller gain

H_p the peripheral controller gain

I_c constant for central drive of ventilation

I_p constant for peripheral drive of ventilation

P_a is the mean arterial blood pressure

$P_{\text{a-CO}_2}$ partial pressure of CO_2 in arterial blood

$P_{\text{a-p}}$ is the mean blood pressure in the arterial region of the pulmonary circuit

$P_{\text{I-CO}_2}$ partial pressure of CO_2 in inspired air

$P_{\text{v-p}}$ is the mean blood pressure in the venous region of the pulmonary circuit

$V_{\text{A-CO}_2}$ is the effective CO_2 storage volume of the alveolar (lung compartment)

V_s is the stroke volume

Φ_{Av} is the alveolar ventilation

$\Phi_p(t)$ is the pulmonary blood flow at instant t

$\tau_{\text{pr-p}}$ is the peripheral resistance in the pulmonary circuit

Γ_{hr} is the heart rate

3.3. Formulation of Carbon Dioxide Exchange in the Body Tissue Compartment

The body tissues are assumed to be lumped together in one compartment. The change of carbon dioxide concentration is determined by a balance between the net rate of carbon dioxide transported to the tissues by the blood and the net rate at which carbon dioxide is produced by the tissues. By assuming that the gas tensions in the body tissues are uniform and at every instant equal to those in the mixed venous blood returning to the lungs, the mass balance equations for the tissue compartment and the related formulas can be written [5-11]:

$$V_{\text{T-CO}_2}\frac{\mathrm{d}C_{\text{v-CO}_2}(t)}{\mathrm{d}t} = \Gamma_{\text{mr-CO}_2} + \phi_{\text{s}}(t)\left[C_{\text{a-CO}_2}(t) - C_{\text{v-CO}_2}(t)\right] \quad (20)$$

$$\phi_{\text{s}}(t) = \frac{P_{\text{a-s}}(t) - P_{\text{v-s}}(t)}{\tau_{\text{pr-s}}} \quad (21)$$

$$C_{\text{v-CO}_2} = -\sigma P_{\text{v-CO}_2} + \frac{C_{\text{bb}}}{2} + \frac{\sqrt{\left[C_{\text{bb}} + 2(0.03+\sigma)P_{\text{v-CO}_2} + 2k\right]^2 - 8\cdot k\cdot C_{\text{bb}}}}{2} - k \quad (22)$$

$$C_{\text{a-CO}_2} = -\sigma P_{\text{a-CO}_2} + \frac{C_{\text{bb}}}{2} + \frac{\sqrt{\left[C_{\text{bb}} + 2(0.03+\sigma)P_{\text{a-CO}_2} + 2k\right]^2 - 8\cdot k\cdot C_{\text{bb}}}}{2} - k \quad (23)$$

$C_{\text{a-CO}_2}$ concentration of bound and dissolved CO_2 in arterial blood

C_{bb} is the buffer base concentrations

$C_{\text{v-CO}_2}$ concentration of bound and dissolved CO_2 in the mixed venous blood

$P_{\text{a-CO}_2}$ is the partial pressure of CO_2 in arterial blood

$P_{\text{a-s}}$ is the mean blood pressure in the arterial region of the systemic circuit

$P_{\text{v-CO}_2}$ partial pressure of CO_2 in the mixed venous blood

$P_{\text{v-s}}$ is the mean blood pressure in the venous region of the systemic circuitt is the time

$V_{\text{T-CO}_2}$ effective tissue storage volume for CO_2

Φ_s the systemic blood flow

$\Gamma_{\text{mr-CO}_2}$ steady state metabolic CO_2 production rate

$\tau_{\text{pr-s}}$ is the peripheral resistance in the systemic circuit

σ and k are constant parameters

Generally, we assume the equality between left cardiac output, right cardiac output, the systemic blood flow and the pulmonary blood flow [11]. The results of those equations will be used to model the production and elimination of acids and bases wastes, subject of the following paragraph.

3.4. Formulations for the Kidney/Dialyser Compartment

To simulate the production and elimination of acids and bases in body fluids, the buffer base concentration has been considered as a further state variable. The change in the buffer content of whole blood is determined as function of the changes in plasma pH and in plasma's bicarbonate $\left(HCO_3^-\right)$ content [16]. The buffer base concentrations in the equations (21) and (22) can be deduced from the following equations [16]:

$$\frac{dC_{\text{bb}}}{dt} = \left[1 - C_{\text{hg}} \times 0.023\right] \times \left[\frac{dC_{\text{HCO}_3^-}}{dt} + \frac{d\text{pH}}{dt}\left(2.3C_{\text{hg}} + 7.7\right)\right] \quad (24)$$

$$\text{pH} = 6.1 + \log\left[\frac{C^{\text{pl}}_{\text{HCO}_3^-}}{0.03P_{\text{a-CO}_2}}\right] \quad (25)$$

C_{bb} is the buffer base concentrations

C_{hg} is the haemoglobin content of whole blood

$C^{\text{pl}}_{\text{HCO}_3^-}$ is the plasma's bicarbonate content

$P_{\text{a-CO}_2}$ is the partial pressure of CO_2 in arterial blood

pH is the plasma pH (potential hydraulic)

The plasma can be considered as a chemical system in which coexist bicarbonate and a non bicarbonate buffer [5-11]. An efficient buffering permits the balance of the net production of acids in the healthy patient. By contrast, due to an insufficient $[H^+]$ ions elimination through the kidneys/dialyser, the buffering term is strongly reduced in the uremic patient. Hemodialysis therapy rests on the premise that the control of blood concentrations for various toxic substances to some maximum value is required for the adequate treatment. We will devise here a general equation system for predicting the concentration profile of any potential toxic substances in the blood during dialysis [1,13,14,17] that can be applied to the urea/creatinine substances:

$$V\frac{dC_{\text{sub}}}{dt} = \Gamma_{\text{ge}} - q_{\text{dc}}C_{\text{sub}} - q_{\text{rc}}C_{\text{sub}} \quad (26)$$

$$C_{t,\text{urea}} = C_{0,\text{urea}}\exp\left[-\frac{q_{\text{dc}}t}{V_{\text{urea}}}\right] + \frac{\Gamma_{\text{ge}}}{q_{\text{rc}}}\left(1 - \exp\left[-\frac{q_{\text{dc}}t}{V_{\text{urea}}}\right]\right) \quad (27)$$

$$\frac{q_{\text{dc}} \times t}{V_{\text{urea}}} = -\ln\left(\frac{C^{\text{post-d}}_{\text{urea}}}{C^{\text{pre-d}}_{\text{urea}}} - 0.008t\right) + \left(4 - 3.5\frac{C^{\text{post-d}}_{\text{urea}}}{C^{\text{pre-d}}_{\text{urea}}}\right)\frac{V_{\text{uf}}}{B^{\text{post-d}}_{\text{W}}} \quad (28)$$

$$\frac{q_{\text{per,dc}}t}{V_{\text{urea}}} = \left(\frac{V_{\text{urine}}}{1.44}\right)\left(\frac{C_{\text{u,urea}}}{C_{\text{b,urea}}}\right)\left(\frac{1440}{V_{\text{urea}}}\right) \quad (29)$$

$$V_{\text{urea,m}} = 2447 - \left(95.16\text{Ag}\right) + \left(107.4B_{\text{H}}\right) + \left(336.2B_{\text{W}}\right) \quad (30)$$

$$V_{\text{urea,w}} = \left(106.9B_{\text{H}}\right) + \left(246.6B_{\text{W}}\right) - 2097 \quad (31)$$

Ag is the age in years

B_{H} is the body height

B_{w} is the body weight

$B^{\text{post-d}}_{\text{W}}$ is the body post-dialysis weight

$C_{\text{b,urea}}$ is the urea concentration in blood

C_{sub} is the toxic substance concentration in the blood to be eliminated

$C_{\text{t,urea}}$ is the final solute concentration at the end of each dialysis session (t).

$C_{\text{u,urea}}$ is the urea concentration in dialysate or in urine

$C^{\text{post-d}}_{\text{urea}}$ is the post-dialysis urea concentration in blood

$C^{\text{pre-d}}_{\text{urea}}$ is the pre-dialysis urea concentration in blood

$C_{0,\text{urea}}$ is the initial solute concentration at the beginning of each dialysis.

q_{dc} is the dialyser clearance for any substance

$q_{\text{per,dc}}$ is peritoneal dialysis clearance for any substance

q_{rc} is the renal clearance for any substance

t is the treatment time

V_{uf} is the ultrafiltrate volume

V_{urea} is the body urea distribution volume

V_{urine} is the dialysate or the urine volume in 24 h

$V_{urea,m}$ is the urea volume for man

$V_{urea,w}$ is the urea volume for woman

Since the total kidney glomerular filtration rate is equal to the sum of the filtration rates in each of the functioning nephrons, it can be used as an index of functioning renal mass. A decrease of the total kidney glomerular filtration rate precedes kidney failure in all forms of progressive disease. Under steady-state conditions and by considering patient's age, gender and ethnicity, the total kidney glomerular filtration rate can be estimated, for the urea clearance and creatinine clearance, from the following equations [12,15]:

$$\Gamma_{gf} = 1.04\left[q_{cr}\right]^{0.751}\left[q_{urea}\right]^{0226} \times (1.109 \text{ if African America}) \quad (32)$$

$$\Gamma^{gf} = 270\left[C^{s}_{cr}\right]^{-1.007}\left[\text{Ag}\right]^{-0.18}\left[C^{B.N}_{urea}\right]^{-0.169} \times\left[0.755 \text{ if female}\right]\times\left[1.178 \text{ if African America}\right] \quad (33)$$

Ag is the patient age

C^{s}_{cr} is the serum creatinine concentration

$C^{B.N}_{urea}$ is the blood urea nitrogen concentration

q_{cr} is the ceatinine clearance

q_{urea} is the urea clearance

Γ_{gf} is the glomerular filtration rate

The equations proposed above permit the calculus of the carbon dioxide arterial pressure, its concentration and the blood flow at each compartment (brain, lung, body tissue). The parameters estimated are directly linked to tissue metabolism, metabolite production and the buffer base concentration. Those later influence the kidney/dialyser function and the status of the hemodialysis Patient. The control role of the model, the function of each compartment and the link between them will be discussed and clarified in the following paragraph.

4. Block Diagram for the Model Functioning

The equations proposed above permit the calculus of the carbon dioxide arterial pressure, its concentration and the blood flow at each compartment (brain, lung, body tissue). The parameters estimated are directly linked to tissue metabolism, metabolite production and the buffer base concentration. Those later influence the kidney/dialyzer function and the status of the hemodialysis patient. This approach is based on an integration over all human body compartments and on the argument that blood pressure is controlled by several physiological mechanisms acting in combination (**Figure 1**).

The parameters indicators of the functioning of each compartment were detailed and explained in every correspondent equations. They represent the input or output quantities of the model depending on its usage sense. So the approach permits to quantify the change that can undergo the carbon dioxide arterial blood pressure, the systolic/diastolic blood pressure, the blood flow and the sympathetic nerve activity in each body compartment when there is any anomaly in the kidney/dialyzer system. To verify the running of the model and because the different compartments detailed above have a linked functioning which means that they work as entity in a closed circuit and every one uses the parameters furnished by the other we have make the calculus by considering a unique system embedding the four compartments and by treating the kidney/dialyzer as regulator of all system (**Figure 1**). So if we change a parameter in the kidney/dialyzer compartment we must record a change for the parameters indicators of the functioning of other compartments (lung compartment, body tissue compartment, brain compartment, cardiac compartment) with a proportional sensibility. The parameter indicator of the kid-ney/dialyser efficiency for which we have made a variation is the glomerular filtration rate. We have chosen values in the band of chronic kidney failure, one near the normal band (Γ_{gf} = 75 ml/min) and the other far distant (Γ_{gf} = 30 ml/min). The parameters indicators of the system functioning that we have chosen are carbon dioxide arterial blood concentration, the systolic/diastolic blood pressure, baroreceptor and sympathetic nerve activities.

5. Results and Discussions

Figure 2 shows the variation of the CO_2 concentration in the arterial blood for the tissue compartment.

We note that at the beginning of the dialysis session and for the two values of glomerual filtration rate the

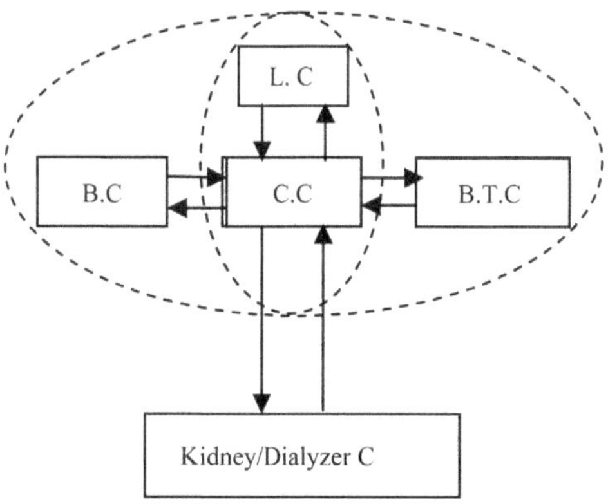

Figure 1. A block diagram for a potential monitoring use of the controlling model (L.C: Lung compartment, C.C: Cardiac compartment, B.C: Brain compartment, B.T.C: Body tissue compartment).

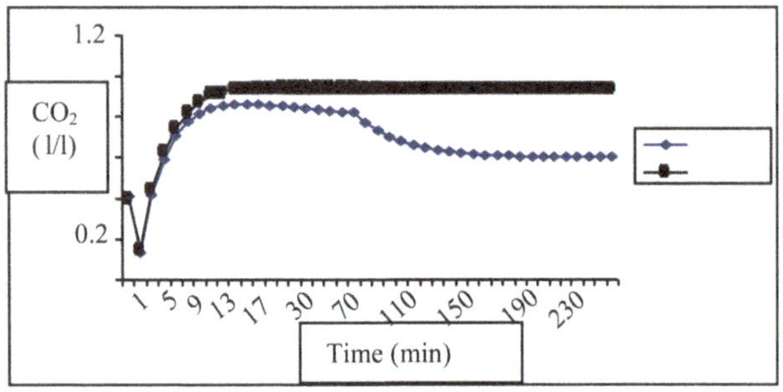

Figure 2. Variation of the CO_2 concentration in B.T.C for (Γ_{gf} = 30ml/min ◆) and (Γ_{gf} = 75 ml/min ■).

CO_2 concentrations are close to each other, decrease slowly in the first five minutes of the session, increase later to reach a threshold closed to 1 l/l for (Γ_{gf} = 75 ml/min) and 0.8 for (Γ_{gf} = 30 ml/min). We record a progressive decrease for the CO_2 concentration corresponding to (Γ_{gf} = 30 ml/min) to reach a value of about 0.6 l/l at the end of the dialysis session but the CO_2 concentration corresponding to (Γ_{gf} = 75 ml/min) retains the threshold value till the finish of the session. The variation of CO_2 arterial blood pressure in the body tissue compartment during the treatment is presented in **Figure 3**. We have noticed the following results:

We record a CO_2 pressure increase in the first 5 min of the dialysis session with a closed values for both situations ((Γ_{gf} = 75 ml/min) and (Γ_{gf} = 30 ml/min). A treshold of about 40mmHg is reached 12 min after the beginning for the curve corresponding to (Γ_{gf} = 75 mmHg) and the pressure maintains the same value till the end. For the trend of the case (Γ_{gf} = 30 ml/min) we remark that the CO_2 arterial blood pressure reaches a maximum of about 18 mmHg at the instant 8 min, it decreases after to attain a minimum of about 20 mmHg at the moment 70 min and it continues with the same value until the end of the session.

In **Figure 4** we present the variation of the barorecepetor activity as average over all the dialysis session.

We can say that, the baroreceptor activities for both (Γ_{gf} = 75 ml/min) and (Γ_{gf} = 30 ml/min) maintains closed values in the first 30 min of the dialysis session with two phases. In the first phase from the beginning of the session till the 10 first minutes there is a decrease of the baroreceptor activity to reach a value of about 0.6. After, the second phase, there is a progressive increase To attain the value 1. From the instant 30 min the slope of increase becomes more important and a net difference between the data of the two curves was recorded. It becomes more significant later. The baroreceptor activity for (Γ_{gf} = 75 ml/min) reaches a treshold of about 1.7 at the moment 90 min while for (Γ_{gf} = 30 ml/min) the threshold was reached at 80 min and is about 1.4.

The mean variation of the sympathetic activity for the two values of the glomerular filtration rate was revolted

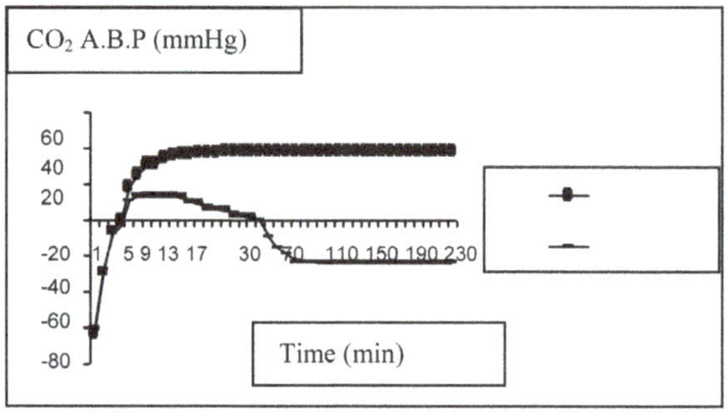

Figure 3. Variation of the CO_2 arterial blood pressure (A.B.P.) in the tissue compartment for (Γ_{gf} = 75 ml/min ▮) and (Γ_{gf} = 30 ml/min ▬).

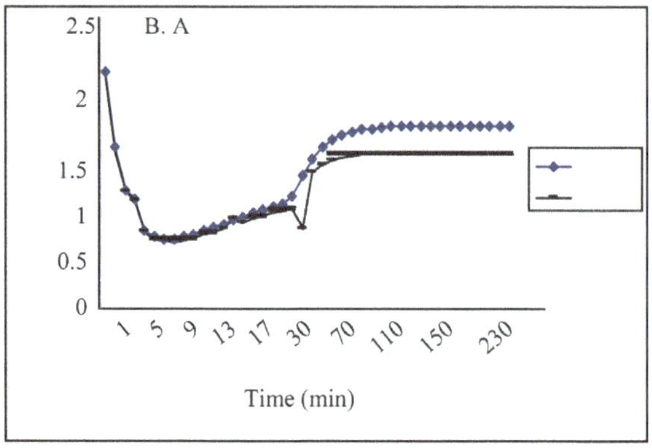

Figure 4. Variation of the baraoreceptor activity (B.A) for two values of glomerular filtration rate (Γ_{gf}) during a dialysis session (Γ_{gf} = 30 ml/min ▬, Γ_{gf} = 75 ml/min ◆).

in the **Figure 5**.

We remark a fast decrease of the sympathetic activities in the first 7 min of session, to attain a minimum of about 1.45, followed by a progressive increase to reach a maximum of about 2.2 (respectively 1.7) for (Γ_{gf} = 75 ml/min) respectively (Γ_{gf} = 30 ml/min)) at 90 min and maintain those values until the end of session. We notice closed sympathetic activity values for both situation (Γ_{gf} = 75 ml/min) and (Γ_{gf} = 30 ml/min) in the first 30 min and after a net difference is recorded which becomes more clear later.

The mean systolic blood pressure variation for the tow values of glomerular filtration rate was mentioned in **Figure 6**. Comparison between calculated and measured systolic blood pressure is in **Figure 7**.

The measured values were taken from the bibliography and represent the mean systolic blood pressure from 32 hemodialysis patients [19]. We remark that the global walk of the three curves is the same A complete validation of the model necessitates to compare the other parameters to those measured, which is a handicap. But, because the systolic blood pressure is considered as the result from the interaction between all the control parameters in the different body compartments of the pa-

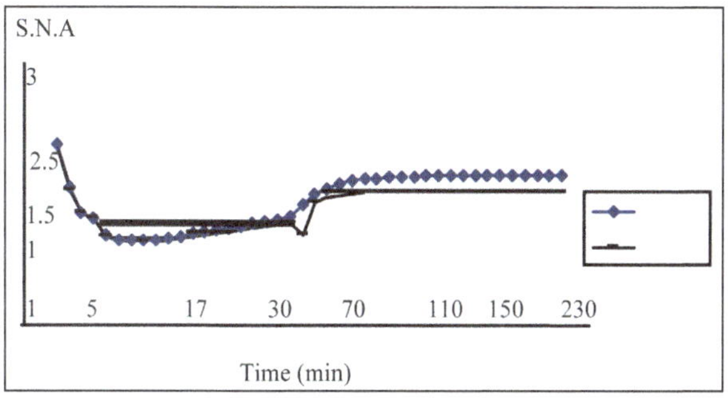

Figure 5. Variation of the sympathetic nerve activity (S.N.A.) for two values of glomerular filtration rate (Γ_{gf}) during a dialysis session (Γ_{gf} = 75 ml/min ◆, Γ_{gf} = 30 ml/min ▬).

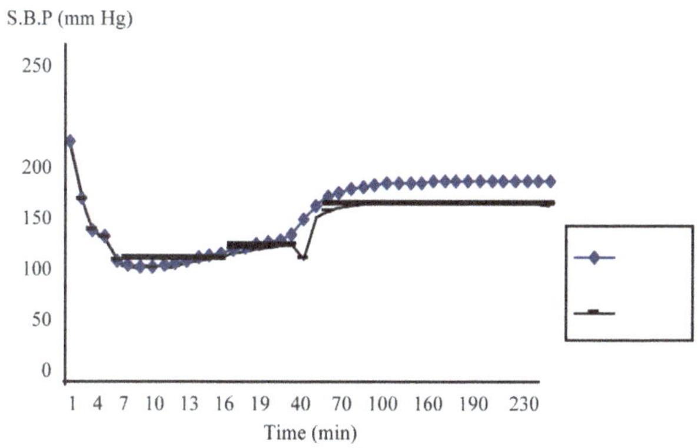

Figure 6. Variation of the systolic blood pressure (S.B.P) for two values of glomerular filtration rate (Γ_{gf}) during a dialysis session (Γ_{gf} = 75 ml/min ◆, Γ_{gf} = 30 ml/min ▬).

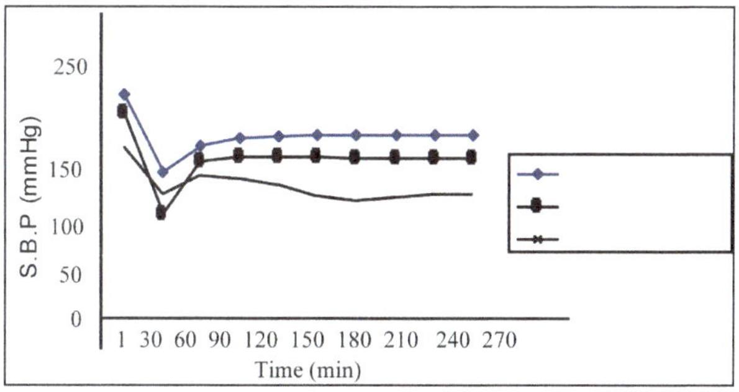

Figure 7. Comparing calculated systolic blood pressure (C.S.B.P) for two values of glomerular filtration rate (Γ_{gf}) to that measured (M.S.B.P) during a dialysis session (Γ_{gf} = 75 ml/min ◆, Γ_{gf} = 30 ml/min ▮, M.S.B.P*).

tient and the mirror of their normal functioning we can say that if validated for the systolic blood pressure the model can be judged satisfying.

A comparison between the trends of the five parameters variations calculated by the model (CO_2 concentration, CO_2 arterial blood pressure, baroreceptor activity, sympathetic activity and systolic blood pressure) shows a certain synchronization between them and that the same tendency is recorded. Which approve the right functioning of the model and its controlling role. In fact that synchronisation can be explained when reminding that blood pressure oscillates with a respiratory cycle, and that ventilation causes changes in intra-thoracic pressure. Those changes affect venous return, cardiac output, blood flows (brain, pulmonary and renal) and arterial pressure. This, in turn, activates both arterial and cardiopulmonary baroreceptors, inhibit vasoconstriction while increasing kidney filtration and renal blood flow. The kidney, help to regulate the blood pressure by increasing or decreasing the blood volume and also by the rennin-angiotension system. The baroreceptors, pressure sensitive nerve endings, are sensitive to the changes in arterial pressure due to changes in blood flow. The arterial baroreceptors are integrated with the autonomic nervous system and stimuli sent from the baroreceptors will either increase sympathetic tone or diminish sympathetic tone of the arterial system when the glomerular filtration rate is not in the normal band. In fact, the renal responses on the reduction of the total renal blood flow are characterized by sodium and water retention and, subsequently, rising cardiac filling pressures. This increase is mediated by sympathetic neurons which release norepinephrine in the vicinity of the authorythmic cells.

So the frequency band corresponding to sympathetic activity is deviated from its normal value. It is assumed that the overall aim of renal auto regulation is to maintain a constant glomerular filtration rate despite large changes in afferent blood pressure. The detection of an eventual imbalance can be done only by detection of the total carbon dioxide concentration and the carbon dioxide arterial blood pressure at any compartment. In fact, changes in oxygen, carbon dioxide and hydrogen ion concentration have very potent effects on blood flow.

When oxygen supply does not meet the metabolism demand, then there is a build-up of metabolites and an increase in its production. This build-up of metabolites will then cause a reflex vasodilation to adjust blood flow and an increase in local flow that accompanies an increase in local metabolism. Which will influence the kidney/dialyser function and the status of the hemodialysis patient. The reason is that an increase in carbon dioxide reacts with water to form hydrogen ions and bicarbonate and an increased carbon dioxide is most probably exerted on the arterial side. Out of their normal boundaries (carbon dioxide concentration and carbon dioxide arterial blood pressure), we can say that there is dysfunction at any level so we must verify the dialysis dose, the time of dialysing, and the state of the membrane. We can also detect the eventual dysfunction if the frequency corresponding to the sympathetic activity of one organ is not in the normal band. In fact, progressive systolic dysfunction of the heart results in a diminishing cardiac output, increasing systemic vacular resistance, smaller effective circulating volume, lower arterial blood pressure, and, ultimately, a reduction of arterial blood pressure below the renal auto-regulatory capacity [20,21].

6. Conclusions

As a way to get better assessment of the hemodialysis patient during dialysis session and to control his status, we have developed a model combining between different levels and systems of the body and describes mathematically processes related to homodynamic, sympathetic nervous system, ventilation, cerebral blood flow, and renal/dialyser clearance. In fact, we have sub-devised the human body into five compartments (brain, lung, cardiovascular system, body tissue, kidney/dialyser) and we have made integration over all of them. This integration is backed by the argument that blood pressure is controlled by several physiological mechanisms acting in combination and that the body compartments work as entity in a closed circuit and every one uses the parameters furnished by the other. So we have applied for every compartment the equations of transfer after considering the usual relationships between flow, concentration and pressure. We have obtained, then, a set of simple differential equations for which we have tried to give analytical solutions. They permit us to quantify the change that can undergo the carbon dioxide arterial blood pressure, the systolic blood pressure, the blood flow and the sympathetic nerve activity in each body compartment when there is any anomaly in the kidney/dialyser system. To test the running of the model, we have made the calculus by considering a unique system embedding the four compartments and by treating the kidney dialyser compartment as regulator of all the system. So if we change a parameter in the kidney/dialyser compartment, we must record a change for the parameters indicators of the functioning of the other compartments with a proportional sensibility.

The parameter indicator of the kidney/dialyser efficiency for which we have made a change is the glomerular filtration rate. We have recorded a clear discard between the two situations for the parameters indicators of the system functioning. The rate of change recorded is about 16.9% for the CO_2 arterial blood concentration, 47.8% for the CO_2 arterial blood pressure, 7.8% for the baroreceptor activity, 4.5% for the sympathetic nerve activity and 4.8% for the systolic blood pressure. To validate the model, we have compared the systolic blood pressures calculated by the model with those measured during a dialysis session taken from bibliography. The global walk of the three curves is the same. A complete validation of the model necessitates to compare the other parameters (sympathetic nerve activity, CO_2 arterial blood pressure and concentration, baroreceptor activity) with those measured, which means we have to propose relationships that govern the measurable parameters, makes predictions, perturbs the object, and observes the fidelity with which the model describes the behaviour of the system.

Finally, we can say that this approach, if completely validated, could prove to be utile to workers undertaking physiologic modeling in a variety of area and it provides practical examples of use in a clinical setting. In fact the model presented can be used to prepare a clinical practice guidelines that present useful equations to the laboratory assessment and as a framework within which clinical phenomena can be better understood by searching ideal biomarkers for acute kidney injury and to evaluate their accuracy as diagnostic tests for the risk identification phase, surveillance phase, diagnosis phase and prognosis phase. Those biomarkers, formulated in the text, would help clinicians and scientists diagnose the most common form of acute kidney injury in hospitalised patients, early and accurately and may aid to risk-stratify patients with acute kidney injury by predicting the need for renal replacement therapy, the duration of acute kidney injury, the length of stay and mortality. So it can participate in differentiating between patients who would benefit from early initiation of renal replacement therapy and those from whom support should be withheld or withdrawn.

REFERENCES

[1] H. J. Kemp, A. Pham and C. R. Tomson, "Urea Kinetic Modelling: A Measure of Dialysis Adequacy," *Annals of Clinical Biochemistry*, Vol. 38, 2001, pp. 20-27.

[2] R. W. De Boer, J. M. Karemaker and J. Strackee, "Hemodynamic Fluctuation and Baroreflex Sensitivity in Humans: A Beat-to-Beat Model," *American Journal of Physiology-Heart and Circulatory Physiology*, Vol. 253, No. 3, 1987, pp. H680-H689.

[3] K. Kotani, K. Takamasu, Y. Ashkenazy, H. Eugene Stanley and Y. Ya mamoto, "Model for Cardio Respiratory Synchronization in Humans," *Physical Review E*, Vol. 65, 2002, pp. 1-9.

[4] H. Seidel and H. Herzel, "Bifurcations in a Non Linear Model of the Baroreceptor Cardiac Reflex," *Physica D: Non Linear Phenomena*, Vol. 115, No. 1-2, 1998, pp. 145-160.

[5] L. Chiari, G. Avanzolini, F. Grandi and G. Gnudi, "A Simple Model of the Chemical Regulation of Acid-Base Balance in Blood," *Proceedings of 16th International Conference IEEE-EMBS*, Baltimore, 3-5 November 1994, pp. 1025-1026.

[6] L. Chiari, G. Avanzolini and M. Ursino, "Acid-Base Regulation in Hemodialysis Patients Modeled by a Simple Respiratory Chemostat," *18th Annual International Conference of the IEEE Engineering in Medicine and Biology Society*, Amsterdam, 1996, pp. 1810-1812.

[7] F. Grandi, L. Chiari, A. Cappello and G. Avanzolini, "A New Analytic Model of UREA Kinetic Oriented to Pa-

rameter Estimation," *International Conference of IEEE-EMBS*, Maryland, 3-5 November 1994, pp. 1164-1165.

[8] M. Ursino, L. Coli, C. Brighenti, L. Chiari, A. De Pascalis and G. Avanzolini, "Predictioof Solute Kinetcis, Acid-Base Status and Blood Volume Changes during Profiled Hemodialysis," *Annals of Biomedical Engineering*, Vol.28, No. 2, 2000, pp. 204-216.

[9] M. C. Khoo, R. E. Kronauer, K. P. Strohl and A. S. Slutsky, "Factors Inducing Periodic Breathing in Humans. A General Model," *Journal of Applied Physiology: Respiratory, Environmental & Exercise Physiology*, Vol. 53, No. 3, 1982, pp. 644-659.

[10] M. C. Khoo, "Modelling and Parameter Estimation in Respiratory Control," Academic Press, New York, 1977.

[11] S. Timischl, "A Global Model for the Cardiovascular and Respiratory System," Ph.D. Thesis, Unversität Graz, Graz, 1998, p. 109.

[12] A. S. Levey, J. P. Bosch, J. B. Lewis, T. Greene, N. Rogers and D. Roth, "A More Accurate Method to Estimate Glomerular Filtration Rate from Serum Creatinine: A New Prediction Equation," *Annals of Internal Medicine*, Vol. 130, No. 6, 1999, pp. 461-470.

[13] J. T. Daugirdas, "Chronic Hemodialysis Prescription: A Urea Kinetic Approach," In: *Handbook of Dialysis*, Little, Brown and CO., Boston, 1994, pp. 92-120.

[14] F. A. Gotch, "*Kt/V* Is the Best Dialysis Dose Parameter," *Blood Purification*, Vol. 18, No. 4, 2000, pp. 276-285.

[15] L. A. Stevens and A. S. Levey, "Clinical Implications of Estimating Equations for Glomerular Filtration Rate," *Annals of Internal Medicine*, Vol. 141, No. 12, 2004, pp. 959-961.

[16] P. A. Stewart, "How to Understand Acid-Base: A Quantitative Acid-Base Primer for Biology and Medicine," Elsevier North Holland Publication, city, 1981.

[17] V. Bordin, C. Catalano, A. Berto, G. Silvestrin, D. Di Landro and F. Fabbian, "Determining the Prescription for Online Predilution Hemofiltration: Our Experience and a Review of the Literature," *Dialysis and Transplantation*, Vol. 31, No. 6, 2002.

[18] B. A. Molitoris, V. Y. Melnikov, M. D. Okusa and J. Himmelfarb, "Technology Insight: Biomarker Development in Acute Kidney Injury: What Can We Anticipate?" *Nature Clinical Practice Nephrology*, Vol. 4, No. 3, 2008, pp. 1-8.

[19] C. Silvio, C. Andrea, S. Stefano, B. Fabio, B. Stefano, G. Andrea, C. Leonardo and S. Antonio, "Model-Based Study of the Effects of the Hemodialysis Technique on the Compensatory Response to Hypovolemia," *Kidney International*, Vol. 65, 2004, pp. 1499-1520.

[20] S. C. Malpas, "Sympathetic Nervous System Overactivity and Its Role in the Development of Cardiovascular Disease," *Physiological Reviews*, Vol. 90, No. 2, 2010, pp. 513-557.

[21] R. Zhang, J. A. H. R. Claassen, S. Shibata, S. Kilic, K. Martin-Cook, R. Diaz-Arrastia and B. D. Levine, "Arterial-cardiac. Baroreflex Function: Insights from Repeated Squat-Stand Maneuvers," *American Journal of Physiology-Regulatory, Integrative and Comparative Physiology*, Vol. 297, 2009, pp. R116-R123.

Mechanical Analysis of Viscous-Elastic Fluid Acting on Residual Oil in the Micro Pore

Lili Liu[1], Chao Yu[1], Lihui Wang[2], Chengchuyue Fu[1], Peixiang Li[3]
[1]Department of Petroleum Engineering, Northeast Petroleum University, Daqing, China
[2]Daqing Petroleum Equipment Group, Daqing, China
[3]Technical Development Department, The 2nd Oil Production Factory of Daqing Oilfield Company Limited, Daqing, China

ABSTRACT

In order to analyze the normal deviatoric stress that viscous-elastic fluid acting on the residual oil under the situation of different flooding conditions and different permeabilities, Viscous-elastic fluid flow equation is established in the micro pore by choosing the continuity equation, motion equation and the upper-convected Maxwell constitutive equation, the flow field is computed by using numerical analysis, the forces that driving fluid acting on the residual oil in micro pore are got, and the influence of flooding conditions, pore width and viscous-elasticity of driving fluid on force is compared and analyzed. The results show that: the more viscous-elasticity of driving fluid increases, the greater the normal deviatoric stress acting on the residual oil increases; using constant pressure gradient flooding, the lager the pore width is, the greater normal deviatoric stress acting on the residual oil will be.

Keywords: Viscous-Elastic Fluid; Residual Oil; Normal Deviatoric Stress; Micro Pore

1. Introduction

In the enhanced oil recovery, the viscous-elastic fluid flooding is a very important part. Many scholars have made a lot of efforts on viscous-elastic fluid enhancing oil recovery. By experimental studies Wang Demin [1] found that viscous-elastic polymer solution can improve flooding efficiency in micro pore; Yin Hongjun [2] studied that the rheology of viscous-elastic polymer solution in the dead end: Xia Huifen [3] discussed the microscopic mechanism of viscous-elastic polymer solution improving flooding efficiency. In the paper, the force of viscous-elastic fluid acting on residual oil is analyzed theoretically under different flooding conditions, different permeabilities, which lays the foundation for further analysis of deformation.

2. Calculation Model

In order to analyze the force acting on residual oil in micro pore of different width, a calculation model is established shown in **Figure 1**, Where H is the pore width, takes 20 μm, 40 μm, 60 μm respectively. The core is oil wet. The film height is 10 μm, the length is 40 μm.

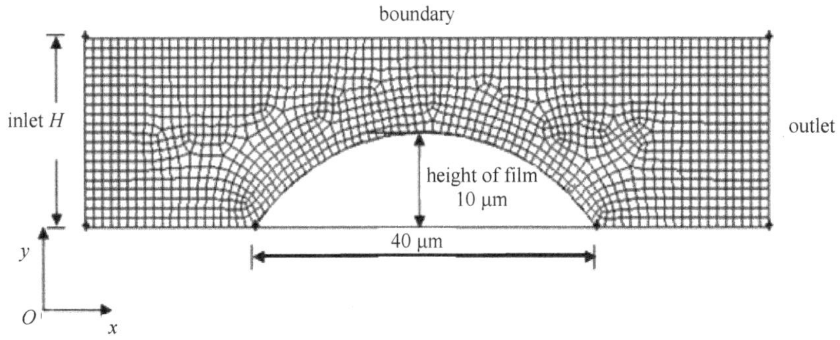

Figure 1. Calculation model.

The boundary conditions are setted: The inlet and an outlet, respectively, use a constant velocity (1e - 5 m/s), constant flow rate (2e - 10 m^3/s) and constant pressure gradient (0.02 MPa/m). The boundaries meet the no-slip condition and assume that the film is stationary.

3. Basic Equations

Viscous-elastic fluids have got complex rheological properties. Based on experiment research, the upper-convected Maxwell constitutive equation is suitable to describe the rheological properties of viscous-elastic fluid. The corresponding continuity equation, motion equation and constitutive equation [4] are:

Continuity equation:

$$\nabla \cdot \boldsymbol{u} = 0 \quad (1)$$

where: $\boldsymbol{u}$ is velocity vector.

Momentum equation:

$$\mathrm{d}\boldsymbol{u}/\mathrm{d}t = \boldsymbol{F} + \mathrm{div}\boldsymbol{P}/\rho \quad (2)$$

where: $\boldsymbol{F}$ is mass force; ρ is density; $\boldsymbol{P}$ is stress tensor.

$$\boldsymbol{P} = -p\boldsymbol{I} + \boldsymbol{T} \quad (3)$$

where $\boldsymbol{I}$ is unit tensor; $-p$ is static pressure; $\boldsymbol{T}$ is deviatoric stress tensor.

Constitutive equation:

$$\boldsymbol{T} + \lambda \overset{\nabla}{\boldsymbol{T}} = \eta \mathbf{A}_1 \quad (4)$$

where: λ is relaxation time, which is the characterization of fluid elasticity; η is shearing viscosity; A_1 is one order Rivlin-Ericksen tensor.

4. Calculation Results

The Weissenberger number is a dimensionless quantity which is used to describe fluid elasticity. The greater its value is, the stronger the fluid elasticity is. The formula is:

$$We = \lambda v/l$$

where We is the Weissenberger number; v is characteristic velocity; l is characteristic length.

4.1. The Change of Normal Deviatoric Stress Acting on Residual Oil under Different Flooding Conditions

For the same pore width (20 μm, for example), using constant velocity (1e - 5 m/s), constant flow rate (2e - 10 m^3/s) and constant pressure gradient (0.02 MPa/m), respectively. The normal deviatoric stress acting on the residue oil is calculated by different viscous-elastic fluid. As is shown in **Figure 2**, for the same pore width, the conditions of constant velocity and constant flow rate are the same.

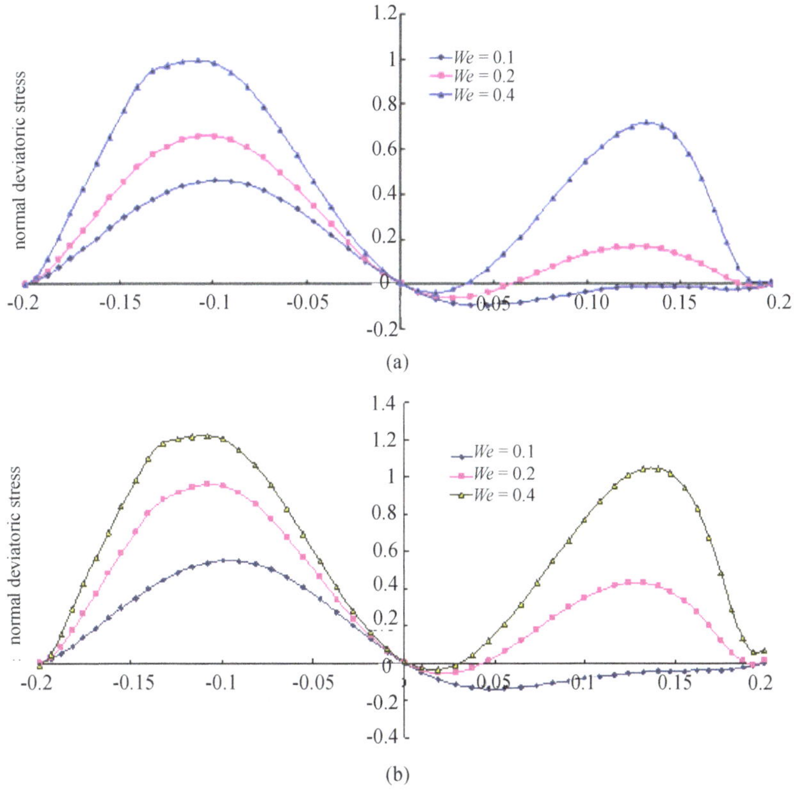

Figure 2. Normal deviatoric stress acting on the residue oil by viscous-elastic fluid of different *We* (pressure is positive, tension is negative); (a) Constant velocity (constant flow rate); (b) Constant pressure gradient.

It can be seen from **Figure 2**, no matter which of the conditions of constant velocity, constant flow rate or constant pressure gradient, the normal deviatoric stress increases as *We* increases. Which shows that the more viscous-elasticity of viscous-elastic fluid is, the more bigger normal deviatoric stress is.

4.2. The Change of the Normal Deviatoric Stress Acting on Residual Oil under Different Permeability Micro Pores

For the same *We*, micro pores of 20 μm 40 μm and 60 μm respectively, the forces acting on residual oil under the flooding conditions of constant velocity (1e - 5 m/s), constant flow rate(2e - 10 m^3/s) and constant pressure gradient (0.02 MPa/m), are calculated, which is shown in **Figure 3**.

It can be seen from **Figure 3**, when flooding with constant pressure gradient, with the increase of the pore width, the normal deviatoric stress increases. On the contrary, when flooding with constant flow rate and constant velocity, the normal deviatoric stress reduces with the pore width increases. Flooding with pressure gradient in the oilfield, the high permeability zone supports more force, residual oil deforms greater relatively and is more

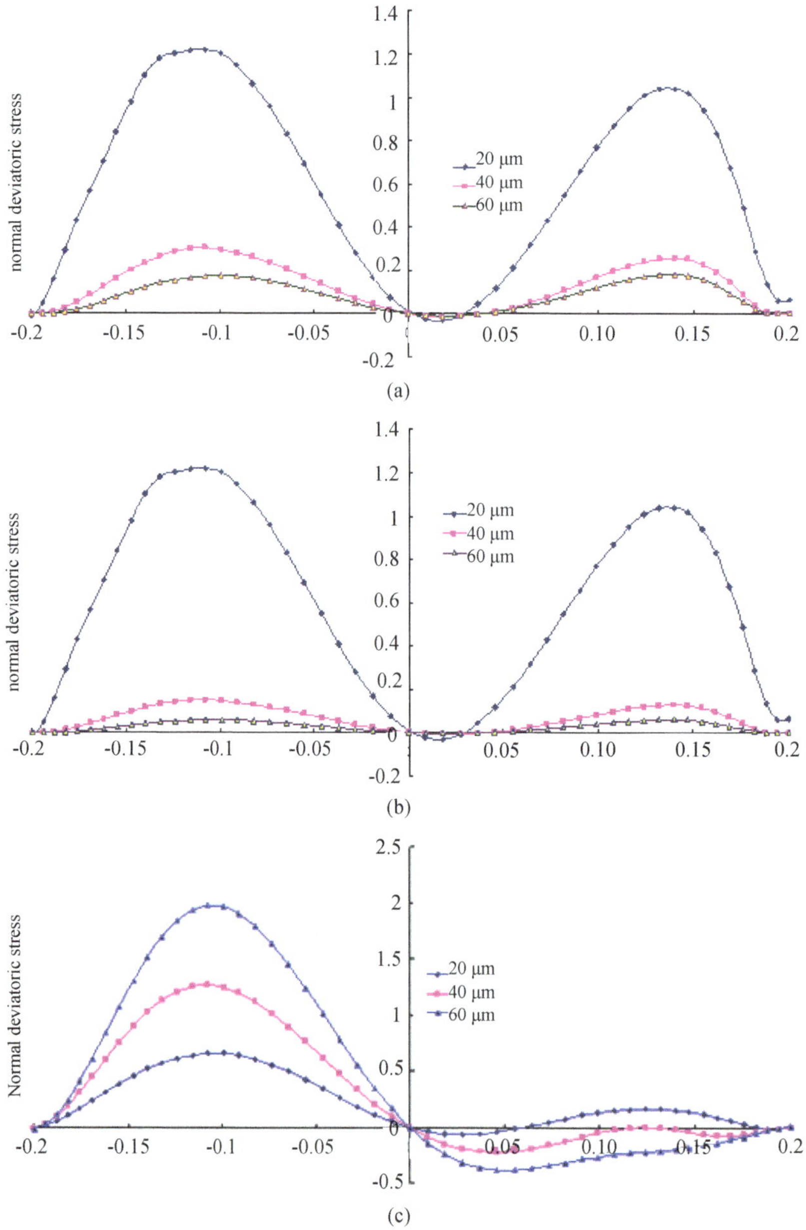

Figure 3. The normal deviatoric stress acting on the residual oil under different permeability micro pores (pressure is positive, tension is negative); (a) Constant velocity (*We* = 0.2); (b) Constant flow (*We* = 0.2); (c) Constant pressure gradient (*We* = 0.2).

susceptible to be displaced.

5. Conclusion

Using constant pressure gradient flooding, the greater the permeability is, the greater the force of residual oil acting by viscous-elastic fluid is; using constant flow rate or constant velocity flooding, the situation is just the opposite; as *We* increases, the force of viscous-elastic fluid acting on residual oil increases.

REFERENCES

[1] D. M.Wang, J. C. Cheng, Q. Y. Yang, *et al.*, "Viscous-Elastic Polymer Can Increase Micro Scale Displacement in Cores," SPE63227, 2000, pp. 2-8.

[2] H. J. Yin, D. M. Wang, H. Y. Zhong, *et al.*, "Study on Flow Behaviors of Viscoelastic Polymer Solution in Micro Pore with Dead End," SPE101950, 2006, pp. 4-15.

[3] H. F. Xia, D. M. Wang, Z. C. Liu, *et al.*, "Mechanism of Viscous-Elastic Polymer Solution to Improve Microscopic Flooding Efficiency," *Acta Petrolei Sinica*, Vol. 22, No. 4, 2001, pp. 60-65.

[4] S. F. Han, "Non-Newtonian Fluid Constitutive Equation and the Analytic Theory," Science Press, Beijing, 2002.

5

Numerical Simulation of Hydraulic Transport of Sand-Water Mixtures in Pipelines

Changhee Kim[1], Cheolheui Han[2*]
[1]Hyundai Engineering & Construction Company, Seoul, South Korea
[2]Department of Aeronautical and Mechanical Design, Korea National University of Transportation, Chungju, South Korea

ABSTRACT

The development of empirical model for the hydraulic transport of sand-water mixtures is important for the design of economical solid-liquid transportation system in chemical and waste-disposal industries. The hydraulic transport characteristics of sand-water mixtures in circular pipelines are numerically investigated by using the FLUENT commercial software. Eulerian granular multiphase (EGM) model with the k-e turbulent model is used for the computation. Present method is validated by the computed values with the measured data. The effect of the concentration and pipe sizes on the relative solid effect is numerically investigated. It is found that the effect of the volumetric delivered concentration on both hydraulic gradient and solid effect increases as the Reynolds number decreases. When the Reynolds number is small, the increase in the volumetric delivered concentration has an effect of decreasing the hydraulic gradient whereas the solid effect increases with the volumetric delivered concentration stepping up. The effect of the pipe diameter is not the critical parameter for deciding the values of the relative solid effect in the sand-water mixture transportation.

Keywords: Computational Fluid Dynamics; Hydraulic Transport; Solid Effect; Sand-Water Mixture

1. Introduction

Solid-liquid flow in pipelines has been a popular mode of transportation in chemical and waste-disposal industries. Circular pipes are normally used for long distance transportation of variety of materials in bulk quantities [1]. Both time and pumping power required for hydraulically transporting huge amounts of soils in a state of slurry (sand-water mixture) are crucial factors for successful dredging and reclaiming projects [2]. Thus, the economical hydraulic transportation of solids to a long distance has been attempted by using several methods [3,4].

During the several past decades, the complex computational scheme describing the microscopic processes in the solid-liquid flow and the computing costs has been a drawback to the development and use of commercial softwares. Ling *et al.* [5] investigated the double slurry flow in the pipe for the fully turbulent flow using Eulerian granular multiphase (EGM) model. Al Araby *et al.* [6] performed a numerical study on the single phase combined free and force convection in the entrance region of a horizontal pipe with its wall temperature fixed as a constant value. Lin and Ebadian [7] used a simplified three-dimensional algebraic slip mixture (ASM) model [8] and the RNC k-e turbulent model [9] while focusing on the developing process of volume of fraction and density distributions, mean velocity profiles.

Kim *et al.* [10] experimentally investigated the hydraulic transport characteristics of sand-water mixtures in circular and square pipelines by changing the Reynolds number and volumetric delivered concentration.

In the present paper, the numerical computation results are presented by first validating the numerical results with the measured data. The present study is also mainly focused on the heterogeneous and homogeneous flow regimes. The effect of the concentration and the geometrical changes on the relative solid effect will be discussed.

2. Numerical Method

2.1. Governing Equations

The continuity equation for the mixture is

$$\frac{\partial}{\partial t}\left(\alpha_q \rho_q\right) + \nabla \cdot \left(\alpha_q \rho_q \boldsymbol{v}_q\right) = \sum_{p=1}^{n} \dot{m}_{pq} \qquad (1)$$

*Corresponding author.

where α_q denotes the volume fraction of single material q among the materials that constitutes the mixture, ρ_q is the density of a material q, $\dot{m}_{pq}$ $\dot{m}_{pq}$ is the mass flow rate transferred from the pth material to the qth material.

The momentum equation for the single material q can be expressed as

$$\frac{\partial}{\partial t}(\alpha_q \rho_q \mathbf{v}_q) + \nabla \cdot (\alpha_q \rho_q \mathbf{v}_q \mathbf{v}_q) = -\alpha_q \nabla p + \nabla \cdot \overline{\overline{\tau}}_q + \sum_{p=1}^{n} (\mathbf{R}_{pq} + \dot{m}_{pq} \mathbf{v}_{pq}) + \alpha_q \rho_q (\mathbf{F}_q + \mathbf{F}_{lift,q} + \mathbf{F}_{vm,q}) \quad (2)$$

where the stress-strain tensor of a material q, $\overline{\overline{\tau}}$, is

$$\overline{\overline{\tau}} = \alpha_q \mu_q (\nabla \mathbf{v}_q + \nabla \mathbf{v}_q) + \alpha_q \left(\lambda_q - \frac{2}{3} \mu_q \right) \nabla \cdot \mathbf{v}_q \overline{\overline{I}} \quad (3)$$

where μ_q and λ_q is the shear and bulk viscosity of the material q, respectively. $\mathbf{F}_q$ denotes the external force, $\mathbf{F}_{lift,q}$ is the lift force and $\mathbf{F}_{vm,q}$ is the virtual mass force. $\mathbf{R}_{pq}$ represents the interaction among the materials, p is the static pressure.

$$\frac{\partial}{\partial t}(\alpha_q \rho_q \mathbf{v}_q) + \nabla \cdot (\alpha_q \rho_q \mathbf{v}_q \mathbf{v}_q) = -\alpha_q \nabla p + \nabla \cdot \overline{\overline{\tau}}_q + \alpha_q \rho_q \mathbf{g} + \alpha_q \rho_q (\mathbf{F}_q + \mathbf{F}_{lift,q} + \mathbf{F}_{vm,q}) + \sum_{p=1}^{n} \left(K_{pq} (\mathbf{v}_p - \mathbf{v}_q) + \dot{m}_{pq} \mathbf{v}_{pq} \right) \quad (4)$$

$$\frac{\partial}{\partial t}(\alpha_s \rho_s \mathbf{v}_s) + \nabla \cdot (\alpha_s \rho_s \mathbf{v}_s \mathbf{v}_s) = -\alpha_s \nabla p - \nabla p_s + \nabla \cdot \overline{\overline{\tau}}_s + \alpha_s \rho_s \mathbf{g} + \alpha_s \rho_s (\mathbf{F}_s + \mathbf{F}_{lift,s} + \mathbf{F}_{vm,s}) + \sum_{l=1}^{n} \left(K_{ls} (\mathbf{v}_l - \mathbf{v}_s) + \dot{m}_{ls} \mathbf{v}_{ls} \right) \quad (5)$$

where K_{ls}, K_{sl} are the coefficients that represent the momentum transfer between the fluid and the soild, and can be denoted as follows

$$K_{ls} = \frac{3(1+e_{ls}) \left(\frac{\pi}{2} + C_{frls} \frac{\pi^2}{8} \right) \alpha_s \rho_s \alpha_l \rho_l (d_l + d_s)^2 g_{o,ls}}{2\pi (\rho_l d_l^3 + \rho_s d_s^3)} |\mathbf{v}_l - \mathbf{v}_s| \quad (5)$$

The shear stress of the solid includes the shear viscosity (μ_s) due to the collision and transfer among the particles and the bulk viscosity (λ_s), and can be written as follows:

$$\mu_s = \mu_{s,col} + \mu_{s,kin} + \mu_{s,fr} \quad (6)$$

where $\mu_{s,col}$ is the component due to collision, $\mu_{s,kin}$ is the component due to motion, and $\mu_{s,fr}$ is the frictional component. In the present study, the k-e turbulence model is used for the eddy viscosity modeling.

2.2. Boundary Conditions

In the present study, the mean velocity inlet boundary condition and pressure outlet boundary condition are imposed on the inlet and the outlet of the pipeline, respectively. The turbulence intensity level is set to 4% for the intermediate velocities in the sand-water mixture flows. The pipe length L was set to 100D. The density of water is 998.2 kg/m^3. The specific gravity (Ss) and diameter (d_{50}) of Jumunsin sands are 2.65 and 0.54 mm, respectively [10]. Water temperature is set to 20°C.

A multi-block unstructured, non-uniform grid system with hexahedral elements is used to discretize the computation domain. Using the symmetry boundary condition, the computational domain is set to the half of the physical domain. Heat transfer is neglected with the steady state slurry flow assumption. The second-order upwind and central difference are selected as the discretization schemes of both convection and diffusion terms, respectively. With the SIMPLEC algorithm for the coupling between the velocity and pressure, under-relaxation technique is applied in the dependent variables.

3. Results and Discussion

In order to validate the present method computed results are compared with measured data. **Figure 1** shows the velocity profiles from the bottom of the circular pipe to the top. Present results are compared with the measured data in [11]. The diameters of the pipe and the sand in the experiment are 51.5 mm and 0.165 mm, respectively. The volumetric delivered concentration (C_{vd}) is 8.4%. Though the present computation slightly over-predicts at the center and under-predicts at the point where the flow velocity has the maximum value, the present results agree with the measured data.

Figure 2 shows the change of the hydraulic gradient due to the changes in the Reynolds number and the volumetric delivered concentration (C_{vd}). The amount of solids delivered inside the pipeline is represented by the volumetric delivered concentration (C_{vd}) that is calculated as the ratio between solid and slurry flow rates (Q_s / Q_m). The Reynolds number is defined as $V_m D_h \rho_w / \mu_w$, where D_h and μ_w represent the hydraulic diameter of the pipes and the viscosity of the water, respectively. Mean velocity (V_m) is a bulk velocity of the mixture defined as the volumetric flow rate of a matter passing through a pipeline, Q_m, divided by the cross-sectional area, A.

$$V_m = \frac{Q_m}{A} = \frac{4Q_m}{D^2} \quad (7)$$

where Q_m equals the sum of Q_s and Q_f that represent the

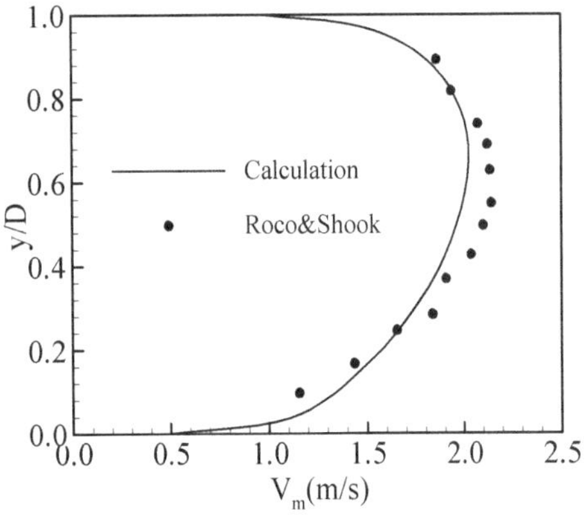

Figure 1. Comparison of velocity profile at C_{vd} = 8.4%, d = 0.165 mm and D = 51.5 mm.

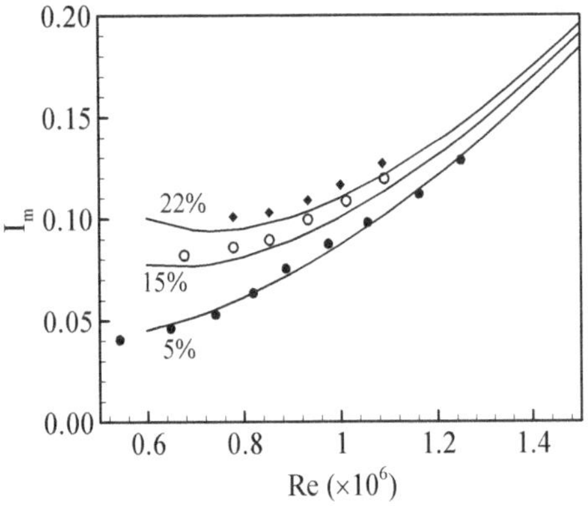

Figure 2. Concentration (C_{vd}) effect to hydraulic gradient (I_m) at d = 0.54 mm and D = 200 mm.

flow rates of the solid particles and the fluid, respectively. In this paper, the definition of the Reynolds number is also used for water and slurry flows together. Present results are compared with the measured data [2]. The diameters of the pipe and the sand in the experiment are 150 mm and 0.54 mm, respectively. It can be seen from the figure that present results agree well with the measured data when the Reynolds are within the range from 0.7×10^6 to 1.3×10^6.

Figure 3 shows the relation of the volumetric delivered concentration with the hydraulic gradient (I_m) in the given Reynolds number. The energy loss due to friction can be assessed by using hydraulic gradient (I_m) that means the pressure gradient for a mixture expressed in height of water per length of line.

$$I_m = \frac{\Delta p}{\rho_f g L} \quad (8)$$

As shown in **Figure 3**, the effect of the volumetric delivered concentration on hydraulic gradient increases as the Reynolds number decreases. When the Reynolds number is small, the increase in the volumetric delivered concentration has an effect of decreasing the hydraulic gradient.

Figure 4 shows the relation of the volumetric delivered concentration with solid effect $(I_m - I_w)$ in the given Reynolds number. In evaluating additional friction loss due to the solid particles, solid effect represented by $(I_m - I_w)$ is used. I_w means the friction gradient for water alone at the flow rate equal to the mixture flow rate (Q_m). **Figure 4** shows that the effect of the volumetric delivered concentration on solid effect increases as the Reynolds number decreases. When the Reynolds number is small, the solid effect increases with the volumetric delivered concentration stepping up. Thus, it can be said that, when the flow speed is low, the friction among the solid particles becomes a dominant factor, which results in the increase in the energy loss due to friction. When the flow speed is high, the solid particle distributions become uniform and the solid particles flow through the pipe with less probability of settling on the bed. Thus, the friction between the solid particles with the wall or deposited solid particles becomes small.

Figure 5 shows the effect of the solid particle size on the relative solid effect as a function of the Reynolds number $(V_m D_h \rho_w / \mu_w)$. It can be seen from the figures that the large size solid particles has the large relative

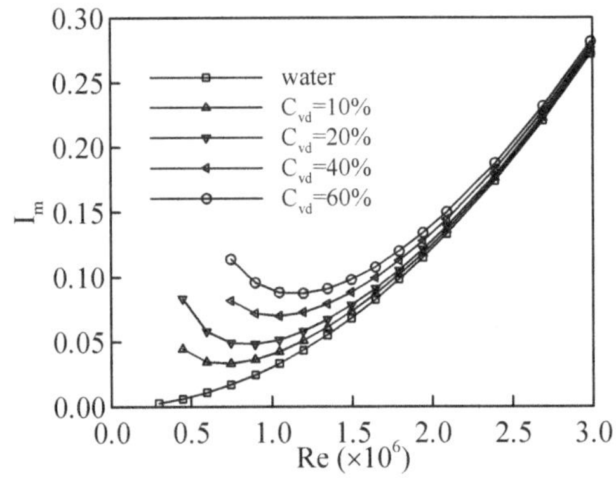

Figure 3. Concentration (C_{vd}) effect on both hydraulic gradient (I_m) hydraulic gradient (I_m).

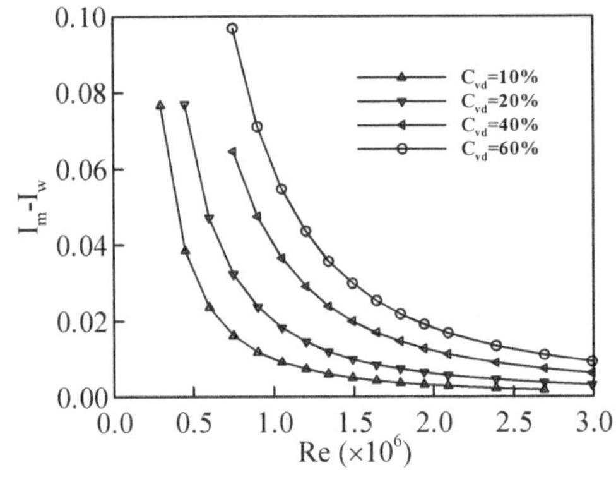

Figure 4. Concentration (C_{vd}) effect on both hydraulic gradient (I_m) and solid effect ($I_m - I_w$).

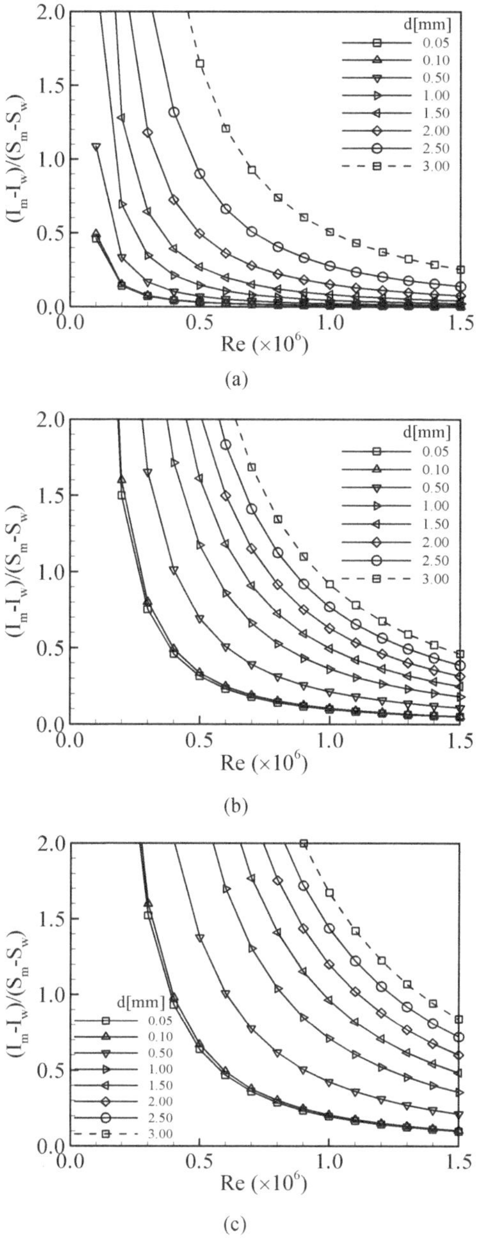

Figure 5. Correlation of data with relative solid effect. (a) *D* = 100 mm; (b) *D* = 400 mm; (c) *D* = 600 mm.

solid effect. However, when the solid particle size is very small ($d < 0.1$ mm), the particle size does not affect the value of the relative solid effect. Wilson *et al.* [12] found the following relationship.

$$\frac{I_m - I_w}{S_m - S_w} \approx 0.22\left(\frac{V_m}{V_{50}}\right)^{-M} \quad (9)$$

where 0.22 represents the value of the relative solid effects when $V_m = V_{50}$. The magnitude of M is a function of normal stress of solids (granular pressure). In Equation (9), $S_w = 1$ because the working fluid is water. Following Equation (9), we can represent the relative solid effect as a function of the Reynolds number.

$$\frac{(I_m - I_w)}{(S_m - S_w)} = A \cdot \mathrm{Re}^{-M} \quad (10)$$

According to the estimation of Clift *et al.* [13], the value of M is set to 1.7. It can be said from the figures that the relative solid effect is affected more by the particle size than the pipe diameter.

The effect of the pipe diameter on the relative solid effect can be represented as follows [14]:

$$\frac{(I_m - I_w)}{(S_m - S_w)} = A \cdot \left(\frac{V_m}{U'_u}\right)^{-M} \quad (11)$$

Table 1 and **Figure 6** show the effect of non-dimensionalized particle size effect on the relative solid effect as a function of the Reynolds number. As shown on **Table 1**, the non-dimensionalized particle size is set to 0.0025 for the investigated 5 cases. The value of M is also set to 1.7, then the value of A is obtained by computing the relative solid effect with the change in the mean velocity. It can be seen from the **Table 1** and **Figure 6**, that, even though the non-dimensionalized particle size has the same value, the relative solid effect has different values. In case the diameters of both pipe and solid particle are large, the values of the relative solid effect are larger than the other cases. Thus, it can be said from the results in **Figure 6** that the size of the pipe diameter compared to the solid particle size is not the dominant factor. The particle size itself is a critical parameter to the performance of the slurry transportation system.

4. Conclusions

In this paper, the effect of the concentration, and both particle and pipe sizes on the relative solid effect is numerically investigated using Eulerian granular multiphase (EGM) model and k-e turbulent model. Present method is validated by the computed values with the measured data.

It is found that the effect of the volumetric delivered concentration on both hydraulic gradient and solid effect increases as the Reynolds number decreases. When the Reynolds number is small, the increase in the volumetric delivered concentration has an effect of decreasing the

Table 1. Correlation coefficient at *d*/*D* = 0.0025.

	D (mm)	*d* (mm)	*d*/*D*	*A*	*M*
Case 1	200	0.5	0.0025	0.1288	1.7
Case 2	400	1	0.0025	0.2217	1.7
Case 3	600	1.5	0.0025	0.2883	1.7
Case 4	800	2	0.0025	0.3535	1.7
Case 5	1000	2.5	0.0025	0.4150	1.7

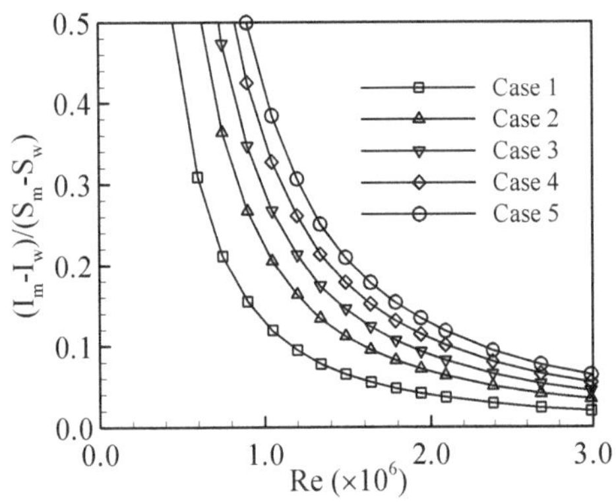

Figure 6. Correlation of data with relative solid effect ($(I_m - I_w)/(S_m - S_w)$).

hydraulic gradient whereas the solid effect increases with the volumetric delivered concentration stepping up.

The effect of the particle size and the pipe diameter on the relative solid effect is calculated and correlated using the previously developed simple model. It is found that the effect of the pipe diameter or the relative size of the particle size with regard to the pipe diameter is not the critical parameter for deciding the values of the relative solid effect in the sand-water mixture transportation. The particle size itself is shown to be the most important parameter when the concentration is fixed.

In future, the effect of the particle size on the deposition limit velocity will be discussed.

REFERENCES

[1] D. R. Kaushal and Y. Tomita, "Comparative Study of Pressure Drop in Multisized Particulate Slurry Flow through Pipe and Rectangular Duct," *International Journal of Multiphase Flow*, Vol. 29, No. 9, 2003, pp. 1473-1487.

[2] V. Matousek, "Flow Mechanism of Sand-Water Mixtures in Pipelines," Delft University Press, Delft, 1997.

[3] M. E. Charles, "Transport of Solids by Pipeline," *Proceedings of Hydrotransport* 1, *BHRA Fluid Engineering*, Cranfield, England, Paper A3.

[4] G. K. Patterson, J. L. Zakin and J. M. Redriguez, "Drag Reduction: Polymer Solutions, Soap Solution and Particle Suspensions in Pipe Flow," *Industrial Engineering of Chemistry*, Vol. 61, No. 1, 1969, pp. 22-35.

[5] J. Ling, C. X. Lin and M. A. Ebadian, "Numerical Investigation of Double-Species Slurry Flow in a Straight Pipe Entrance," *ASME* 2002 *International Mechanical Engineering Congress and Exposition, Heat Transfer, Vol.* 4, New Orleans, 17-22 November 2002, pp. 145-154.

[6] M. A. Al Araby, M. K. Shaban, A. S. Salem and M. M. Mahmoud, "Laminar Combined Free & Forced Convection Heat Transfer in the Entry Length of a Horizontal Pipe," *Proceedings of the ASME Heat Transfer/Fluid Engineering Summer Conference, HT/FED* 2004, *Vol.* 1, Charlotte, 11-15 July 2004, pp. 541-551.

[7] C. X. Lin and M. A. Ebadian, "A Numerical Study of Developing Slurry Flow in the Entrance Region of a Horizontal Pipe," *Computers and Fluids*, Vol. 37, No. 8, 2008, pp. 965-974.

[8] M. Manninen, V. Taivassalo and S. Kallio, "On the Mixture Model for Multiphase Flow," Technical Research Center of Finland, Espoo, VIT Publications 288, 1996.

[9] B. E. Launder and D. B. Spalding, "The Numerical Computation of Turbulent Flows," *Computer Method in Applied Mechanical Engineering*, Vol. 3, No. 2, 1974, pp. 269-289.

[10] C. H. Kim, M. S. Lee and C. Han, "Hydraulic Transport of Sand-Water Mixtures in Pipelines, Part I Experiments," *Journal of Mechanical Science and Technology*, Vol. 22, No. 12, 2008, pp. 2534-2541.

[11] M. V. Roco and C. A. Shook, "Modeling of Slurry Flow: The Effect of Particle Size," *Canadian Journal of Chemical Engineering*, Vol. 61, No. 4, 1983, pp. 494-503.

[12] K. C. Wilson, "Influence of Particle Properties on Solid Effects," *Proceedings of the* 10*th International Kol. Massenguttransport Druch Rohrleitungen*, University. GH Paderborn, Meschede, Germany, 1992.

[13] R. Clift, K. C. Wilson, G. R. Addie and M. R. Cartens, "A Mechanistically-Based Method for Scaling Pipeline Tests for Settling Slurries," *Proceedings Hydrotransport* 8. *BHRA Fluid Engineering*, Cranfield, 1982, pp. 91-101.

[14] G. R. Addie and J. R. Hammer, "Pipeline Head Loss Test of Settling Slurries at Low Velocities," *International Mechanical Engineering Congress*, Perth Western, 1994, pp. 225-230.

6

Performance Evaluation of Heat Exchangers in OTEC Using Ammonia/Water Mixture as Working Fluid

Takafumi Morisaki, Yasuyuki Ikegami
Institute of Ocean Energy, Saga University, Saga, Japan

ABSTRACT

The ocean thermal energy conversion (OTEC) system is a promising solution to provide stable electricity supply. Although the available temperature difference in OTEC systems is small, an ammonia/water mixture as working fluid is expected to decrease irreversible losses in the heat exchangers and to improve system performance. However, in actual heat exchangers, an adequate temperature crossing does not occur in the condenser but in the evaporator. Therefore, clarification of this characteristic is important. To date, the logarithmic temperature difference (LMTD) method is used in performance evaluations of OTEC heat exchangers. This method is of limited use if physical properties of fluids vary. A generalized mean temperature difference (GMTD) method is introduced to perform this evaluation. As changes in fluid property values can be considered in the GMTD method, method dependencies on heat exchanger characteristics, effectiveness, and system characteristics can be studied. In particular, GMTD and LMTD using a pure substance were found to be almost equal. Mean temperature differences using mixtures as working fluid were higher in the evaporator, but lower in the condenser, from the GMTD method than from the LMTD method. For higher ammonia concentrations in ammonia/water mixtures, the mean temperature differences from both methods are different.

Keywords: Ammonia/Water Mixture; Evaporation and Condensation Pressure; OTEC; GMTD; LMTD

1. Introduction

Energy and environmental problems are becoming more critical; the development of natural energy is desired. Ocean thermal energy conversion (OTEC) using temperature differences between warm surface seawater and cold deep seawater can supply stable electrical power [1] integrated into a variety of applications, such as seawater desalination, residential cooling, and aquaculture [2-6]. In addition, the beneficial effects of OTEC systems on global warming and energy shortage problems seem promising. Oceans have a huge amount of stored thermal energy, although, the energy density is low. As an OTEC system operates in conditions of small temperature differences, an appropriate working fluid must have a low boiling point if it is to be used in a closed-cycle OTEC. The OTEC system requires warm water, cold seawater, and a working fluid pump power; the ratio of the warm and cold water pump power to output power is comparatively higher than that of conventional fossil power plants.

Research is continuing on the Rankine cycle which is the conventional closed-cycle process exploiting low-boiling-point working fluids such as ammonia. Using a pure substance as working fluid, the thermal efficiency of the OTEC cycle increases with decreasing irreversible losses in the cycle or increasing effective temperature difference, specifically the difference between evaporating and condensing temperatures of the working fluid. For that reason, it is necessary to decrease energy losses in the heat exchangers, namely to decrease the temperature difference between seawater and working fluid. This is done by increasing the heat transfer area or the overall heat transfer coefficient. Using non-azeotropic mixtures such as ammonia/water as working fluids in the OTEC cycle was proposed by Kalina in 1982 [7]. Uehara *et al.* [8] indicated that reducing the irreversible losses in heat exchangers using non-azeotropic mixtures leads to improvements in the thermal efficiency of the cycle. Furthermore, they focused on the Kalina cycle, and presented a new cycle using the ammonia/water mixture [9].

Several theoretical and experimental investigations have reported that thermal efficiency of such Kalina cycles is higher than that of the Rankine cycle using pure ammonia [10-15]. Other studies have reported that the effective temperature difference and heat transfer coefficient decrease with changes in concentration in the boundary layer during evaporation and condensation [16-18]. Previous studies have investigated the operating conditions and equipment of an experimental apparatus using ammonia/water mixtures to improve system performance. Ikegami *et al.* [19] performed experiments to investigate the influence of system performance on ammonia concentration in the mixture. The experiments were performed to assess the stability of the ammonia/water mixture power cycle [20]. The results from these experiments showed that it is critically important to enhance the performance of the condenser in increasing the performance of the total OTEC system. Furthermore, the characteristics of the ammonia/water mixture cycle have not as yet been thoroughly clarified experimentally.

To enhance the ammonia/water mixture cycle, clarification of the heat exchanger performance is extremely important. The traditional and effectual evaluation method of heat exchanger performance is the logarithmic mean temperature difference (LMTD) method but application of this method is limited to heat exchangers using working fluids having constant heat transfer coefficient and thermal properties. The generalized mean temperature difference (GMTD) method however enables variations in thermal properties in heat exchange processes to be included. Utamura *et al.* [21] first proposed the GMTD method, and applied it to evaluate the heat exchange process between hot water and supercritical carbon dioxide as a working fluid. As a result, validity of the GMTD method was proved. The GMTD is believed applicable in the evaluation of heat exchangers using ammonia/water although properties of the mixture are quite distinct. It is not clear whether the GMTD method is valid for heat exchangers using non-azeotropic mixtures as working fluid.

In this paper, clarification is obtained on the characteristics of heat exchanger using ammonia/water mixtures in improving cycle performance, achieved with the available methods of evaluation heat exchangers. A comparison is made between LMTD and the recently developed GMTD calculation methods as performed on the heat exchanger using ammonia/water mixture and the applicability of the GMTD method is assessed.

2. Temperature Change in Heat Exchangers with Ammonia/Water Mixture

In **Figures 1(a)** and **(b)**, the distributions of the fluid temperature within the heat exchanger is given as a function of distance (Δx) and heat flow rate (ΔQ), respectively. The temperature variation of the ammonia/water mixture changes in the heat exchanger owing to the large difference between its boiling point and its dew point. Consequently, an increase in available exergy of the system is expected with this mixture as working fluid. From **Figure 1(a)**, the improvement in heat transfer performance and the ratios of the working fluid temperature to the heat source temperature in the evaporator and to the cold In **Figures 1(a)** and **(b)**, the distributions of the fluid

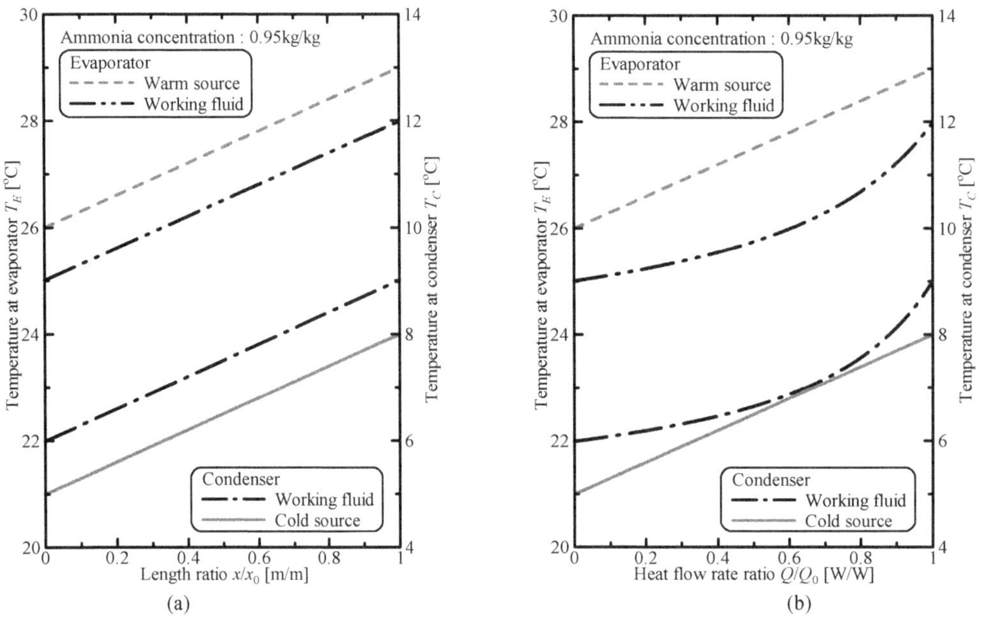

Figure 1. Temperature change in heat exchangers. Relationships between temperature and (a) length ratio and (b) heat-flow-rate ratio for evaporator and condenser.

temperature within the heat exchanger is given as a function of distance (Δx) and heat flow rate (ΔQ), respectively. The temperature variation of the ammonia/water mixture changes in the heat exchanger owing to the large difference between its boiling point and its dew point. Consequently, an increase in available exergy of the system is expected with this mixture as working fluid. From **Figure 1(a)**, the improvement in heat transfer performance and the ratios of the working fluid temperature to the heat source temperature in the evaporator and to the cold source temperature in the condenser are important when considering enhancing cycle performance. From **Figure 1(b)**, the working fluid temperature is a convex curve of the heat flow rate. The working fluid temperature approaches the warm water temperature in the inlet and outlet of evaporator but approaches the cold water temperature in the middle of the condenser. The pinch-point in particular is not well determined by the inlet and outlet temperatures in condenser. In addition, the traditional evaluation method underestimates heat transfer performance. Hence, the GMTD method, which accounts for the variation in the thermal properties, has been introduced for this performance evaluation. In contrast, for a pure ammonia working fluid, temperatures are constant in these heat exchangers because the thermal properties are stable.

2.1. Traditional Evaluation Method

The LMTD method has traditionally been used to evaluate the temperature difference between the heat source and the working fluid in a heat exchanger. The LMTD method is used only when the thermophysical properties of the local fluid in the heat exchanger and the overall heat transfer coefficient are constant. Hence, the fluid in the heat exchanger undergoes temperature changes as shown in **Figure 2** with the horizontal axis conceptually representing the length. The LMTD method is expressed as follows:

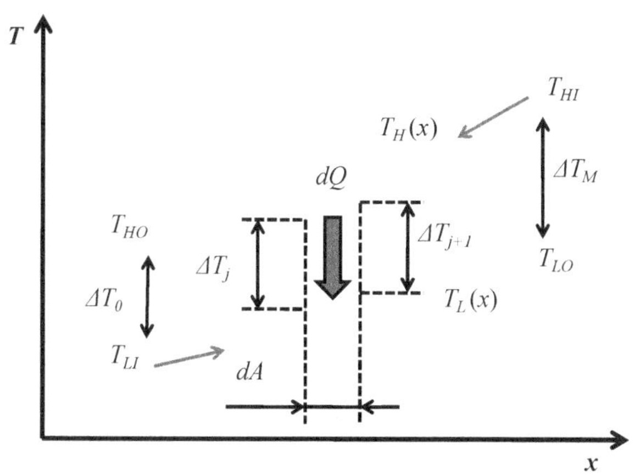

Figure 2. Fluid temperatures in the counter-flow heat.

$$\Delta T_{LMTD} = \frac{(T_{HI} - T_{LO}) - (T_{HO} - T_{LI})}{\ln\left[(T_{HI} - T_{LO})/(T_{HO} - T_{LI})\right]}. \quad (1)$$

where T_{HI}, T_{HO}, T_{LI} and T_{LO} are the high and the low temperature fluids heat exchanger inlet and outlet temperatures.

2.2. Innovated Evaluation Method

For an improved evaluation, it is necessary to consider variations in the thermophysical properties that better reflect conditions within actual heat exchangers. Therefore, the calculation results of LMTD are compared with those of GMTD and assessed. Using **Figure 2**, the heat flow rate for the heat exchanger, dQ from x to $x + dx$, is calculated using:

$$dQ = U_i dA_i \Delta T. \quad (2)$$

where U_i represents the overall heat transfer coefficient for the ith segment, mass flow rate of the warm or cold source, dA_i is the heat transfer surface area, and ΔT is the temperature difference between high- and low-temperature fluids. Integrating equation over all segments of the exchanger yields

$$\int_0^{Q_0}\left[\mathrm{d}Q/\Delta T(Q)\right] = \int_0^{A_i} U_i \mathrm{d}A_i. \quad (3)$$

Then, the GMTD ΔT_{GMTD} is defined as follows [21]:

$$1/\Delta T_{\text{GMTD}} = (1/Q_0)\int_0^{Q_0}\left[\mathrm{d}Q/\Delta T(Q)\right]. \quad (4)$$

The temperature integral is approximated as:

$$\int_0^{Q_0}\left[\mathrm{d}Q/\Delta T(Q)\right] = \sum_{j=1}^{M}\left[2q/(\Delta T_j - \Delta T_{j+1})\right]. \quad (5)$$

Therefore, the GMTD is calculated by:

$$\Delta T_{\text{GMTD}} = M\Big/\sum_{j=1}^{M}\left[2/(\Delta T_j - \Delta T_{j+1})\right]. \quad (6)$$

3. Comparison between Pure Ammonia and Ammonia/Water Mixture

Figure 3 shows the correlation between GMTD and LMTD for the evaporator and the condenser calculated for both pure ammonia and ammonia/water mixture as working fluid. The evaporator and the condenser are counter-flow-type heat exchangers; both warm and cold sources are water; for the warm water, the inlet temperature is 28°C, and the outlet temperature is 25°C; for the cold water, the inlet temperature is 5°C, and the outlet temperature is 8°C. For pure ammonia, the evaporation temperature varied from 27.5°C to 26.0°C; the condensation temperature varied from 5.5°C to 7.0°C. For the ammonia/water mixture, the temperature at the evaporator inlet varied from 27.5°C to 26.0°C, and at the evaporator

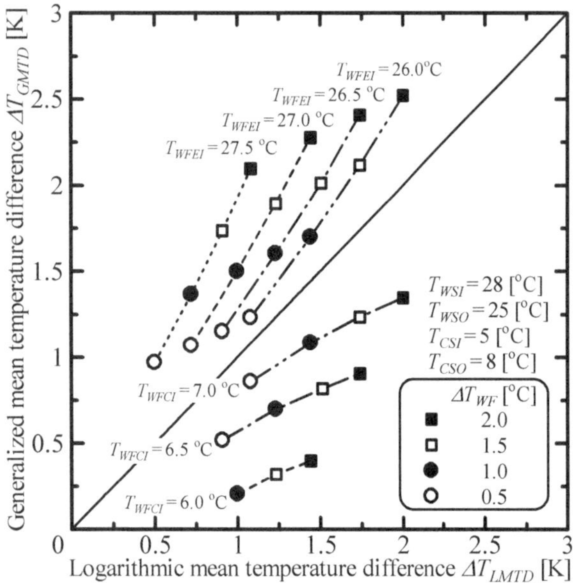

Figure 3. Temperature change in heat exchangers.

the outlet temperature varied from 24.5 to 23.0; the temperature at the condenser inlet varied from 6.0°C to 7.0°C, and at the condenser outlet varied from 8.5°C to 10.0°C; the ammonia concentration is 0.95 kg/kg. The working fluids are assumed to be saturated liquid at the evaporator inlet and outlet, and saturated vapor at the condenser outlet and inlet. The physical properties of ammonia are evaluated using the P-PROPATH database, and those of ammonia/water mixture are evaluated using the M-PROPATH database (Ibrahim and Kelein) [22].

The mean temperature differences between pure ammonia and heat source from the LMTD method and GMTD method are almost equal, as seen in **Figure 3**. In contrast, the mean temperature difference between ammonia/water mixture and heat source is higher in the evaporator from the GMTD method than from the LMTD method, but lower in the condenser. The variation of working fluid temperature with heat-flow-rate ratio produces a convex downward curve, as seen in **Figure 1(b)**. For the evaporator, the working fluid temperature approaches the warm water temperature at the evaporator inlet and outlet. The mean temperature difference is higher in the evaporator from the GMTD method than from the LMTD method. For the condenser, the working fluid temperature approaches the cold water temperature of the condenser. The mean temperature difference with the GMTD is lower than that with the LMTD method.

4. Mass-Flow-Rate Ratio Dependencies from GMTD and LMTD Models

Figure 4(a) shows the mean temperature difference between ammonia/water mixture and warm water from the LMTD and GMTD methods and the evaporation pressure P_E of the working fluid, in the evaporator with the mass-flow-rate ratio of working fluid to warm source. Here P_E is defined as the maximum evaporation pressure of the working fluid, P_{max} for these calculation settings. **Figure 4(b)** shows the mean temperature difference between the ammonia/water mixture and cold water for both LMTD and GMTD methods as well as the conden sation pressure of the working fluid, P_C, in the condenser against the mass-flow-rate ratio of working fluid to cold source. The condensation pressure, P_C is defined as the minimum condensation pressure of the working fluid, P_{min} for these calculation settings. The same settings for the inlet and outlet temperatures for evaporator and condenser described in Section 3 apply here. The ammonia concentration in the mixture is 0.95 kg/kg; the mass-flow-rate ratio of working fluid to heat source, γ varied from 1.0×10^{-3} to 5.0×10^{-2}. The temperature difference at the pinch-point is 0.5°C. The working fluid is also assumed to be saturated liquid at the evaporator inlet and saturated vapor at the condenser outlet. The physical properties of ammonia/water mixture are evaluated using the M-PROPATH database [22].

The mean temperature difference is higher in the evaporator from the GMTD method than from the LMTD method at low flow-rate ratios (**Figure 4(a)**). The pinch-point appears at the evaporator inlet and outlet, reflecting the convex downward nature of the variation of the working fluid temperature compared with the warm water temperature. The convexity of the working fluid temperature changes strongly when the temperature difference of the working fluid between the evaporator inlet and outlet increases. The maximum pressure decreases with decreasing mass-flow-rate ratio at low flow-rate ratio, that is, 1.64×10^{-2} or less under these calculation conditions, because of the increasing working fluid temperature difference between evaporator inlet and outlet. In contrast, the mean temperature differences from GMTD and LMTD methods are almost equal at the high flow-rate ratios. The convexity of the working fluid temperature is not strong at the high flow rate ratio, that is, 5.0×10^{-2} or more under these calculation settings, because the working fluid temperature difference between evaporator inlet and outlet is small.

Figures 5(a) and **(b)** shows an example where there are variations in the working fluid and warm water temperatures in the evaporator at heat-flow-rate ratio of 0.80 kg/kg and 0.95 kg/kg ammonia concentration, respectively, and the temperature difference at the pinch-point is 0 K. The working fluid temperature curve against heat flow rate is convex downward. Its slope rises gradually as the ammonia concentration in the mixture increases. The path between the working fluid outlet temperature and warm source inlet temperature is defined as "State A". The mass-flow-rate ratio for "State A" is under 1.64

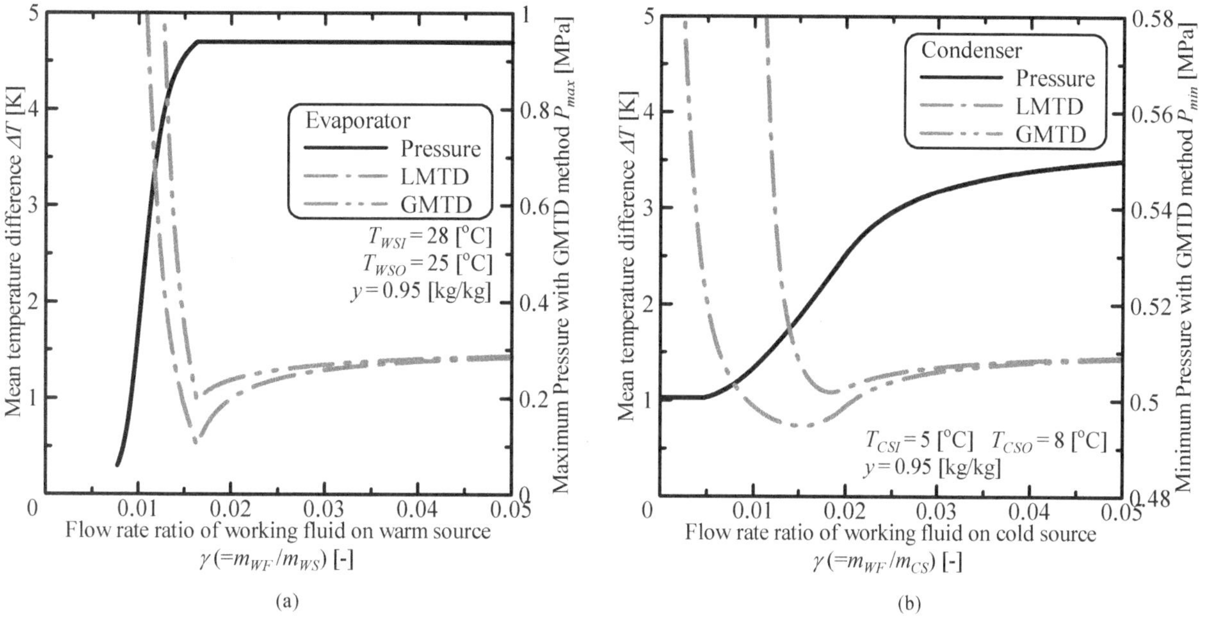

Figure 4. Temperature change in heat exchangers. (a) Evaporator; (b) Condenser.

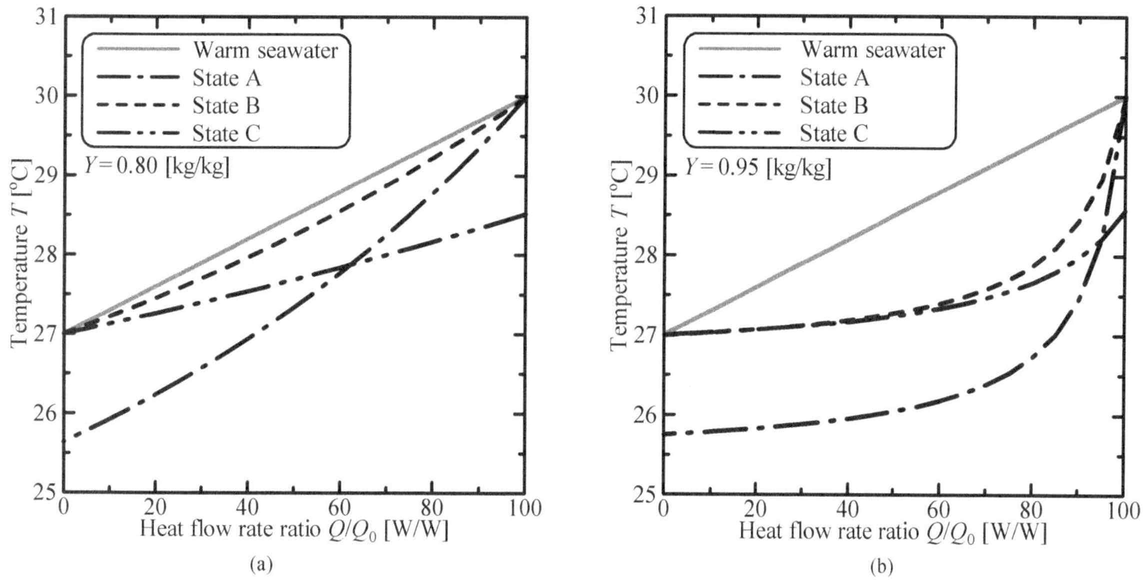

Figure 5. Temperature change in the evaporator. (a) Y = 0.80 kg/kg; (b) Y = 0.95 kg/kg.

$\times 10^{-2}$. A similar path at the evaporator inlet and outlet, defined as "State B", has mass-flow-rate ratio of 1.64×10^{-2}, and the mean temperature differences from the GMTD and LMTD methods are minimal. The path between the working fluid inlet temperature and warm source outlet temperature is defined as "State C". The variation in the maximum pressure is constant because at evaporator inlet the working fluid is saturated liquid. The mass-flow-rate ratio for "State C" exceeds 1.64×10^{-2}.

The mean temperature difference is lower from the GMTD method than from the LMTD method at low flow-rate ratios in the condenser (**Figure 4(b)**). For the condenser, the working fluid is high temperature fluid, and the cold water is low temperature fluid. A pinch-point appears as a consequence of convex downward variation in the working fluid temperature compared with that for the cold water temperature. Similar to the convexity in the working fluid temperature in the evaporator, the temperature in the condenser changes when the temperature difference of the working fluid between the condenser inlet and outlet increases. Therefore, the minimum pressure decreases gradually with decreasing mass-flow-rate ratio. The mean temperature differences from the GMTD and LMTD methods are almost equal at high-flow-rate ratios, that is, 5.0×10^{-2} or more with these calculation settings, because the working fluid temperature difference between evaporator inlet and outlet is small.

Figures 6(a) and **(b)** show as a further example with property variations in the working fluid, the cold water temperatures in the condenser against heat-flow-rate ratio at ammonia concentration of 0.80 kg/kg and 0.95 kg/kg, respectively, when the temperature difference of the pinch-point is 0 K. The path between the working fluid outlet temperature and cold source inlet temperature is defined as "State A". The variation of the minimum pressure is constant from the assumption that the working fluid is saturated liquid at the condenser outlet. The mass-flow-rate ratio for "State A" is 4.4×10^{-3} or less. Similarly, the path between the working fluid outlet temperature and warm source temperature in the condenser is defined as "State B". The pinch-point for "State B" is transferred from the cold water condenser outlet to the inlet when the mass-flow-rate ratio decreases. The mass-flow-rate ratio for "State B" is from 4.5×10^{-3} to 1.96×10^{-2}. The mean temperature difference from the GMTD method is a minimum when the mass-flow-rate ratio is 1.53×10^{-2}, but the minimum from the LMTD method occurs when the mass-flow-rate ratio is 1.85×10^{-2}. The path between the working fluid inlet temperature and warm source outlet temperature is defined as "State C". From the GMTD method, the mean temperature difference is lower for "State B" in the condenser. The mass-flow-rate ratio for "State C" is 1.97×10^{-2} or more.

5. Ammonia Concentration and Cold Source Temperature Dependencies from GMTD and LMTD Models

Obtained using the GMTD method, **Figures 7(a)** and **(b)** show the dependence of the maximum pressure in the evaporator on the mass-flow-rate ratio of working fluid to warm source. Similarly, **Figures 8(a)** and **(b)** show the dependence of minimum pressures for the condenser on the mass-flow-rate ratio of working fluid to cold source. The same calculation settings as describe above in Section 3 are used. The mass-flow-rate ratio of the working fluid with heat source varied from 1.0×10^{-3} to 5.0×10^{-2}. Again, the working fluid is saturated liquid at the evaporator inlet and saturated vapor at the condenser outlet. The physical properties of ammonia/water mixture are evaluated using the M-PROPATH database [8]. The ammonia concentration in the mixture varied from 0.99 to 0.80 kg/kg in **Figures 7(a)** and **8(a)**. The warm water temperature at the evaporator outlet varied from 27°C to 24°C in **Figure 7(b)**. The cold water temperature at the condenser outlet varied from 6°C to 9°C in **Figure 8(b)**.

The convexity of the working fluid temperature changes strongly when the ammonia concentration in the mixture increases (**Figures 5(a)** and **(b)**). For "State A", the maximum pressure decreases gradually at low ammonia concentrations compared with high ammonia concentrations as mass-flow-rate ratio decreases. In contrast, the mass-flow-rate ratio for "State B" increases with increasing warm water temperature difference between evaporator inlet and outlet.

Like convexity for the working fluid temperature in the evaporator, that in the condenser changes when the ammonia concentration in the mixture increases, (**Figures 6(a)** and **(b)**). Therefore, the mass-flow-rate ratio for "State B" increases with decreasing ammonia concentration. The slope of the cold water temperature curve is steep when the cold water temperature difference between the condenser inlet and outlet increases. The range for "State B" increases as the cold water temperature difference increases because the pinch-point at "State B"

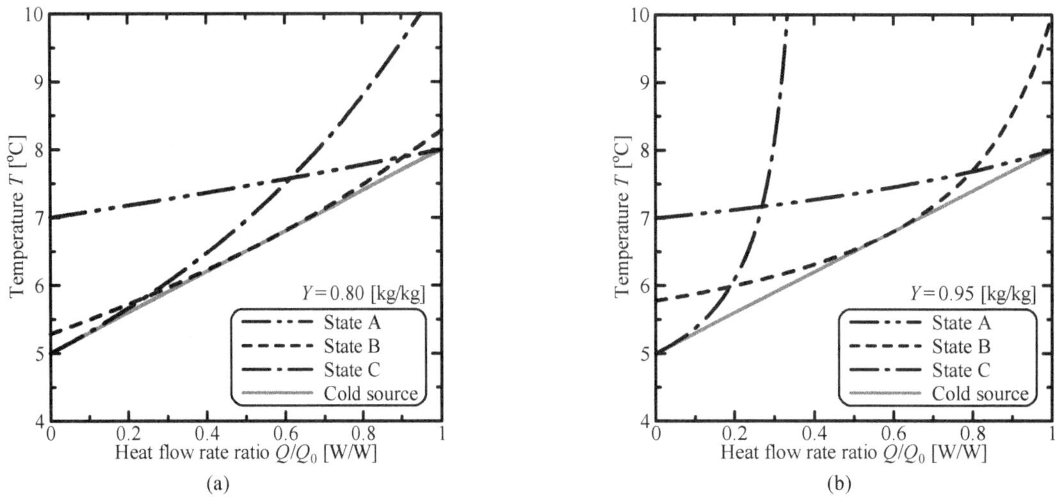

Figure 6. Temperature change in the condenser. (a) Y = 0.80 kg/kg; (b) Y = 0.95 kg/kg.

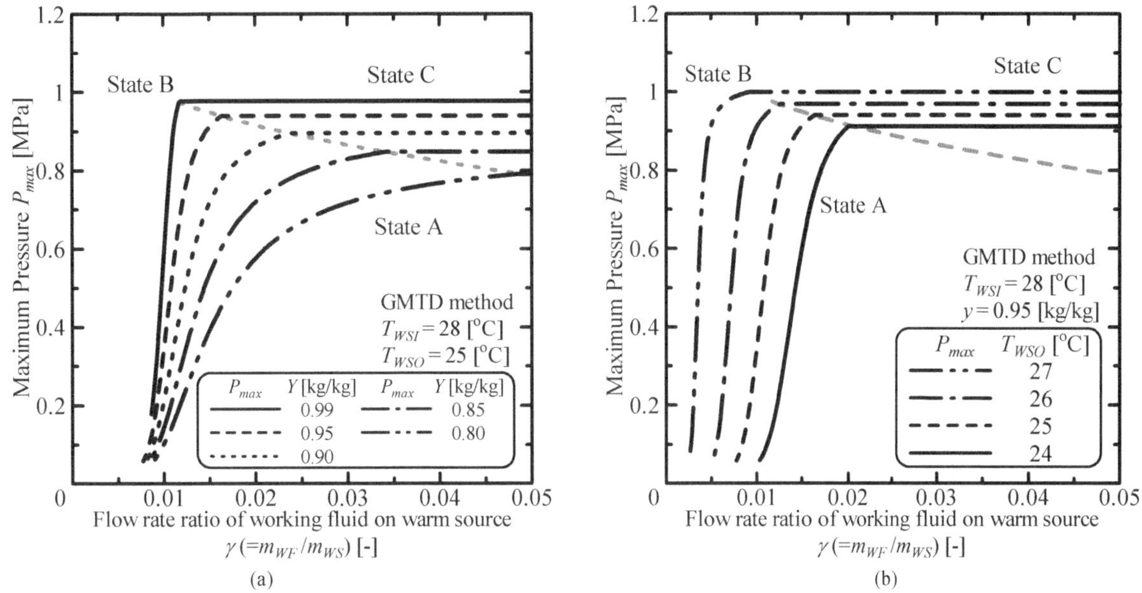

Figure 7. Maximum evaporation pressure. (a) Ammonia concentration; (b) Warm source outlet temperature.

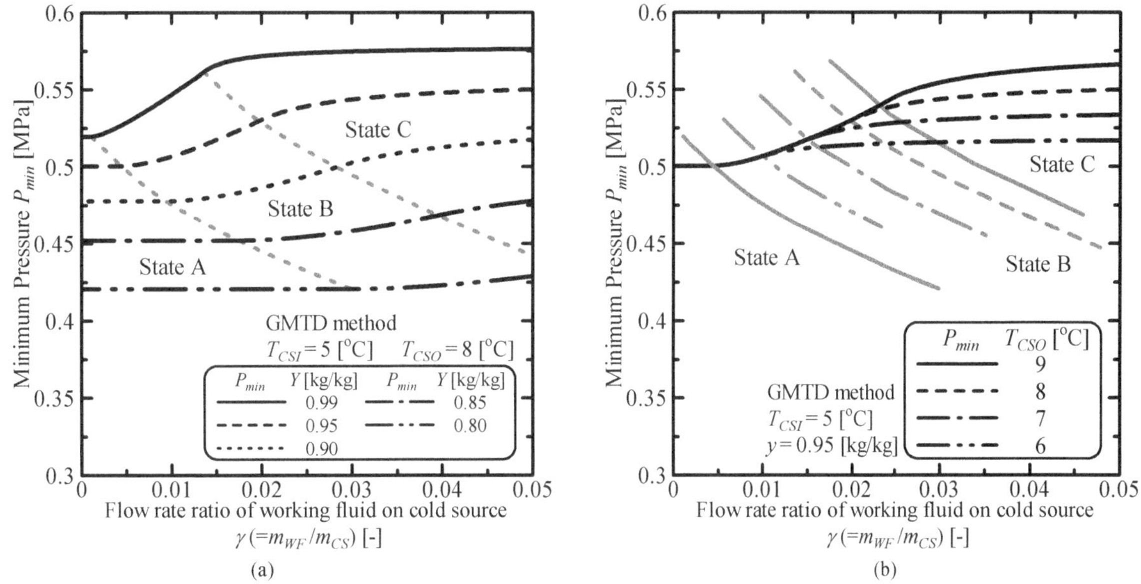

Figure 8. Maximum condensation pressure. (a) Ammonia concentration; (b) Cold source outlet temperature.

is transferred from the cold water condenser outlet to the inlet.

The correlation between the mean temperature differences from both GMTD and LMTD methods plotted in **Figures 7(a)** and **8(a)** is presented in **Figure 9**; the broken line refers to an ammonia concentration of 0.99 kg/kg, the chain line to 0.90 kg/kg, and the two-dot chain line to 0.80 kg/kg; the circles refers to a mass-flow-rate ratio of 5.0×10^{-2}, the black circle 3.0×10^{-2}, and the rectangle 2.0×10^{-2}. The convexity of the working fluid temperature changes strongly when ammonia concentration in the mixture increases (**Figures 5** and **6**). Therefore, the mean temperature difference differs between the GMTD and LMTD methods when the ammonia concentration in the mixture increases.

6. Conclusions

To obtain reliable data on the characteristics of a heat exchanger that uses ammonia/water mixture as working fluid, the recently developed GMTD method, which is to replace the traditional LMTD method, was assessed for validity in application. The conclusions are as follows:

1) Mean temperature differences between pure ammonia and water obtained with the LMTD and GMTD methods are almost equal. In contrast, the evaluation of

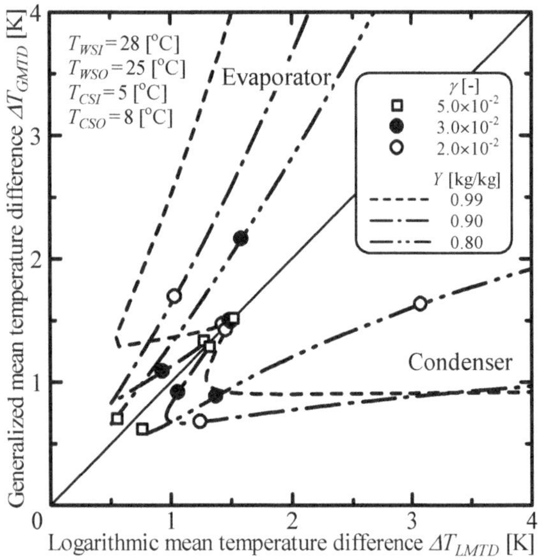

Figure 9. Relationship between GMTD and LMTD.

heat exchange using ammonia/water mixtures using the LMTD method was unsatisfactory. Temperatures in the ammonia/water mixture differ from warm water temperatures in the evaporator, and only approach cold water temperatures in the condenser.

2) Mean temperature differences are higher in the evaporator and lower in the condenser from the GMTD method than from the LMTD method, because the convexity of ammonia/water mixture temperature changes strongly as ammonia concentration increases. From the GMTD method, the mean temperature difference is lower when the working fluid temperature approaches the cold water temperature in the condenser.

3) For "State A", the maximum pressure in the evaporator decreases with decreasing mass-flow-rate ratio of ammonia/water mixture to water, owing to an increase in the temperature difference between the evaporator inlet and outlet. In contrast, the variation in the maximum pressure is constant at "State C" because the working fluid is assumed to be saturated liquid at evaporator inlet. The maximum pressure decreases with decreasing ammonia concentration in the mixture.

4) The variation in the minimum pressure in the condenser is constant assuming that the working fluid at condenser outlet is saturated liquid. The pinch-point for "State B" is transferred from the cold water condenser outlet to the inlet as the mass-flow-rate ratio decreases. Therefore, the minimum pressure for "State B" decreases with decreasing mass-flow-rate ratio. At low mass-flow-rate ratios, the minimum pressure for "State C" decreases gradually with decreasing the mass-flow-rate ratio that reflects the convex downward behavior of the working fluid temperature compared with the cold water temperature. The slope of the cold water temperature is steep when the cold water temperature difference between the condenser inlet and outlet increases. The range of "State B" increases with increasing the cold water temperature difference because the pinch-point for "State B" is transferred from the cold water condenser outlet to inlet.

Hence, the GMTD method can be applied to cycle calculations of power plants and refrigeration systems using ammonia/water mixture. Further research into compatibility conditions and accuracy of the GMTD method will be carried out.

REFERENCES

[1] A. H. A. William and W. Chih, "Renewable Energy from the Ocean: A Guide to OTEC," Oxford University Press, New York, 1994.

[2] Y. Ikegami and H. Uehara, "Conceptual Design of Ocean Thermal Energy Conversion (OTEC) Power Plants in Sri Lanka," *Proceedings of the 8th International Society of Offshore and Polar Engineers*, 1998, pp. 140-148.

[3] Y. Ikegami and H. Uehara, "Optimum Design Point for a Closed-Cycle OTEC System," *Proceedings of the 4th International Society of Offshore and Polar Engineers*, 1994, pp. 383-389.

[4] F. Sun, Y. Ikegami, B. Jia and H. Arima, "Optimization Design and Exergy Analysis of Organic Rankine Cycle in Ocean Thermal Energy Conversion," *Applied Ocean Research*, Vol. 35, 1994, pp. 383-389.

[5] N. Yamada, A. Hoshi and Y. Ikegami, "Performance Simulation of Solar-Boosted Ocean Thermal Energy Conversion Plant," *Renewable Energy*, Vol. 34, No. 7, 2009, pp. 1752-1758.

[6] F. Sun, Y. Ikegami and B. Jia, "A Serformance Simulation of Solar-Boosted Ocean Thermal Energy Conversion Plant," *Renewable Energy*, Vol. 41, 2012, pp. 220-219.

[7] A. I. Kalina, "Generation of Energy by Means of a Working Fluid, and Regeneration of a Working Fluid," United States Patent 4346561, 1982.

[8] H. Uehara and Y. Ikegami, "Parametric Performance Analysis of Otec Using Kalina Cycle," *ASME*, 1993, pp. 203-207.

[9] H. Uehara, Y. Ikegami and T. Nishida, "Performance Analysis of OTEC System Using a Cycle with Absorption and Extraction Processes," *Transactions of the Japan Society of Mechanical Engineers, Series B*, 64-624, 1998, pp. 384-389.

[10] G. Wall, C. C. Chuang and M. Ishida, "Exergy Study of the Kalina Cycle," *American Society of Mechanical Engineers (ASME)*, 10-15, 1989.

[11] Y. M. El-Sayed and M. A. Tribud, "Theoretical Comparison of the Rankine and Kalina cycles," *ASME Publication*, 1, 1985, pp. 97-102.

[12] V. P. Pall and L. Eliasson, "Factors Influencing the Economics of the Kalina Power Cycle and Situations of Superior Performance," *International Geothermal Conference*, 2003, pp. 32-40.

[13] M. D. Mirolli, "Cementing Kalina Cycle Effectiveness, the Kalina Cycle for Cement Kiln Waste-Heat-Recovery Power Plants," *IEEE Industry Applications Magazine*, Vol. 12, No. 4, 2006, pp. 60-64.

[14] U. R. Nasruddin, M. Rifaldi and A. Noor, "Energy and Exergy Analysis of Kalina Cycle System (KCS) 34 with Mass Fraction Ammonia-Water Mixture Variation," *Mechnical Science and Technology*, Vol. 23, No. 7, 2009, pp. 1871-1876.

[15] Z. Guzovic, D. Loncar and N. Ferdelji, "Possibilities of Electricity Generation in the Republic of Croatia by Means of Geothermal Energy," *Energy*, Vol. 35, No. 8, 2010, pp. 3429-3440.

[16] K. E. Starling, D. W. Johnson, H. Hafezzadeh, L. W. Fish, H. H. West and K. Iqbal, "Use of Mixtures as Working Fluids in Ocean Thermal Energy Conversion Cycles—Phase I," US DOE Report, 1974.

[17] K. E. Starling, D. W. Johnson, H. Hafezzadeh, L. W. Fish, H. H. West and K. Iqbal, "Use of Mixtures as Working Fluids in Ocean Thermal Energy Conversion Cycles—Phase II," US DOE Report, 1978.

[18] C. B. Panchal, D. L. Hillis, J. J. Lorenz and D. T. Yung, "OTEC Performance Tests of the Trane Plate-Fin Heat Exchanger," Ocean Thermal Energy Conversion Program Argonne National Laboratory, 1981.

[19] Y. Ikegami, H. Goto, T. Morisaki and T. Furukawa, "Effect of Working Fluid Flow Rate and Ammonia Concentration on OTEC Using Ammonia/Water Mixture as Working Fluid," *13th Asian Congress of Fluid Mechanics*, 2010, pp. 1026-1029.

[20] Y. Ikegami, K. Urata, J. Inadomi, H. Goto, T. Morisaki, K. Inoue and S. Goto, "Investigation on the Stability of OTEC System Using Ammonia/Water Mixture as Working Fluid by Continuous Operation for Two Weeks," *Proceedings of the 20th International Society of Offshore and Polar Engineers*, 2010.

[21] M. Utamura, K. Nikitin and Y. Kato, "Generalization of Logarithmic Mean Temperature Difference Method for Heat Exchanger Performance Analysis," *Thermal Science & Engineering*, Vol. 15, No. 3, 2007, pp. 163-172.

[22] PROPATH Group, "A Program Package of Thermo-Physical Properties of Fluids, Version 11.1," PROPATH Group.

7

Performance of an Axial Cascade

Basharat Salim
King Saud University, Riyadh, KSA

ABSTRACT

A cascade is defined as an infinite row of equidistant similarly aerofoil bodies. The cascade is used to divert a flow stream with a minimal loss. The flow over an axial cascade presents a complicated intra blade fluid dynamic interaction that causes the flow to behave differently than the flow over a single aerofoil blade. A cascade tunnel in which different shapes of cascades could be tested for performance evaluation has been build. An axial flat plate cascade of 70 mm chord, 215 mm span and 0.55 pitch chord ratio has been tested in this tunnel for evaluating its aerodynamic performance. The results depict that the Reynolds number, incidence and blade angle of the cascade control the aerodynamic performance of the axial cascade.

Keywords: Axial; Cascade; Performance; Deflection; Incidence

1. Introduction

A cascade is a row of geometrically similar blades arranged at equal distance from each other and aligned to the flow at the same angle. It is an infinite row of equidistant similarly aerofoil bodies. The cascade is used to divert a flow stream with a minimal loss. It forms the basic block for the design and development of turbomachinery, particularly the axial compressor and axial turbine. The turbine usually shows tolerance to the blade design and alignment errors because blades of a turbine stage perform under a favorable pressure gradient whereas compressor blades are prone to aerodynamic losses because these have to work under adverse pressure gradients due to diffusing nature of the flow field. The bladings of an axial compressor and axial turbine have high solidity which makes the flow structure in these machines highly complicated as the flow around any blade in these machines is affected by the presence of the adjacent blades. These blades are said to form a cascade. The pressure ratio developed by a cascade depends on its aerodynamic characteristics. There are various types of shapes used in compressor blading. These range from C series, NACA series, DCA series and other special profiles. Similarly, aerofoil shapes used in wind turbines are of NACA series, thin aerofoil shape and of special arc shapes. The aerodynamic parameters for compressor blades are stagger, solidity, camber, camber line shapes, thickness chord ratio and thickness distribution. The peculiar geometry of their blades causes the flow to be three dimensional. The passage flow in these machines is grossly affected by passage pressure gradient, secondary flows, boundary layer effects, cross flows, tip clearances, complicacies of stator rotor aerodynamic interaction, shock boundary layer interaction and inlet flow conditions. Therefore, understanding of cascade flow results in systematic improvement of the aerodynamic art for design of such passages.

Specification of a blade of an axial compressor or turbine is at the very heart of the design of an axial turbomachinery. There have been many attempts to find solution of these problems. Roundbash [1], Scholz [2] and Gostalow [3] have discussed the summary of these works. Compusty [4] has reported the work of Felix and Emery [5] in which it has been suggested that the shape of a compressor blade has an insignificant effect on its performance at low mach numbers. Their tests on C4 and NACA65 series cascades revealed same results. Bacur [6] has studied the flow development in S shaped profiles in a cascade tunnel. Andrews [7] has found out that the camber line shape, leading edge radius, and thickness chord ratio have small effect on the performance of a cascade. Serovy, *et al.* [8] have carried out detailed investigation of inter-passage flow in a baseline and modified versions of a two-stage axial compressor and have observed similarities in their aerodynamic performance. Pullen and Harrvey [9] have arrived at a loading parameter for an axial flow turbine cascade by taking sweep of the blade into account. They have validated their results with the experimental results of linear cascade tests of low pressure ratio axial turbine. Kalpatrick and Barrows [10] have tested cascades of varying aspect ratio from 1.5 to 3. Mustaphe, *et al.* [11] have presented the results of an off design performance of a turbine cascade at its mid span with varying Mach numbers, Reynolds numbers and incidences. Nagasaki and Yamsaki [12] have developed a CFD code to compute unsteady aerodynamic forces on a vibrating annular cascade and have compared it with the linearized theory and have found the two to be almost similar.

2. Experimental Setup and Methodology

The experimental setup for testing the cascade was constructed and it is a blow down, close jet, low speed cascade tunnel shown in **Figure 1**. The tunnel has a radial blower of 3.5 kW power, pressure rise of 3.5 kPa and a flow rate of 0.83 cubic meters per second. The blower speed is 3450 rpm. The blower is fitted with an inlet duct of 200mm internal diameter and 300 mm length. The inlet duct has a cup and cone valve at the free end to control the flow rate through the cascade tunnel. The cup can be moved into the cone by an axial movement system. The cup can fully close the valve. A circular duct of diameter 270 mm is fitted at the exit of the blower. The duct is connected to a settling chamber. The settling chamber has three parts. The inlet part is of diffusing section with inlet of diameter 270 mm where as its exit is rectangular of size 680 mm × 600 mm. This diffusing section is connected to the middle part which is further divided into three portions of size 680 mm × 600 mm and length 190 mm. These parts are connected tightly and wire gauze is fitted at each connection. These gauzes reduce the turbulence of the flow emerging out of the blower. The exit of the settling chamber is a rectangular converging duct of inlet size 680 mm × 600 mm and exit of size 220 mm × 210 mm. The converging section is connected to a small rectangular duct of size 220 mm × 210 mm and length 200 mm. This duct is in turn connected to the rectangular test section of size 220 mm × 210 mm and length 300 mm. The test section is fabricated from Perspex sheet of 12 mm thickness. All the ducts etc are made from sheet metal. The test section discharges the air to the atmosphere. The cascade row is fitted at the end of the test section. The cascade row was made to rotate from −10° to +10° by a moving mechanism which kept the pitch of the blades same all through. The axial cascade had a chord of 70 mm, span of 215 mm and pitch chord ratio of 0.55. The velocity traverses upstream and downstream of the cascade row were taken by using two Pitot Static probes, one upstream and other downstream, which was fitted on traverse systems. The traverses were taken at two flow rates corresponding to the two free stream Reynolds numbers for different blade angle setting. These traverses yielded the total velocity, V, and its flow angle, α, at the central location. The blades were oriented to the incoming flow through cascade blade angle changing mechanism that was fitted to the cascade tunnel. It could orient flow at any angle simultaneously.

3. Results and Discussion

The main concern of this study was to build a cascade tunnel in which different shapes of cascades could be tested for performance. This paper presents the results of the investigation that were carried out for finding out the performance of an axial flat plate cascade in that purpose build experimental setup. Karanja and Sayers [13] and Wolf and Jhonston [14] have documented that presence of a presence of a diffuser downstream of a straight duct reconfigures the flow upstream of it in that duct. This becomes more urgent in case where the straight duct is blocked by the cascade row. Hence it becomes necessary to determine a position which can represent free stream location. To achieve this traverse of velocity probe were carried out upstream of the cascade row and a location

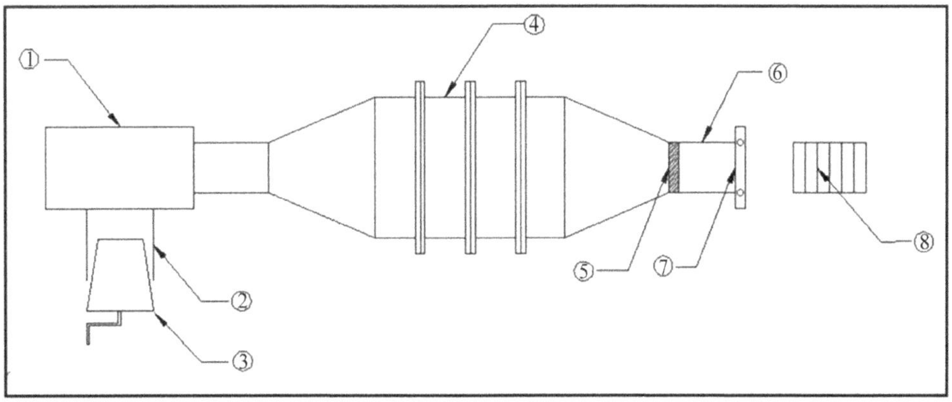

Figure 1. Experimental set up; (1) Blower; (2) Inlet duct; (3) Cup and cone; (4) Settling chamber; (5) Distorter; (6) Test section; (7) Cascade blade angle changing mechanism (8) Cascade row.

was found where the velocity profile would match the normal velocity profile in a rectangular duct. The free stream condition of symmetrical flow behavior in the straight duct was found at a distance of 1.75 times the width of the straight duct upstream of the leading edge of the cascade. This was taken as a free stream location for all the test conditions. The investigations were carried out at two flow rates corresponding to free stream Reynolds numbers of 2.26×10^5 and 2.52×10^5. It may be said that the free stream Reynolds number was based on duct dimensions. The inlet velocity profiles for free stream of the cascade were measured at about six chords distance upstream of the cascade. These profiles were measured with a NPL standard Pitot static probe using an inclined water manometer for the measuring the pressure. **Figure 2** shows the free stream velocity profiles at two Reynolds numbers for the axial cascade. The results show uniform velocity profile for the most of the passage width of the tunnel except near the walls. The nonuniformity at the walls can be attributed to the wall effect of the tunnel and related boundary layer phenomenon. Therefore it can be concluded that the present investigations present the results of an axial cascade with a uniform upstream flow.

Cascade Flow: The flow over an axial cascade presents a complicated intra blade fluid dynamic interaction that causes the flow to behave differently than the flow over a single aerofoil blade. But if the aspect ratio of the cascade blade row is high then two dimensionality of the flow prevails. In most of the flow analysis problems of the cascade the flow is usually assumed to be two dimensional. The most important aerodynamic flow parameters that govern the two dimensional flow behavior in a cascade are angle of incidence, angle of deflection, angle of deviation, static pressure rise coefficient, total pressure loss coefficient, drag coefficient and lift coefficient. These flow parameters were measured by a precaliberated three hole probe near the centre of the span of the cascade at half chord distance upstream and downstream of the central blade of the cascade. The probe output was measured with the help of an inclined water manometer. The different flow and performance variables for the cascade were calculated using two dimensional cascade aerodynamics (3). The incidence angle, i, was calculated as, $i = \alpha - \beta$. All other calculation were based on the measurements of the probe though the use of two dimensional theory. The relevant relationships for each parameter is given in the definitions of these parameters in the nomenclature.

Figure 3 reveals the variation the angle of incidence with the blade angle. The angle of incidence is defined as the difference between the flow angle and blade angle at the inlet of the cascade. The blade angle was varied between −10 degrees to 10 degrees. These depict that the angle of incidence decreases with the blade angle at both Reynolds numbers for the cascade. All other flow parameters have been shown as a combined variation of the blade angle and the angle of incidence. This is due to the fact that the variation in a cascade is dependent on blade angle whereas in an isolated blade or aerofoil the dependence is mainly on the angle of incidence or angle of attack. **Figure 4** shows the variation of deviation angle with the blade angle and the incidence angle at two Rey-

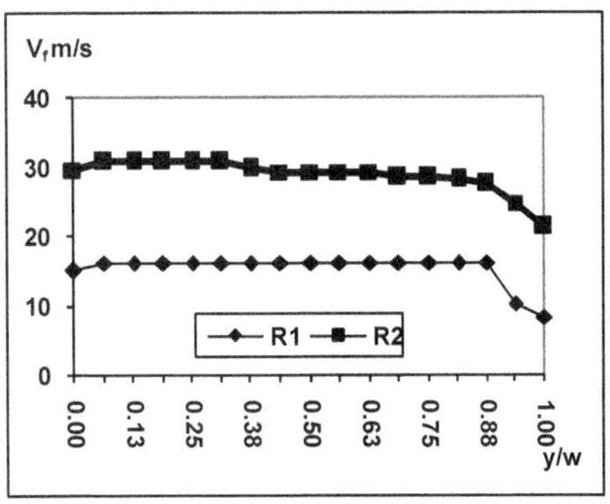

Figure 2. Variation of free stream velocity in the Cascade tunnel R1 = 2.26×10^5 and R2 = 2.52×10^5.

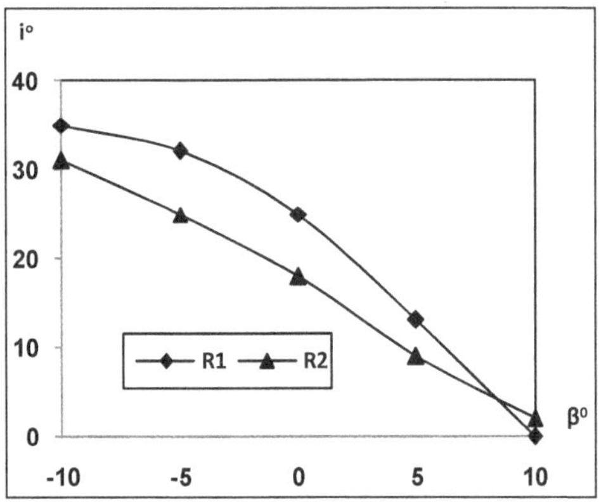

Figure 3. Variation of angle of incidence with the blade angle R1 = 2.26×10^5 and R2 = 2.52×10^5.

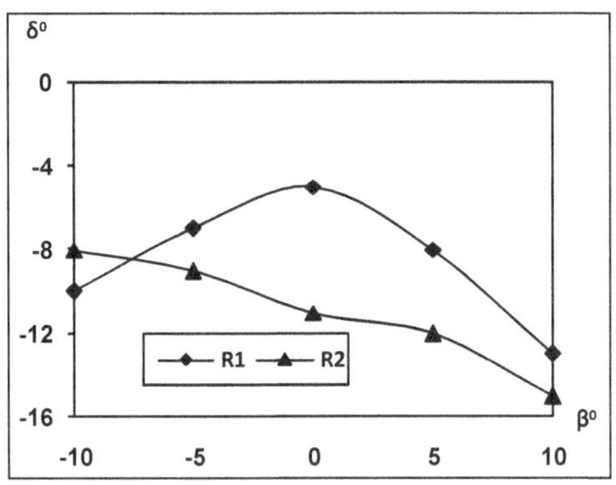

Figure 4. Variation of deviation angle with blade angle and angle of incidence. R1 = 2.26×10^5 and R2 = 2.52×10^5.

nolds numbers. The deviation angle is the difference between the flow angle and the blade angle at the exit of cascade row. At lower Reynolds number the deviation angle decrease up to zero blade angle and there after it increases again. At higher Reynolds number it progressively decreases with the blade angle. The deviation at lower Reynolds number is seen to increase with the incidence up to a blade angle of zero and then decreases, A maximum incidence of 35° is observed at lower Reynolds number. At higher Reynolds number the deviation increases with incidence and achieves a maximum value of 20° for a deviation angle of −8°. The deviation of the flow is observed to decrease with the increase in Reynolds number and the blade angle. **Figure 5** shows the variation of deflection angle with the blade angle and angle of incidence at two Reynolds numbers. The deflection angle is the difference between the flow angles at the inlet and the exit of cascade row. The deflection angle depicts a continuous decreasing trend with the increase in the blade angle and it increases with the increase in the incidence. Further the deflection decreases at higher Reynolds number. Thus flow is found to deflect less at higher blade angles and flow rate.

Variation of static pressure rise coefficient through the cascade is presented in **Figure 6**. It shows that the variation at lower Reynolds number is of a wavy form showing lowest and highest values at blade angles of ±5°. The value of Cp at +5° blade angle is 0.77 where at −5° blade angle it is −0.11. At higher Reynolds number a small change is observed in the values of Cp with the change in the blade angle whereas almost no changes are observed with the angle of incidence. The variation of total pressure loss coefficient along the passage through the cascade is presented in **Figure** 7. At lower Reynolds number the total pressure loss coefficient is minimum at zero blade angle where as it increases both for the positive and negative values of the blade angle. At higher Reynolds number the total pressure loss coefficient is almost invariant with the blade angle. The least changes in the values total pressure loss coefficient is an indicator of stable behavior within the passage that creates such fluid dynamic conditions that give rise to lower flow losses in the cascade passage. A minimum value of total pressure loss coefficient of 0.06 at zero blade angle is observed at lower Reynolds number where as very small loss is depicted at higher Reynolds number. The total pressure loss coefficient is observed to have a minimum value of 0 - 6 at incidences below 10 degrees and it shows two values for the same incidence corresponding to negative and positive values of the blade angle.

Figure 8 presents variation of lift coefficient at two Reynolds numbers for the cascade. The lift coefficient is

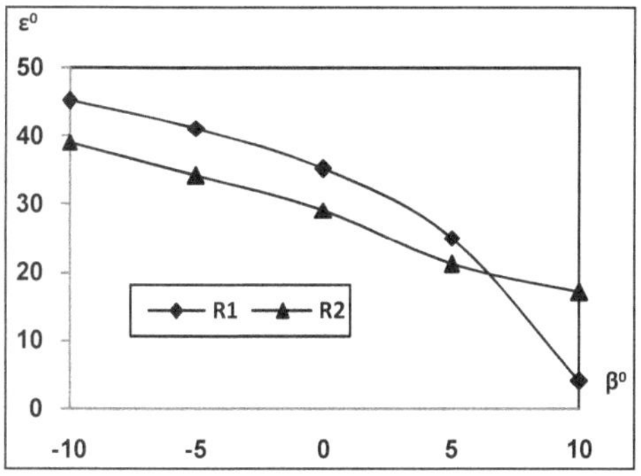

Figure 5. Variation of deflection angle with blade angle and angle of incidence. R1 = 2.26 × 10^5 and R2 = 2.52 × 10^5.

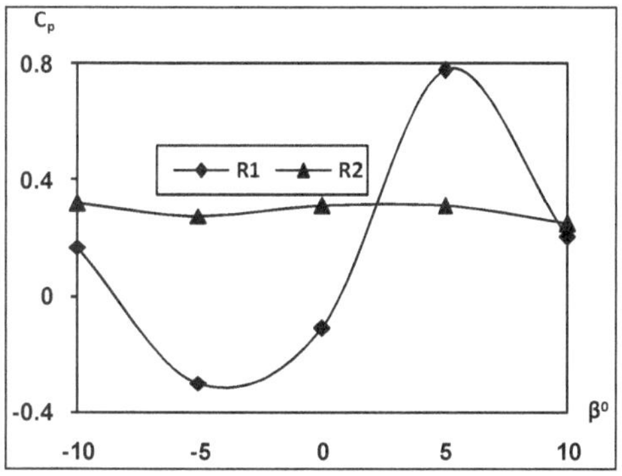

Figure 6. Variation of static pressure rise coefficient with blade angle and angle of incidence. R1 = 2.26 × 10^5 and R2 = 2.52 × 10^5.

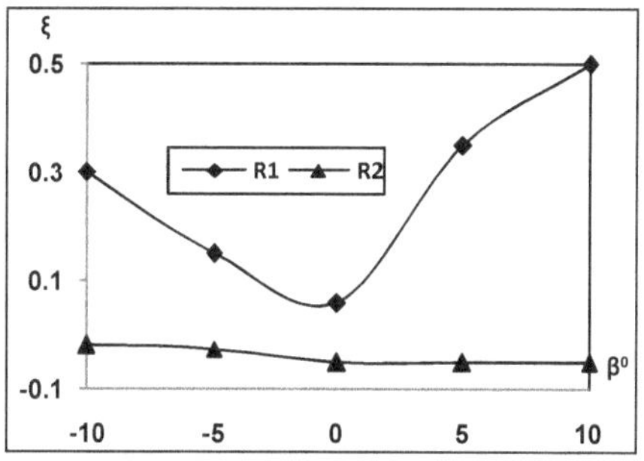

Figure 7. Variation of total pressure loss coefficient with blade angle and angle of incidence. R1 = 2.26 × 10^5 and R2 = 2.52 × 10^5.

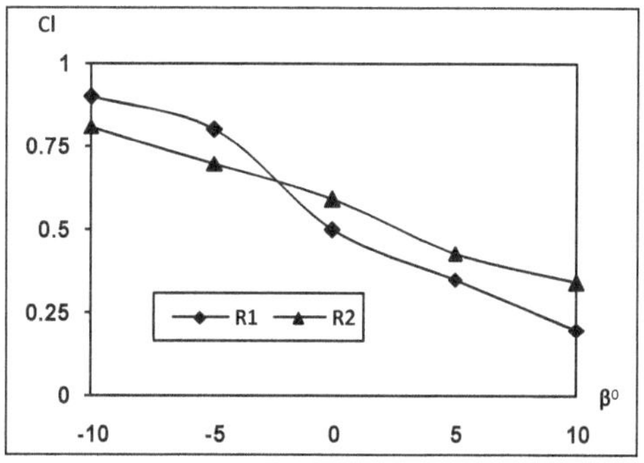

Figure 8. Variation of lift coefficient with blade angle and angle of incidence. R1 = 2.26 × 10^5 and R2 = 2.52 × 10^5.

observed to decrease with the blade angle at both the Reynolds numbers. Compared to this the lift coefficient is observed to increase with the incidence. This variation is typical of a linear cascade. Even in case of a flat plate the lift coefficient directly increases with the increase in the incidence. At negative blade angles the lift coefficient is seen to be more at lower Reynolds number this corresponds to the higher values of incidence. The drag coefficient variations of the cascade, shown in **Figure 9**, indicate that at lower Reynolds number, the drag for the cascade is minimum at zero blade angle and increases as the blade angle gets altered from that blade angle in a parabolic manner. At that instance the incidence is about 14°. As the incidence increases the drag coefficient achieves twin relations. First in the range of decreasing negative blade angles where drag coefficient increases with the incidence and the next where drag coefficient decreases with the incidence which corresponds to the increasing positive blade angles. But at higher value of Reynolds number the drag coefficient is almost invariant with blade angle for all the configurations. It is obvious that drag will be minimum at zero blade angle because at that time the blocked area in the tunnel is minimum and as blocked area is altered the drag coefficient gets changed. Thus the drag coefficient blade angle and the incidence are interrelated to each other.

A comparison of the main performance parameters like lift coefficient and drag coefficient was carried out with different published empirical results for a flat plate with no distortion case. Lanzafame, *et al.* [13] have obtained the following empirical formula for the lift coefficient and the drag coefficient.

$Cd = Cd\max\sin^2\alpha$. For a flat plate $Cd\max = 2.0$, ref [10]

$Cl = Cl\max\sin\alpha\cos\alpha$ where $Cl\max = 2\pi\alpha$, ref [10]

Using this empirical result the comparison between achieved results and theoretical results is presented in **Figure 10** for drag coefficient. The results depict that the measured drag is more than that obtained from the theory. It is due to the fact that the theory considers only a single

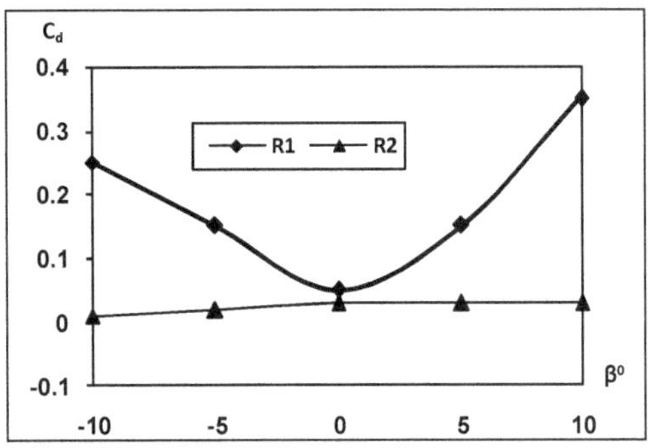

Figure 9. Variation of drag coefficient with blade angle and angle of incidence. R1 = 2.26 × 10^5 and R2 = 2.52 × 10^5.

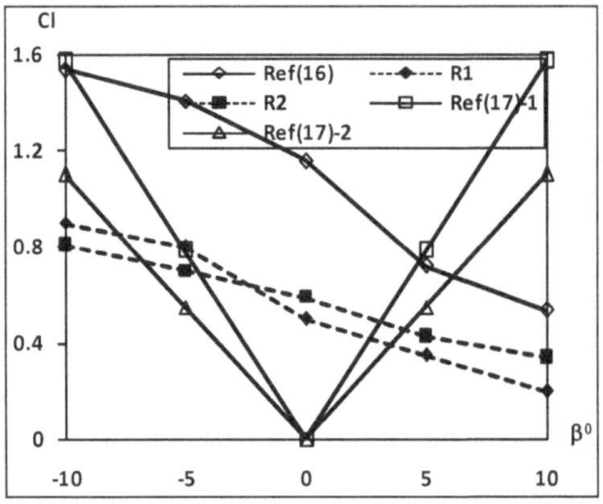

Figure 10. Comparison of experimental and theoretical lift coefficient R1 = 2.26 × 10^5 and R2 = 2.52 × 10^5.

plat in a infinite control volume that is unaffected by the presence of other elements of the cascade and the tunnel wall boundaries. Comparative lift coefficient variations are presented in **Figure 11**. This figure also shows that the experimental values of lift coefficient are lesser than the theoretical values of the lift coefficient. The reasoning for this result is also that the theoretical results represent a flat plat aerodynamics in an infinite space which is not true for the cascade. A comparison of the results has also been made with the momentum theory and lifting line theory [14] for a flat plate. The lift coefficient and drag coefficient by these theories are given as under.

Lifting Line theory

$$Cl = 0.11\times\alpha^0$$

Momentum theory

$$Cl = \pi AR\alpha$$

The drag in both cases is given as

$$Cd = \frac{Cl^2}{\pi AR}$$

where AR is aspect ratio of the cascade = b/c.

The comparison of these results also shows that the theoretical values of lift coefficient are higher than the experimental values whereas theoretical drag coefficients are lower than the experimental values. This difference is because of the fact that a flat plate behaves differently when it is placed in infinite surroundings rather than as a part of a cascade. It may be added here that the lift coefficient and drag coefficient are basically dependent both on the blade angle and the angle of incidence. \this is illustrated in the **Figures 12** and **13**. It may be seen that these figure show similarity with the **Figures 8** and **9**. It is because the angle of incidence is dependent on blade angle through the flow angle.

4. Conclusion

A blow down cascade tunnel facility for testing cascades

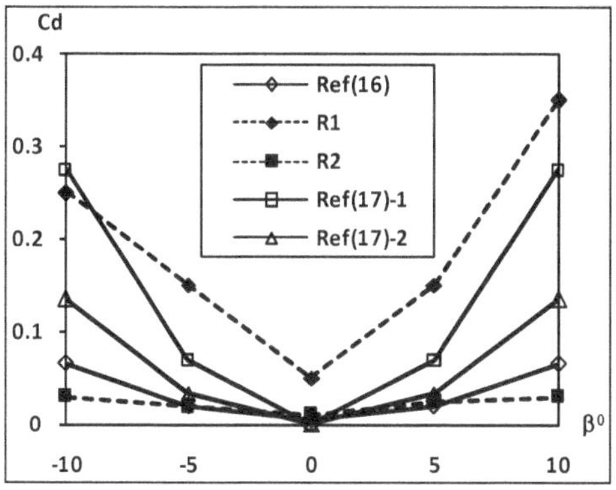

Figure 11. Comparison of experimental and theoretical drag coefficient R1 = 2.26 × 10^5 and R2 = 2.52 × 10^5.

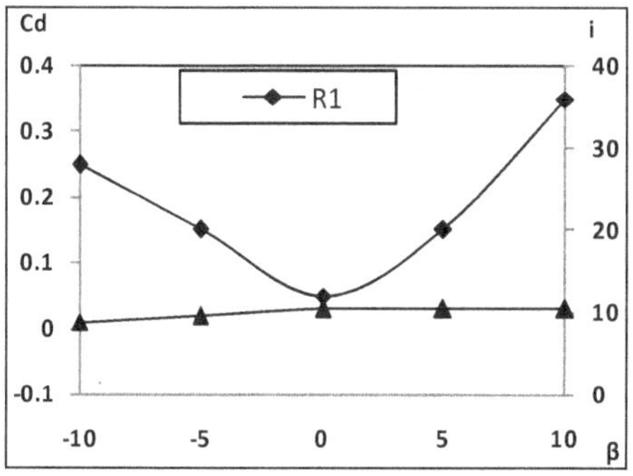

Figure 12. Variation of drag coefficient with Blade angle and angle of incidence. R1 = 2.26 × 10^5 and R2 = 2.52 × 10^5.

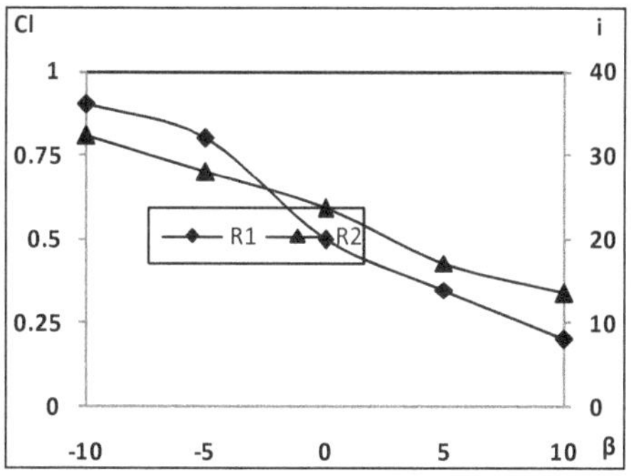

Figure 13. Variation of lift coefficient with Blade angle and angle of incidence. R1 = 2.26 × 10^5 and R2 = 2.52 × 10^5.

of different shapes at low speed has been built. Aerodynamic performance of an axial cascade of chord 70 mm, span 215 mm and pitch chord ratio of 0.55 has been evaluated. From the investigations, it has been found that under the conditions of the test the angle of incidence decreases with the blade angle. The deviation of the flow is observed to decrease with the increase in Reynolds number and the blade angle. The flow is found to deflect less at higher blade angles and flow rate. At higher Reynolds number, a small change is observed in the values of Cp with the change in the blade angle, whereas almost no changes are observed with the angle of incidence. At higher Reynolds number, the total pressure loss coefficient is almost invariant with the blade angle. The lift coefficient is observed to decrease with the blade angle. The drag coefficient, blade angle and the incidence are interrelated to each other. Further the theoretical values of both coefficients are seen to be different than the measured values. Measured drag is higher and the lift is lower.

5. Acknowledgements

This project was supported by Deanship of Scientific Research and Research Center College of Engineering, King Saud University, Riyadh, KSA.

REFERENCES

[1] W. H. Roundebush, "Potential Flow in Two Dimensional Cascades, Chapter IV of 'Aerodynamic Design of Axial Flow Compressors'," NASA Sp 36, US Government Printing Office, Washington DC, 1965.

[2] N. Scholz, "Aerodynamics of Cascades," AGARDograph No. 220, Nato Science and Technology Organization, Amsterdam, 1977.

[3] J. P. Gostlow, "Cascade Aerodynamics," Pergaman Press, Oxford, 1984.

[4] N. A. Cumpsty, "Compressor Aerodynamics," Longman Scientific and Technical, England, 1989.

[5] A. Felix and J. Emery, "A Comparison of Typical NGTE and NACA Axial Blade Compressor Blade Sections in Cascades at Low Speeds," NACA TN 3937, US Government Printing Office, Washington DC, 1957.

[6] I. D. Baciu, "A S Profile Axial Cascade Tested in a Wind Tunnel," *International Conference on Hydraulic Machinery and Equipment*, Timosoars, 16-17 October 2008, pp 45-48.

[7] S. J. Andrew, "Tests Related to the Effect of Profile Shape and Camber-Line on Compressor Cascade Performance," Aeronautical Research Council R&M 2743, Aeronautical Research Council, England, 1949.

[8] G. Serovy, P. Kavanagh and T. Okishi, "Aerodynamics of Advanced Axial Turbomachinery," NTIS Report ADA 131360, US Government Printing Office, Washington DC, 1983.

[9] G. Pullen and N. Harrvey, "Influence of Sweep on the Axial Flow Turbine Cascade at Mid Span," *Transactions of ASME, Journal of Turbomachinery*, Vol. 129, No. 3, 2007.

[10] D. Kalpatrik and R. Burrow, "Aspect Ratio Effects on Compressor Cascade Blade Flutter," ARC Technical Report, R&M No. 3103, Aeronautical Research Council, England, 1958.

[11] H. Mustaphe, D. Jourim and S. A. Scolander, "Aerodynamic Performance of a Transonic Turbine Cascade at Off Design Conditions," *Journal of Turbomachines*, Vol. 123, No. 3, 2001, pp. 510-519.

[12] T. Nagasaki and N. Yamasaki, "Linear Unsteady Aerodynamic Forces on a Vibrating Annular Cascade Blade,"

Journal of Thermal Sciences, Vol. 12, No. 2, 2003, pp. 138-143.

[13] R. Lanzafame and M. Messina, "Fluid Dynamic Wind Turbine Design: Critical Analysis, Optimization and Application of BEM Theory," *Renewable Energy*, Vol. 32, No. 14, 2007, pp. 2291-2305.

[14] Mcormik and Barnes "Aerodynamics, Aeronautics and Flight Mechanics," John Wiley and Sons, New York, 1979.

Nomenclature

AR: Aspect ratio of the cascade = *b*/*c*;
c: Chord of the cascade blades;
Cd: Drag coefficient.

$$Cd = \frac{(P_1 - P_2)\, s \cos \propto_m}{\frac{1}{2}\rho V_m^2 c}$$

*Cd*max: Maximum drag coefficient;
Cl: Lift coefficient.

$$Cl = 2\frac{s}{c}\cos \propto_m (\tan \propto_1 - \tan \propto_2) - Cd \tan \alpha_m$$

*Cl*max: Maximum lift coefficient;
Cp: Static pressure rise coefficient.

$$Cp = \frac{p_1 - p_2}{P_1 - p_1}$$

i: Incidence angle $i = \alpha_1 - \beta_1$;
l: Blade span;
P: Total pressure;
p: Static pressure;
s: Pitch;
V: Total velocity;
α: Flow angle;
β: Blade angle;
δ: Deviation angle $\delta = \alpha_2 - \beta_2$;
ε: Deflection angle $\varepsilon = \alpha_1 - \alpha_2$;
ξ: Total pressure loss coefficient $\xi = \frac{P_1 - P_2}{P_1 - p_1}$;
ρ: Air density.

Subscript

1: Cascade inlet;
2: Cascade outlet;
m: Mean;
f: Free stream.

8

Heat Transfer Enhancement of Cu-H_2O Nanofluid with Internal Heat Generation Using LBM

Mohammad Abu Taher[1], Yeon Won Lee[2], Heuy Dong Kim[1*]
[1]School of Mechanical Engineering, Andong National University, Andong, South Korea
[2]School of Mechanical Engineering, Pukyong National University, Busan, South Korea

ABSTRACT

Fluid flow and heat transfer analysis of Cu-H_2O nanofluid in a square cavity using a Thermal Lattice Boltzmann Method (TLBM) have been studied in the present work. The LBM has built up on the D2Q9 model and the single relaxation time method called the Lattice-BGK (Bhatnagar-Gross-Krook) model. The effect of suspended nanoparticles on the fluid flow and heat transfer analysis have been investigated for different non dimensional parameters such as particle volume fraction (φ) and particle diameters (d_p) in presence of internal heat generation (q) of nanoparticles. It is seen that flow behaviors and the average rate of heat transfer in terms of the Nusselt number (Nu) as well as the thermal conductivity of nanofluid are effectively changed with the different controlling parameters such as particle volume fraction ($2\% \leq \varphi \leq 10\%$), particle diameter ($d_p$ = 5 nm to 40 nm) with fixed Rayleigh number, Ra = 10^5. The present results of the analysis are compared with the previous experimental and numerical results for both pure and nanofluid and it is seen that the agreement is good indeed among the results.

Keywords: Nanofluid; Lattice-Boltzmann; Volume Fractions; Particle Diameter; Heat Generation

1. Introduction

The term nanofluid is envisioned to describe a solid-liquid mixer which consists of nano-sized solid particles and a base liquid and this is one of new challenges for thermo-sciences provided by the nanotechnology. It has potential applications in the microelectromechanical systems (MEMS) and electronics cooling industries. Therefore, in the recent years, the micro and nano systems have become of great interest due to their important and promising applications in various fields. As a result, the research topic of nanofluids has been receiving increased attention worldwide. In fact, numerous theoretical and experimental studies of the effective thermal conductivity of suspensions that contains solid particles have been conducted since Maxwell's theoretical work was published more than 100 years ago. All of the studies on thermal conductivity of suspensions have been considered with millimeter to micrometer sized particles. The use of the conventional millimeter and micrometer-sized particles in heat transfer fluids in practical devices is greatly limited by the tendency of such particles to settle rapidly and to clog mini and micro channels. Until now, it is very difficult to solid particles from eventually settling out of suspension because of particle size. However, nanoparticles appear to be ideally suited for applications in which fluids flow through small passages, because the nanoparticles are stable and small enough not to clog flow passage. Therefore, the main goal of nanofluids is to achieve the maximum thermal properties with the minimum possibility of particle volume fractions (less than 1%) and size of particles (less than 10 nm) by uniform dispersion and stable suspension of nanoparticles in base fluids/liquids. The performance of heat transfer of nanofluid depends on more factors such as shape of particles, the dimension of particles, the volume fractions of particle in the suspensions, and the thermal properties of particle materials [1,2]. However, significant amounts of experimental and theoretical work have been performed on buoyancy induced flow in conventional fluid [3,4]. The mechanism of nanoparticles, enhancement of heat transfer characteristics are described more detailedly in the books [5,6] and Yu *et al.* [7].

From the microscopic point of view, classical mechanics has no insight into the microstructure of the substance; however, statistical mechanics calculated the

*Corresponding author.

properties of state on the basis of molecular motions in a space, and on the basis of the intermolecular interactions. The Lattice Boltzmann equation (LBE) is one of the methods available to deal with such problems by Shan and Chen [8]. In general case, most of authors considered only mass and momentum conservation in LBM [9]. The macroscopic equations of these models correspond to the Navier-Stokes (NS) equations with ideal gas equation of state and a constant temperature. However, sometimes it is important to simulate thermal effects simultaneously with the fluid flows [10]. The more detailed thermal Lattice-Boltzmann method (TLBM) with some examples is discussed in the books written by Mohammad [11] and Succi [12].

A lattice Boltzmann model has been developed by Shan and Chen [13] to simulate the fluid flows containing multiple phases and components. It is also known as Shan-Chen (S-C) multi-component model. Using the two-component LBM, the Rayleigh-Benard convection in two and three dimensions for different cases have been studied by many researchers [14-17]. However, Buick and Gretaed [18] introduced a body force into LBM by modifying the collision function. To overcome the limitations of multi-component flows in the existing traditional computational methods, many researchers were interested to use multi-component LBM for multi-phase flows like solid-fluid mixtures [19-21]. In the present work, multi-component thermal Lattice-Boltzmann method (TLBM) is used for simulating natural convection H_2O-Cu nanofluid with Boussinesq approximation in a square cavity considering the internal heat generation effect of Cu nanoparticles. As far as we know, there is no work on nanofluid heat transfer under natural convection using TLBM with considering internal heat generation effect as well. The results of the analysis are compared with experimental and numerical data both for pure and nanofluids, and shown a relatively good agreement.

2. Formulation of the Problem

2.1. Mathematical Analysis

To ensure the model satisfies the N-S equations for a fluid under the influence of body force, multi-component Lattice-Boltzmann equation (LBE) can be written as [19]

$$\begin{aligned} & f_i^\sigma\left(x+e_i\Delta t, t+\Delta t\right)-f_i^\sigma\left(x,t\right) \\ & =-\frac{1}{\tau_m^\sigma}\left(f_i^\sigma\left(x,t\right)-f_i^{\sigma,eq}\left(x,t\right)\right) \\ & +\frac{2\tau_m^\sigma-1}{2\tau_m^\sigma}\frac{D}{A_i c^2}\vec{e}_i\cdot\vec{F}^\sigma \end{aligned} \tag{1}$$

The function $f_i^\sigma(x,t)$ is the particle distribution function of σ = 1, 2 components with lattice velocity vectors e_i, A_i is the adjustable coefficient, D is the dimension, F^σ is applied force, and $\omega^\sigma = 1/\tau_m^\sigma$ is the relaxation parameter that depends on the local macroscopic variables ρ and $\rho\boldsymbol{u}$. These variables should satisfy the following laws of conservation:

$$\rho^\sigma\left(x,t\right)=\sum_i m^\sigma f_i^\sigma\left(x,t\right)$$

and

$$\rho^\sigma(x,t)\boldsymbol{u}^\sigma = m^\sigma\sum_i \boldsymbol{e}_i f_i^\sigma\left(x,t\right) \tag{2}$$

where m^σ is the molecular mass. For two dimensional D2Q9 model, the equilibrium distribution function can be defined as

$$f_i^{\sigma,eq} = \rho^\sigma w_i\left[1+\frac{3}{c^2}\boldsymbol{e}_i.\boldsymbol{u}^{\sigma,eq}+\frac{9}{2c^4}\left(\boldsymbol{e}_i.\boldsymbol{u}^{\sigma,eq}\right)^2-\frac{3}{2c^2}u^{\sigma,eq2}\right] \tag{3}$$

where w_i is the lattice weighting factors. Therefore the equilibrium velocity becomes

$$\boldsymbol{u}^{eq} = \frac{\sum_\sigma \rho^\sigma\boldsymbol{u}^\sigma/\tau_m^\sigma}{\sum_\sigma \rho^\sigma/\tau_m^\sigma}+\frac{\boldsymbol{F}^\sigma}{2\rho^\sigma} \tag{4}$$

Simultaneously, the lattice Boltzmann energy equation without viscous dissipation for nanofluid defined as [19]

$$\begin{aligned} & g_i^\sigma\left(x+e_i\Delta t, t+\Delta t\right)-g_i^\sigma\left(x,t\right) \\ & =-\frac{1}{\tau_\theta^\sigma}\left(g_i^\sigma\left(x,t\right)-g_i^{\sigma,eq}\left(x,t\right)\right)+\Delta t\, w_i\, G^\sigma \end{aligned} \tag{5}$$

The energy distribution function can be written as

$$g_i^{\sigma,eq} = \varepsilon^\sigma w_i\left[1+\frac{3}{c^2}\boldsymbol{e}_i.\boldsymbol{u}^{\sigma,eq}+\frac{9}{2c^4}\left(\boldsymbol{e}_i.\boldsymbol{u}^{\sigma,eq}\right)^2-\frac{3}{2c^2}u^{\sigma,eq2}\right] \tag{6}$$

$\varepsilon^\sigma(x,t)=\sum_i g_i^\sigma(x,t)$ is the internal energy variable and $G^\sigma = q^\sigma/\rho^\sigma C_p^\sigma$ is the source or force term per unit time, which is related to the heat generation (q^σ) per unit volume. In the present study, we consider the heat is generated from nanoparticles only, not from base fluid. The mean velocity, temperature, viscosity and thermal diffusivity of the nanofluid can be written as

$$\begin{aligned} & \nu=\left(\sum_\sigma c_m^\sigma\tau_m^\sigma-\frac{1}{2}\right)Cs^2\Delta t, \alpha=\left(\sum_\sigma c_\theta^\sigma\tau_\theta^\sigma-\frac{1}{2}\right)Cs^2\Delta t \\ & \boldsymbol{u}=\frac{\sum_\sigma m^\sigma\sum_i f_i^\sigma(x,t)\boldsymbol{e}_i}{\sum_\sigma \rho^\sigma}, T=\frac{\sum_\sigma \varepsilon^\sigma(x,t)}{\sum_\sigma \rho^\sigma C_p^\sigma} \end{aligned} \tag{7}$$

where c_m^σ and c_θ^σ are the concentration of viscosity and diffusivity of each component respectively. Solving the Equations (1) and (5) with other approximations, we get all information that we interested in our study.

2.2. Numerical Analysis

The physical configuration with boundary conditions for the present study is shown in **Figure 1**. A closed square cavity of length H is considered here. The horizontal walls are assumed to be insulated whereas the vertical walls are maintained at constant but different temperatures T_h (hot) and T_c (cold). For natural convection, the momentum and energy equations are coupled and the flow is driven by temperature or mass gradient, Under Boussinesq approximation, the force term per unit mass can be written as

$$F(x,t) = \rho(x,t)\, g\, \beta\left(T(x,t) - T_{ref}\right)$$

The ratio of the buoyancy force to the product of viscous force and heat diffusion rates defines the Rayleigh number, $\mathrm{Ra} = \mathrm{Pr} \times \mathrm{Gr} = g\beta\Delta T H^3 / \alpha v$, where Gr is the Grashof number and Pr is the Prandtl number. The boundary conditions are defined as:

$$u = v = \frac{\partial T}{\partial y} = 0, \text{ at } y = 0, H \text{ and } 0 \le x \le L$$

$$T = T_h, u = v = 0, \text{ at } x = 0 \text{ and } 0 \le y \le H$$

$$T = T_c, u = v = 0, \text{ at } x = L \text{ and } 0 \le y \le H$$

2.3. Code Validation

To verify our results with the conventional benchmark and with experimental results, first we consider the cavity is filled with pure fluid like air (Pr = 0.71). It is shown in **Figure 2**. It can be seen from these figures that our present numerical simulation is in very good agreement with numerical results of [3] and experimental solutions of [4]. Moreover, the present LBM also applied for nanofluids and it is necessary to verify our method with other published work. A good agreement is obtained between the present solution and bench mark solution of [3] as illustrated in **Figure 3**. This confirms that our method is correct both for pure and nanofluids.

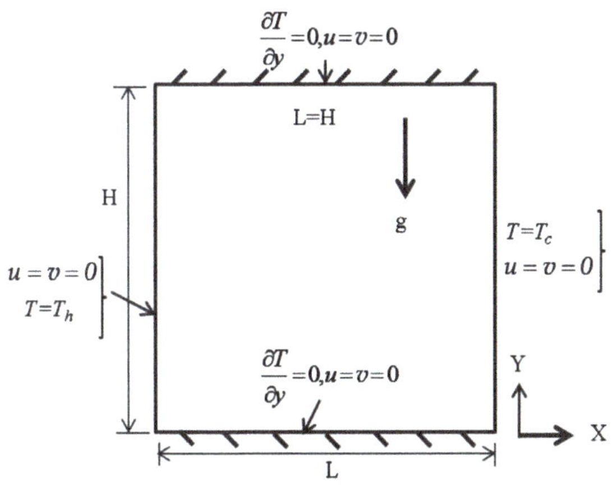

Figure 1. The configuration of the problem under consideration.

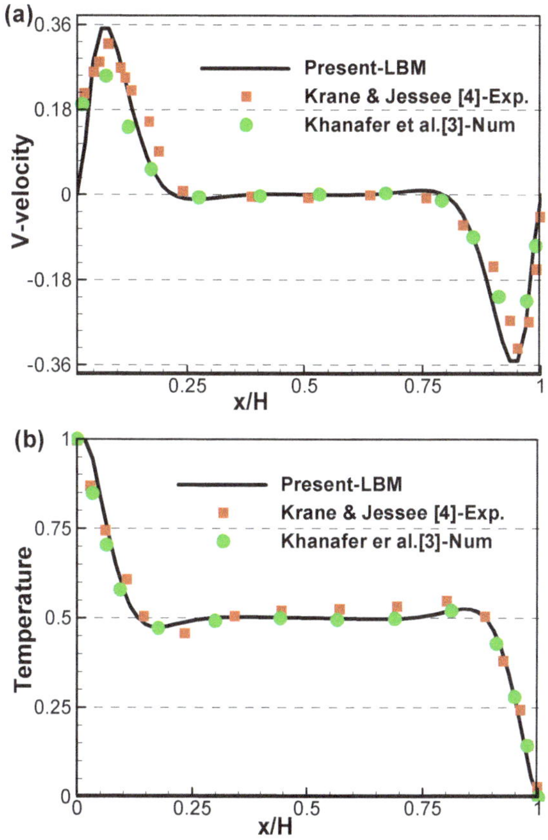

Figure 2. Comparison of dimensionless (a) velocity and (b) temperature profiles for pure fluid (Pr = 0.71, Ra = 1.89 × 10^5).

3. Results and Discussions

Equations (1) and (5) are solved simultaneously on a uniform 2D grid system along with boundary conditions and other equations described in the above sections. Each numerical time step consists of three stages (1) collision, (2) streaming, and (3) boundary conditions steps followed by the LBM approaches. Fluid flow and heat transfer enhancement of nanofluid for a wide range of controlling parameters are discussed in this section. It is assumed that the nanoparticles of Cu are uniformly suspended in water; there is no aggregation of nanoparticles in the fluid medium.

The computed velocity field in the buoyancy driven cavity flow for pure and nanofluids are seen in **Figures 4(a)** and **(b)** respectively. The arrows denote the velocity vectors in both magnitude and direction and the solid lines represent the stream lines of H_2O-Cu nanofluid. The general magnitude of the velocity can be seen to increase as the hot fluid, near the hot wall, flow upward and the

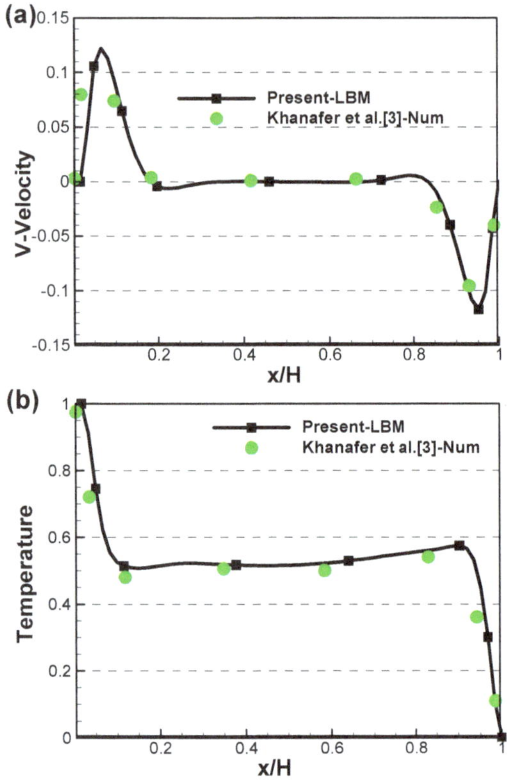

Figure 3. Comparison of velocity and temperature profiles for nanofluid (Pr = 6.2, Gr = 10^5, ϕ = 10%).

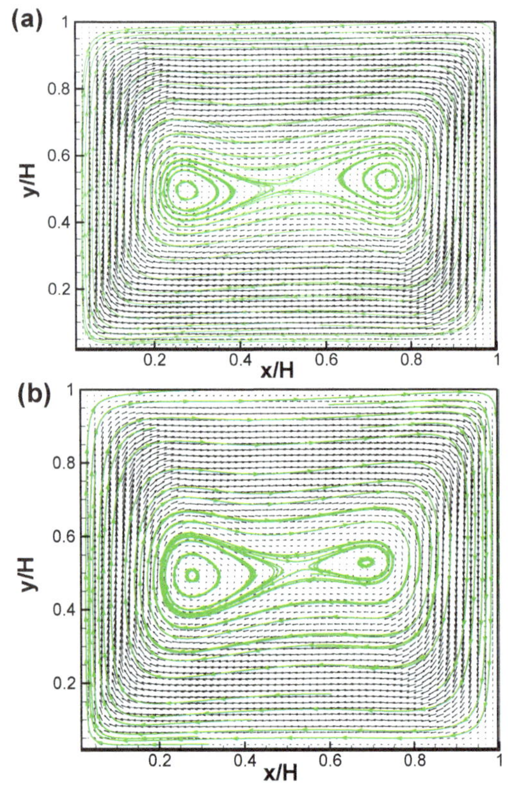

Figure 4. Velocity vectors and stream lines for (a) pure fluid (H_2O) and (b) nanofluid (H_2O-Cu).

cold fluid, near the cold wall, flow downward due to the effect of buoyancy force for both pure and nanofluids. It is one of the main criteria of natural convection in a differently heated cavity. Thus it is clear that the nanofluids flow behavior like a pure fluid.

The effect of internal heat generation of nanoparticles on the velocity and temperature fields at the mid section of the cavity for different particle diameters and volume fractions are illustrated in **Figures 5** and **6**.

It is seen from **Figure 5(a)** that the velocity profiles of nanofluid decrease remarkably with increasing the particle diameter near hot wall and increase near the cold wall. But, in presence of internal heat generation of nanoparticles (dashed lines, **Figure 5(a)**), the velocity profiles near the hot and cold walls have shown infinitesimal changed, decreased (zoom viewed) near hot wall. It is also seen that the location of the local maximum and minimum for all cases have tend to moved to right from hot wall and left from cold wall with increasing the particle diameter. For dp = 5 nm, the local maximum and minimum are seen at approximately x/H = 0.06 and x/H = 0.85, whereas for dp = 10 and 20 nm, the locations are observed at x/H = 0.09 and x/H = 0.93; and x/H = 0.10 and x/H = 0.90 respectively. These represented that the velocity boundary layer become thicker with increasing particle diameters.

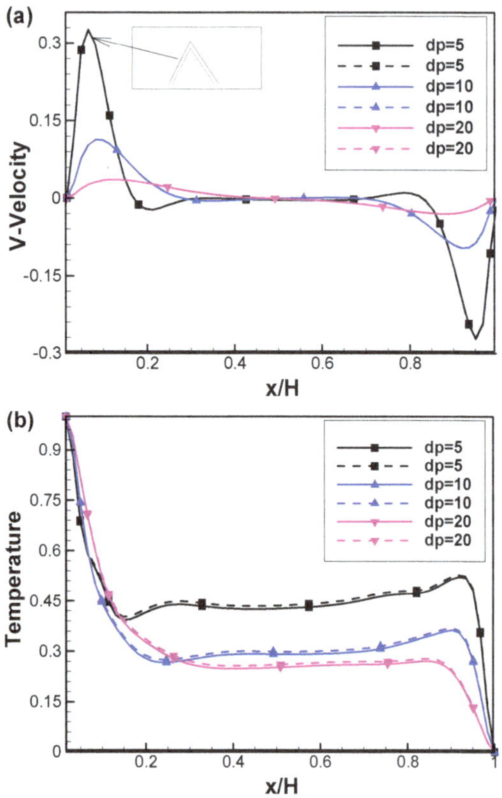

Figure 5. (a) Dimensionless velocity and (b) temperature profiles for different particle diameters (nm) (solid lines: no heat generation, dashed lines: with heat generation).

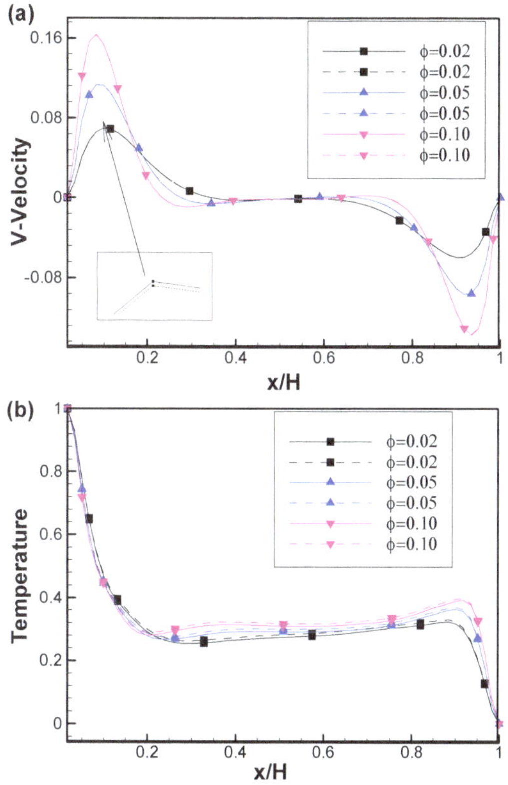

Figure 6. (a) Dimensionless velocity and (b) temperature profiles for different particle volume fractions (%) (solid lines: no heat generation, dashed lines: with heat generation).

From **Figure 5(b)**, it is observed that, the temperature profiles increase slightly near the hot wall but decrease significantly throughout the cavity with increasing the particle diameters and finally satisfied the boundary conditions. Moreover, in presence of heat generation, dased lines in **Figure 5(b)**, the temperature profiles slightly increased compare to without heat generation. It means that, if the nanoparticles generate the heat, the particle lost some energy to the fluid and consequently the average temperature of nanofluid increased. From **Figure 5 (b)**, the significant temperature gradient observed in the vicinity of the heated surface approximately, $0 \leq x/H \leq 0.15$, and unheated surface $0.95 \leq x/H \leq 1$ for $dp = 5$ nm; $0 \leq x/H \leq 0.22$ and $0.92 \leq x/H \leq 1$ for $dp = 10$ nm; $0 \leq x/H \leq 0.28$ and $0.88 \leq x/H \leq 1$ for $dp = 20$ nm. Therefore, it is seen that thermal stratification and increase in temperature in the direction of heat flow in the core region respectively $0.15 < x/H < 0.95$, $0.22 < x/H < 92$ and $0.28 < x/H < 0.88$ for $dp = 5$ nm, 10 nm and 20 nm. This indicates that thermal boundary layer increased with increasing particle diameters.

The effects of volume fractions with internal heat generation of nanoparticles on fluid flow are shown in **Figure 6**. It is shown that the velocity profiles of nanofluid increases remarkably with volume fractions of nanofluid near hot wall and decreased near cold wall. In more details, the velocity along the vertical walls of the cavity show a higher level of activity as predicted by thin layer of hydrodynamic velocity boundary layers. Whereas, the variation of the velocity at the center of the cavity for all volume fractions of nanoparticles are negligible compared with those of boundaries in all cases. This is expected as the maximum fluid motion at the boundaries and almost stagnant in the core region. The locations of the local maximum and minimum velocities for all cases are approximately at the same position of $X(=x/H)$. The effect of dimensionless temperature profiles with increase of volume fractions at the centerline of the cavity, which is perpendicular direction of the heated walls of the cavity, as shown in **Figure 6(b)**. The variation of temperature near hot and cold walls versus dimensionless horizontal length is linear for all volume fractions, which is the characteristic of heat transfer by convection. For all values of ϕ, a significant temperature gradient is observed in the vicinity of the heated surface approximately, $0 \leq x/H \leq 0.22$, and unheated surface $0.92 \leq x/H \leq 1$. Moreover, it is seen that thermal stratification and increase in temperature in the direction of heat flow in the core region $0.22 < x/H < 0.92$. In physical meaning, the temperature should decrease in the direction of heat flow, but for natural convection in a cavity, the rate of cooling is expected to be higher near the heated and unheated walls due to the fluid motion and hydrodynamics effects. This phenomenon is attributed to buoyancy-induced cellular flows in the boundary layer adjoining the heated and cold walls. The fluid motion increases the rate of heat transfer.

A usual means of characterizing heat transfer is to calculate the Nusselt number. Actually the Nusselt number Nu is a dimensionless form of the heat transfer coefficient. To investigate the heat transfer performance in terms of effective thermal conductivity of the nanofluid, the local Nusselt number Nu_L and the averaged Nusselt number Nu over the flow channel are respectively defined as follows:

$$Nu_L = \frac{hH}{k_f} \tag{8}$$

The heat transfer coefficient h is defined as

$$h = \frac{q_w}{T_h - T_c}, \quad q_w = -k_{nf}\frac{\partial T}{\partial x} \tag{9}$$

Here T_h and T_c are the temperature of heated and cooled walls respectively. The heat flux, q_w, of nanofluid can be expressed in terms of thermal conductivity K_{nf} of nanofluid, k_f are the thermal conductivity of the base fluid and H is the channel height. The Nusselt number and the average Nusselt number in dimensionless form on the left heated wall are defined as

$$Nu = \int_0^1 Nu_L(Y)\,\mathrm{d}Y,\; Nu_L(Y) = -\frac{k_{nf}}{k_f}\left(\frac{\partial\theta}{\partial X}\right)_{wall} \quad (10)$$

where, θ, is the non-dimensional temperature. On the basis of the definition [5], the effective thermal conductivity for a two-component mixture (Cu-H_2O) is defined as

$$k_{nf} = \alpha_{nf}\left(\rho C_p\right)_{nf} \quad (11)$$

The average Nusselt number and the thermal conductivity ratios for constant Rayleigh number, Ra = 10^5 at 100,000 time steps with various volume fractions and particle diameters as well as the internal heat generation of nanoparticles are shown in **Figures** 7 and **8**.

The dimensionless rate of heat transfer called Nusselt number (Nu) is significantly decreased with increasing the particle diameters from *dp* = 5 nm to 40 nm, as described in **Figure 7(a)**, though it is increased with particle volume fractions. Moreover, it is observed that in presence of nanoparticles heat generation, the Nusselt number slightly decreases (dashed lines) compare to without heat generation (solid lines). Similar phenomenon is seen for thermal conductivity ratio. It is seen

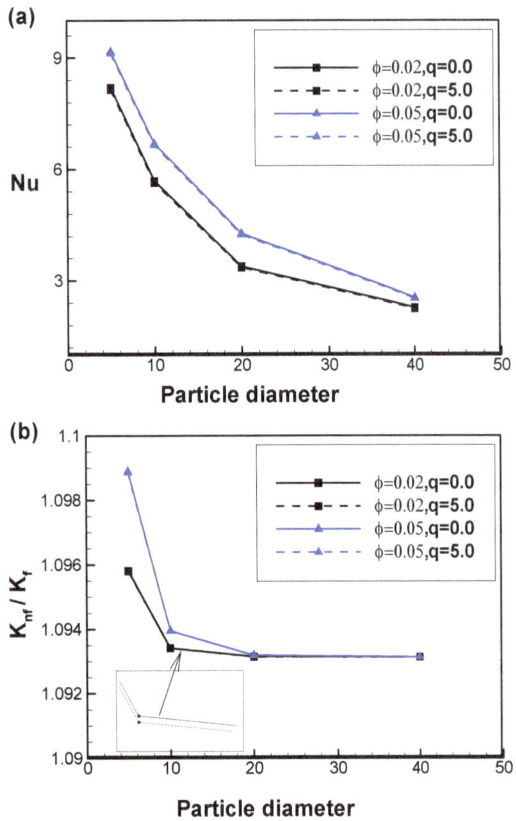

Figure 7. (a) The average rate of heat transfer (Nusselt number, Nu); (b) Thermal conductivity ratios for different particle diameters (solid lines: no heat generation, dashed lines: with heat generation).

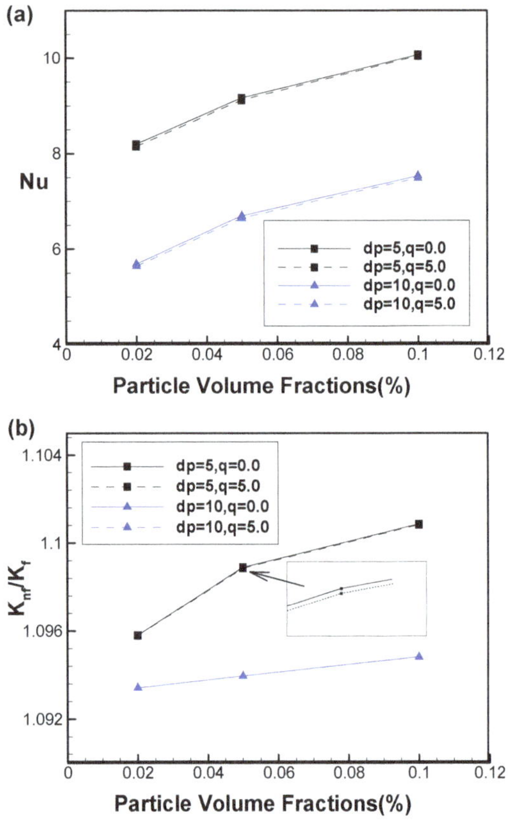

Figure 8. (a) The average rate of heat transfer (Nusselt number, Nu); (b) Thermal conductivity ratios for different particle volume fraction s (%) (solid lines: no heat generation, dashed lines: with heat generation).

almost constant for *dp* > 30 nm for all cases. This is because constant Rayleigh number effect. In natural convection flow, the thermal conductivity is strongly dependent on the Rayleigh number.

It can be seen from **Figure 8** that both of the measured dimensionless rate of heat transfer in terms of Nusselt number and the thermal conductivity are increased with increased of volume fractions. This increased is significantly depending on particle size. For increasing particle diameter both of the heat transfer characteristics are dramatically decreased. With the heat generation of nanoparticles, it is also observed that both of heat transfer rate and conductivity ratio decreases very slightly.

Actually, in the above cases, the particle size is important factor. The large size of particle diameter means that the less number of particles. Therefore, the surface area of nanoparticle decreases, giving less heat transfer area between the phases. In natural convection flow, only buoyancy force is a dominant force, therefore, particle movement is very important. The main problem for large size of particle is the rapid settling in the fluid. Other problems are abrasion and clogging. These problems are highly undesirable for many practical cooling industries.

4. Conclusions

The Thermal Lattice-Boltzmann Method (TLBM) is successfully applied in this work to simulate buoyancy-driven heat transfer characteristics and flow performance of Cu-H_2O nanofluid in a square cavity. Throughout our calculation, it is assumed that constant Rayleigh number, Ra = 10^5. So the fluid flow and heat transfer analysis are depending on particle size and volume fractions. Moreover, it is considered that the internal heat generation effect of nanoparticles to the base fluid, and then finally, we have shown our result of the resultant fluid called nanofluid. The present results indicate that the following statements:

- Both of the thermal boundary layer and velocity boundary layer thickness increased with increasing the particle diameter. However, the change of thickness of thermal and velocity boundary layers are negligible with increasing the particle volume fractions within the range up to 10%.
- The heat transfer features, the heat transfer rate and the thermal conductivity ratio, of a nanofluid are increased with increase of the nanoparticle volume fractions.
- However, the rate of heat transfer and thermal conductivity ratio are significantly decreased with increase of particle diameter.
- In addition, all of the above features are very slightly changed (decreased) with internal heat generation effect of nanoparticle due to the constant Rayleigh number and Prandtl number at the present study.
- For small particle diameter, the nanofluid behaves more like a fluid than the conventional solid-fluid mixer. The results from our analysis have shown quite good and this temperature characteristic enhancement plays a significant role in engineering applications such as in the electronic cooling industries or MEMS devices.

REFERENCES

[1] S. Lee, S. U.-S. Choi, S. Li and J. A. Eastman, "Measuring Thermal Conductivity of Fluids Containing Oxide Nanoparticles," *Journal of Heat Transfer*, Vol. 121, No. 2, 1999, pp. 280-289.

[2] Y. Xuan and Q. Li, "Heat Transfer Enhancement of Nanofluids," *International Journal of Heat and Fluid Flow*, Vol. 21, No. 1, 2000, pp. 58-64.

[3] K. Khanafer, K. Vafai and M. Lightstone, "Buoyancy-Driven Heat Transfer Enhancement in a Two-Dimensional Enclosure Utilizing Nanofluids," *International Journal of Heat and Mass Transfer*, Vol. 46, No. 19, 2003, pp. 3639-3653.

[4] R. Krane and J. Jessee, "Some Detailed Field Measurement for a Natural Convection Flow in a Vertical Square Enclosure," *Proceedings of the First ASME-JSME Thermal Engineering Joint Conference*, Honolulu, 20-24 March 1983, pp. 323-329.

[5] S. K. Das, S. U.-S. Choi, W. Yu and T. Pradeep, "Nanofluids, Science and Technology," Wiley Interscience, Hoboken, 2007.

[6] M. A. Taher, "Fluid Flow and Heat Transfer Analysis of Pure and Nanofluids Using Lattice-Boltzmann Method," Ph.D. Thesis, Pukyong National University, Busan, 2009.

[7] W. Yu, D. M. France, J. L. Routbort and S. U.-S. Choi, "Review and Comparison of Nanofluid Thermal Conductivity and Heat Transfer Enhancements," *Heat Transfer Engineering*, Vol. 29, No. 29, 2008, pp. 432-460.

[8] X. Shan and H. Chen, "Lattice Boltzmann Model for Simulating Flows with Multiple Phases and Components," *Physical Review E*, Vol. 47, No. 3, 1993, pp. 1815-1820.

[9] M. A. Taher and Y. W. Lee, "Numerical Study on Reduction of Fluid Forces Acting on a Circular Cylinder with Different Control Bodies Using Lattice-Boltzmann Method," *International Journal of Energy & Technology*, Vol. 4, No. 18, 2012, pp. 1-7.

[10] M. A. Taher, S. C. Saha, Y. W. Lee and H. D. Kim, "Numerical Study of Lid-Driven Square Cavity with Heat Generation Using LBM," *American Journal of Fluid Dynamics*, Vol. 3, No. 2, 2013, pp. 40-47.

[11] A. A. Mohammad, "Applied Lattice Boltzmann Method for Transport Phenomena, Momentum, Heat and Mass Transfer," The University of Calgary, Calgary, 2007.

[12] S. Succi, "The Lattice Boltzmann Equation for Fluid Dynamics and beyond," Oxford University Press, Oxford, 2001.

[13] X. Shan, "Simulation of Rayleigh-Bernard Convection Using Lattice Boltzmann Method," *Physical Review E*, Vol. 55, No. 3, 1997, pp. 2780-2788.

[14] M. A. Taher, K.-M. Li and Y.W. Lee, "Numerical Study of H_2O-Cu Nanofluid Using Lattice-Boltzmann Method," *Journal of the Korean Society of Marine Engineering*, Vol. 34, No. 1, 2010, pp. 53-61.

[15] A. K. Santa, S. Sen and N. Chakraborti, "Study of Heat Transfer Augmentation in a Differentially Heated Square Cavity Using Copper-Water Nanofluid," *International Journal of Thermal Sciences*, Vol. 47, No. 9, 2008, pp. 1113-1122.

[16] J. Cai, X. Huai, R. Yan and Y. Cheng, "Numerical Simulation on Enhancement of Natural Convection Heat Transfer by Acoustic Cavitation in a Square Enclosure," *Applied Thermal Engineering*, Vol. 29, No. 10, 2009, pp. 1973-1982.

[17] S. Sivasankaran, T. Asaithambi and S. Rajan, "Natural Convection of Nanofluids in a Cavity with Linearly Varying Wall Temperature," *Maejo International Journal of Science and Technology*, Vol. 4, No. 3, 2010, pp. 468-482.

[18] J. M. Buick and C. A. Greated, "Gravity in Lattice Boltzmann Model," *Physical Review E*, Vol. 61, No. 5, 2000, pp. 5307-5320.

[19] Y. Xuan and Z. Yao, "Lattice Boltzmann Model for Nanofluids," *Heat Mass Transfer*, Vol. 41, No. 3, 2005, pp. 199-205.

[20] N. S. Martys and H. Chen, "Simulation of Multicomponent Fluids in Complex Three-Dimensional Geometries by the Lattice Boltzmann Method," *Physical Review E*, Vol. 53, No. 1, 1996, pp. 743-750.

[21] Y. M. Xuan, K. Yu and Q. Li, "Investigations of Flow and Heat Transfer of Nanofluids by the Thermal Lattice-Boltzmann Model," *Progress in Computational Fluid Dynamics*, Vol. 5, No. 1-2, 2005, pp. 13-19.

Numerical Analysis of the Losses in Unsteady Flow through Turbine Stage

Sławomir Dykas, Włodzimierz Wróblewski, Dawid Machalica
Institute of Power Engineering and Turbomachinery, Silesian University of Technology, Gliwice, Poland

ABSTRACT

This paper presents an analysis of the operation of a stage of an aircraft engine gas turbine in terms of generation of flow losses. The energy loss coefficient, the entropy loss coefficient and an additional pressure loss coefficient were adopted to describe the losses quantitatively. Distributions of loss coefficients were presented along the height of the blade channel. All coefficients were determined based on the data from the unsteady flow field and analyzed for different mutual positioning of the stator and rotor blades. The flow calculations were performed using the Ansys CFX commercial software package. The analyses presented in this paper were carried out using the URANS (Unsteady Reynolds-Averaged Navier-Stokes) method and two different turbulence models: the common Shear Stress Transport (SST) model and the Adaptive-Scale Simulation (SAS) turbulence model, which belongs to the group of hybrid models.

Keywords: Turbine Stage; Stator; Rotor; Loss Coefficients

1. Introduction

Aerodynamic losses in the flow through a turbine stage have an adverse impact on energy conversion efficiency. Obviously, due to the turbine stage specificity, the nature of these losses is unsteady. This is mainly related to the impact of the rotating rotor blade ring with the stationary stator blade ring and the intense turbulent phenomena which are responsible for blade losses [1]. These losses include as follows [2]:

- Profile losses which result from the friction of the viscous fluid against the blade surface, as well as from the finite thickness of the trailing edge, behind which the so-called aerodynamic wake is formed.
- Boundary losses which result from the effect that the near-wall layer and the medium flowing in the blade channel have on each other. The medium velocity in the near-wall boundary layer area, on the surfaces limiting the channel from top and bottom, is lower than in the remaining part of the channel. This leads to an imbalance of the forces resulting from the distribution of pressure and the deflection of the medium stream. Consequently, streamlines get deflected towards the base of the blade channel as well as in the direction of the convex (sucking) blade surface. The intensity of the described phenomena increases as the blade channel relative height becomes smaller.
- Losses resulting from incomplete feed. These losses occur in some steam turbines fed with high-parameter steam and in small gas turbines. They result from the fact that the stages used in such structures are not fed at full perimeter. Losses related to incomplete feed manifest themselves by gas being pushed out of the rotor channels which, due to the rotation of the rotor, found themselves before the fed nozzle segment and by the gas being mixed in areas at the edge of the feeding segment.
- Losses related to the cooling of the first stages of the gas turbine.

The energy dissipation phenomena occurring in a turbine stage are more and more often analyzed by means of the computational fluid dynamics (CFD) tools [3-5]. This paper presents the methodology and the results of CFD analyses comprising the unsteady flow in the turbine stage of an aircraft engine [6]. The use of the Ansys CFX commercial software package makes it possible to identify places where losses arise and in many cases allows their physical interpretation. The energy loss coefficient, the entropy loss coefficient and an additional pressure loss coefficient [1,2] were adopted to describe the losses quantitatively. The analyses presented in this paper were carried out using the URANS method and two different turbulence models: the common Shear Stress Transport (SST) model and the Adaptive-Scale Simulation (SAS) turbulence model, which belongs to the group of hybrid models. The aim of the comparison of results obtained from analyses conducted with different turbulence models was to show whether the application of the SAS model, which is much more demanding in terms of equipment, is justified in the type of simulation under discussion.

2. Definitions of Loss Coefficients and Their Physical Interpretation

The expansion process in the turbine reaction stage (**Figure 1**) occurs both in the stator and rotor. It is an irreversible process accompanied by energy dissipation caused mainly by aerodynamic, thermodynamic and leakage-related losses.

Three types of loss coefficients [2] were used in this work to make a quantitative assessment of the losses arising in blade channels:

2.1. Entropy Loss Coefficient

The form of the entropy loss coefficient is derived directly from the definition of isentropic efficiency. This efficiency is defined for static parameters before and after the blade ring, where the difference between the values of real and isentropic enthalpy of the end of the expansion process is replaced with an increase in entropy, according to the second law of thermodynamics.

$$\zeta_{S\,st} = 1-\eta_{s0-1} = 1-\frac{h_0-h_1}{h_0-h_1+T_1\left(s_1-s_0\right)} = \frac{T_1\left(s_1-s_0\right)}{h_0-h_1+T_1\left(s_1-s_0\right)} \quad (1)$$

$$\zeta_{S\,ro} = 1-\eta_{s1-2} = 1-\frac{h_1-h_2}{h_1-h_2+T_2\left(s_2-s_1\right)} = \frac{T_2\left(s_2-s_1\right)}{h_1-h_2+T_2\left(s_2-s_1\right)} \quad (2)$$

This coefficient is based on the theorem that the only reliable representation of losses occurring in the flow where an adiabatic process takes place is the increase in entropy [1]. This increase may have several sources. One of them is flow resistance resulting from movement of any viscous fluid. Apart from that, an increment in entropy may be caused by heat transfer at a finite temperature difference and by the unbalanced nature of some processes.

2.2. Energy Loss Coefficient

According to the energy conservation law (the first law of thermodynamics), the energy in a system remains constant; only its form may change. Therefore, the concept of energy loss is impossible from the physical point of view. However, it is common practice to use the energy loss coefficient in order to define the value of energy which does not participate in work generation. The

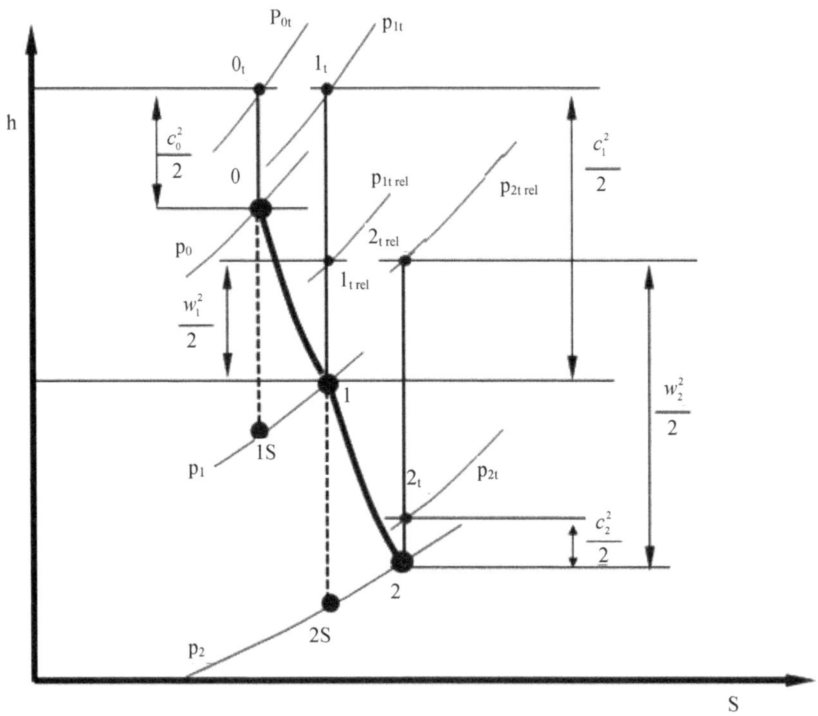

Figure 1. Expansion process in the turbine stage in the h-s chart.

form of this coefficient was derived from the energy conservation law, which states—for the stator blade ring—that total enthalpy is constant, while constant rothalpy is assumed for the rotor blade ring. For perfect gas it may eventually be written using total and static pressures.

$$\zeta_{En\,st} = 1 - \frac{h_{1t} - h_1}{h_{0t} - h_{1s}} \xrightarrow{ideal\ gas} 1 - \frac{1 - \left(\frac{p_1}{p_{1t}}\right)^{\frac{\varkappa-1}{\varkappa}}}{1 - \left(\frac{p_1}{p_{0t}}\right)^{\frac{\varkappa-1}{\varkappa}}} \quad (3)$$

$$\zeta_{En\,ro} = 1 - \frac{h_{2t\,rel} - h_2}{h_{1t\,rel} - h_{2s}} \xrightarrow{ideal\ gas} 1 - \frac{1 - \left(\frac{p_2}{p_{2t\,rel}}\right)^{\frac{\varkappa-1}{\varkappa}}}{1 - \left(\frac{p_2}{p_{1t\,rel}}\right)^{\frac{\varkappa-1}{\varkappa}}} \quad (4)$$

2.3. Pressure Loss Coefficient

This is one of the most common loss coefficients in use. Its popularity results from the fact that determination of the parameters needed to find the coefficient is very easy. This feature is essential in experimental testing, where static and total pressure measurements are commonly applied and relatively easy to perform.

$$\zeta_{P\,st} = \frac{p_{0t} - p_{1t}}{p_{0t} - p_1} \quad (5)$$

$$\zeta_{P\,ro} = \frac{p_{1t\,rel} - p_{2t\,rel}}{p_{1t\,rel} - p_2} \quad (6)$$

The pressure loss coefficient relates a loss of total pressure in the blade ring to the theoretical value of dynamic pressure that the medium would feature at the rotor outlet if loss p_t did not occur. This loss may be caused by friction, presence of shock waves, etc.

3. Turbine Stage Geometry

This work presents an analysis of the unsteady flow in a gas turbine intended for the aircraft industry and described in [6]. The characteristic feature of the turbine is that it has just one stage, which distinguishes it from other turbines used in turbofan engines. Another important factor in the selection of the turbine stage was the publication of complete geometry of the blades and operating parameters of the analyzed stage in [6]. The stage under consideration is composed of 36 38.1 mm high stator blades located at an average radius of 469.9 mm and 64 rotor blades with the same height and radius of location. Due to the lack of data concerning the geometry of the tip seal of the rotor blades, a geometry was selected that is typical for this stage type. The geometry of the stage under analysis is presented in **Figure 2**.

An unstructured hybrid-type mesh composed of approximately 550 k nodes for each blade channel was used to discretize the computational domain. One channel for the stator blade ring and two channels for the rotor blade ring were assumed for the computations. This gives 1.7 M mesh nodes in total. The mesh of the tip seal of the rotor blades was generated as fully structured. It is composed of approximately 300 k nodes. While making the mesh, special care was taken to achieve accurate discretization in the near-wall boundary layers ensuring $y^+ \approx 1$.

The boundary conditions used in the analysis are presented in **Table 1**. The very high drop in pressure between the stage inlet and outlet makes the flow field much more complex. Two turbulence models—the Shear Stress Transport (SST) and the Scale-Adaptive Simulation (SAS) [7] were used in the performed numerical simulations.

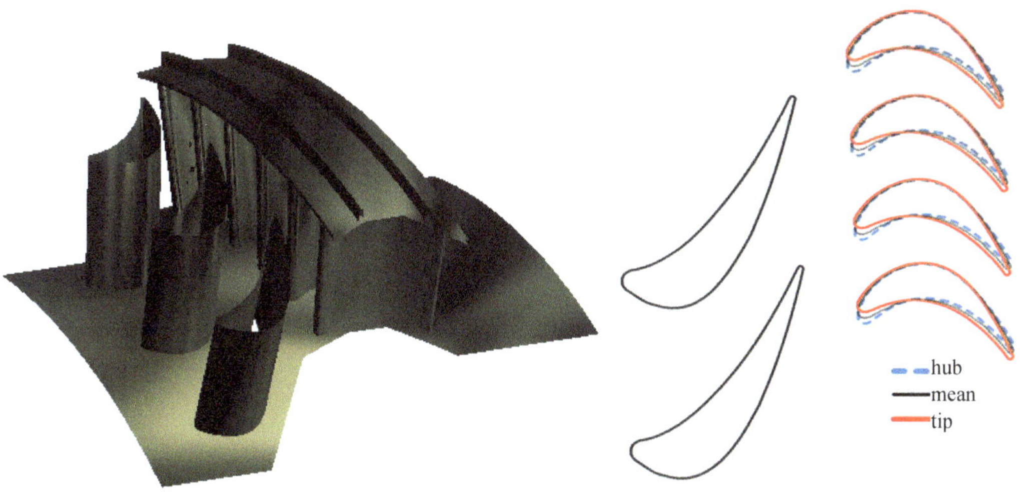

Figure 2. Geometry of the analyzed turbine stage.

Table 1. Data assumed for the reference analysis.

		Steady analysis	Unsteady analysis
	Medium	Air as perfect gas	
	Turbulence model	SST	SST-SAS
	Rotor rotational speed	8081 rev/min	
	Time step		1×10^{-5} s
	Stator/rotor interface type	Stage	Transient rotor-stator
Inlet	Total pressure	101,325 Pa	
	Static temperature	288.2 K	
	Turbulence intensity	5%	
Outlet	Static pressure	26,000 Pa	
	Radial equilibrium	Yes	

The stage-type interface between the stator and rotor was used for steady computations. Its characteristic feature is "averaging" parameters in the circumferential direction. It is also better than an alternative frozen rotor-type interface. For the unsteady analysis, the transient rotor-stator-type interface was applied.

4. Calculation Results

The performed analyses of the unsteady flow field make it possible to determine aerodynamic losses and the impact of the mutual positioning of the stator and rotor on their magnitude. The application of the Ansys CFX package for numerical analyses made it possible to obtain curves of loss coefficients depending on the relative height of the flow channel. In this case, the individual parameters of the medium taken for the analysis were averaged in terms of mass in the circumferential direction. While finding loss coefficients, the parameters with subscript 2 denoting the medium after the rotor were determined in the channel extension. This is to make it possible to take account of the losses resulting from the presence of the blade wake and shock waves. The mutual positioning of the stator and rotor blades for individual steps is presented in **Figure 3**.

4.1. Entropy Loss Coefficient Analysis

Figure 4 presents distributions of the entropy loss coefficient along the channel height. In the case of the stator blade (**Figures 4(a)** and **(c)**) the value of $z_{S,st}$ for almost 80% of the blade height is less than 0.05 (5 percentage points). This proves that the energy conversion level in the stage under analysis is high. The entropy loss coefficient reaches the highest values on the surfaces limiting the channel from top and bottom. This is caused by the occurrence of boundary losses in these areas. The values of these losses at the channel top are smaller than at the bottom, which results from the bigger width of the blade channel. Moreover, higher values of the Mach number occur in the bottom part of the channel than in the remaining parts. The blade channel under analysis features a confusor section and therefore, to reach the velocity of Ma > 1, the medium has to be subjected to Prandtl Meyer expansion. Theoretically, this is an isentropic process. However, the change in the flow direction related to it leads to the creation of shock waves which are the source of losses.

Comparing **Figure 4(a)** with **Figure 4(c)**, which present the distribution of the entropy loss coefficient for the stator channel, it can also be seen that the coefficient values are higher for calculations performed using the SST turbulence model. The differences reach 0.005, which, for example in a new unit designing process, is a significant value.

The minimum and maximum values of the entropy loss coefficient are achieved for both turbulence models used for the same time step. However, the values of coefficient $z_{S,st}$ obtained for the two models differ from each other substantially. For intermediate time steps the discrepancies are even greater.

The distribution of the entropy loss coefficient for the rotor channel is more complex (**Figures 4(b)** and **(d)**). The high value of $z_{S,ro}$ in the upper part of the channel is the effect of, among others, the seal operation. The medium leaving the seal cavity generates secondary flows [2] which lead to a considerable increase in the entropy loss coefficient in the area of 0.8 - 1 of the channel height (**Figure 5**). In the case of the stator, the decrease in the entropy loss coefficient in this region was much more abrupt and small values of the coefficient were obtained already at 0.95 of the channel height. The data presented in **Figures 4(b)** and **(d)** also

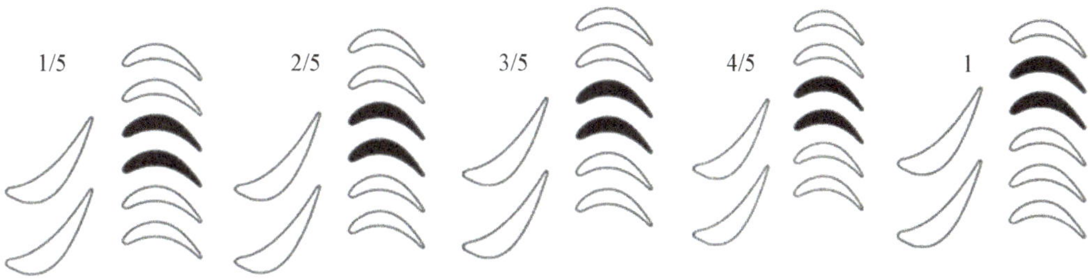

Figure 3. Mutual positioning of the stator/rotor blades for individual time steps.

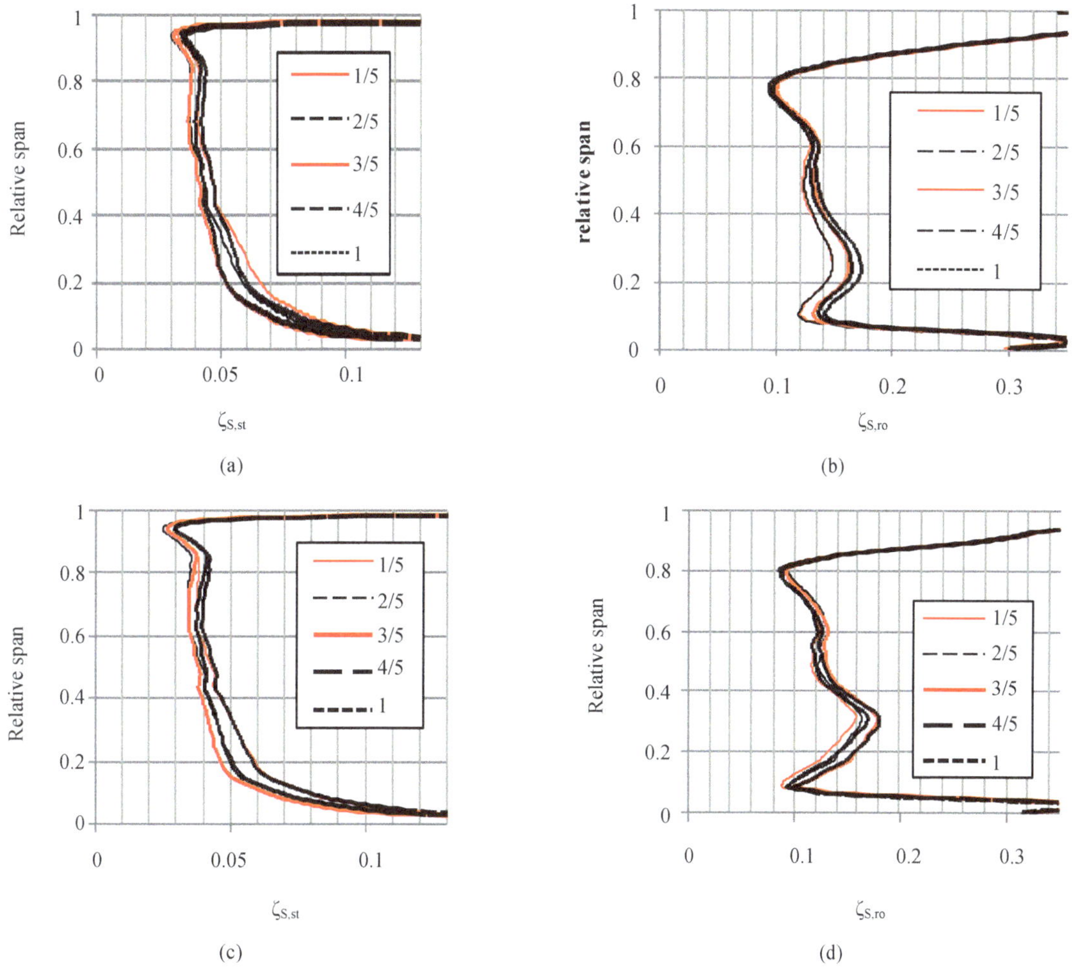

Figure 4. Distribution of the entropy loss coefficient for reference analyses (low temperatures): (a) stator, SST; (b) rotor, SST; (c) stator, SAS; (d) rotor, SAS.

indicate that the entropy loss coefficient for almost the entire height of the channel assumes values higher than 0.1. This proves that the efficiency of the analyzed rotor blade ring is much lower than that of the stator.

At 0.3 of the relative height of the channel coefficient $z_{S,ro}$ reaches values bigger than 0.15. This is related to the substantial increment in the entropy of the medium in the trailing edge area. At this height the increment is clearly bigger than in the remaining part of the channel (**Figure 6**). Also in the area of ~0.3 of the relative height of the channel the biggest changes in the loss coefficient occur depending on the mutual stator/rotor positioning. These changes probably result from the unsteady nature of the blade wake vortices and from the shock waves arising in the trailing edge region.

In the case of the rotor blade ring channel, a big discrepancy is observed between the entropy loss coefficient values calculated for the results obtained using different turbulence models. The difference is especially noticeable in the area of 0.2 - 0.3 of the relative height of the

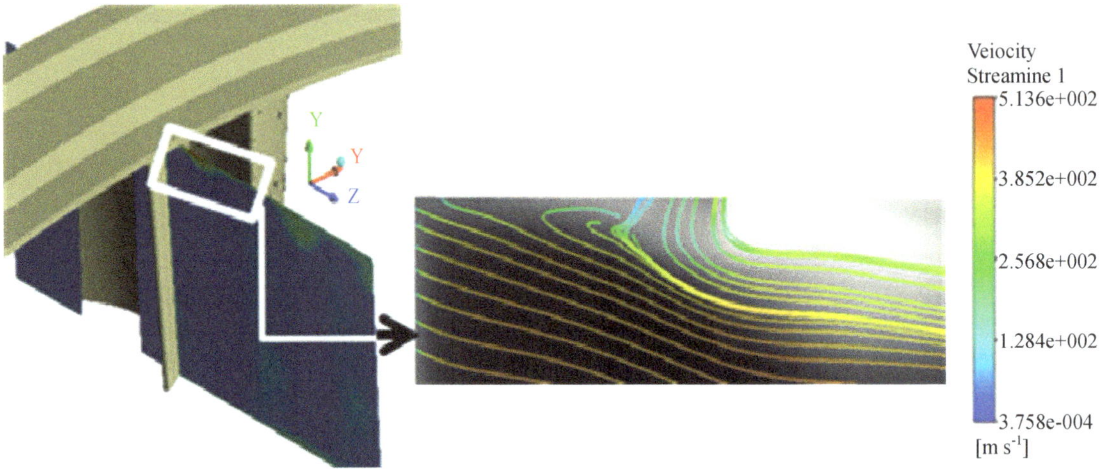

Figure 5. Generation of reverse flows in the area where the medium leaving the seal mixes with the main flow.

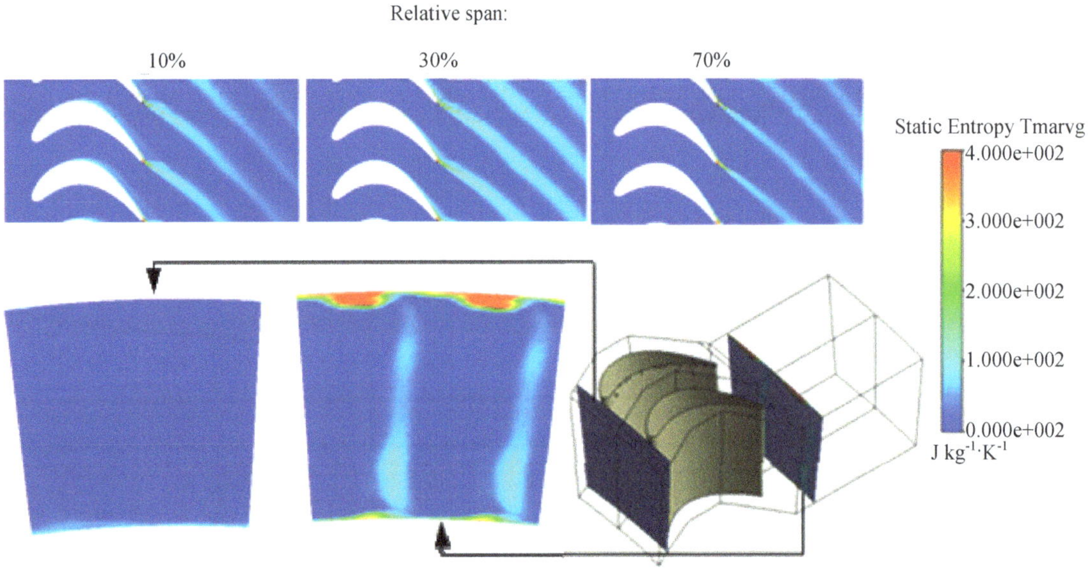

Figure 6. Distribution of static entropy in the rotor channel; reference analysis; SAS turbulence model.

channel, where the values assumed by coefficient $z_{S,ro}$ are high. The SAS model located the mentioned area at a bigger height of the channel compared to the SST model. Moreover, for the SST turbulence model, the values of $z_{S,ro}$ in this area diverge less from the values assumed in other regions of the blade. The fluctuations in the values of $z_{S,ro}$ are in this region higher for the SST model and reach 0.03.

4.2. Energy Loss Coefficient Analysis

Figures 7(a) and **(c)** present distributions of the energy loss coefficient along the height of the stator blade. The amplitude of fluctuations in the values of $z_{En,st}$ decreases rather uniformly with the blade height. Above 0.97 of the channel relative height the curves formed for individual time steps coincide. The biggest changes in the energy loss coefficient occur at 0.2 of the channel relative height, where they reach 0.035.

In the case of the SST turbulence model, an intersection of $z_{En,st}$ curves occurs for subsequent time steps 2/5 and 3/5. This proves that there is a phase shift in changes in the energy loss coefficient for areas above and below the point of intersection. As this point is situated in the upper part of the blade, where amplitudes of fluctuations in the values of $z_{En,st}$ are slight, the phase shift mentioned above will have little effect on "smoothing" the amplitudes of changes in the loss coefficient calculated for the entire stator blade ring. It is the big amplitude of the changes in the bottom part of the channel that has a decisive impact on how the coefficient evolves.

Figures 7(b) and **(d)** show distributions of values of the energy loss coefficient calculated for the channel

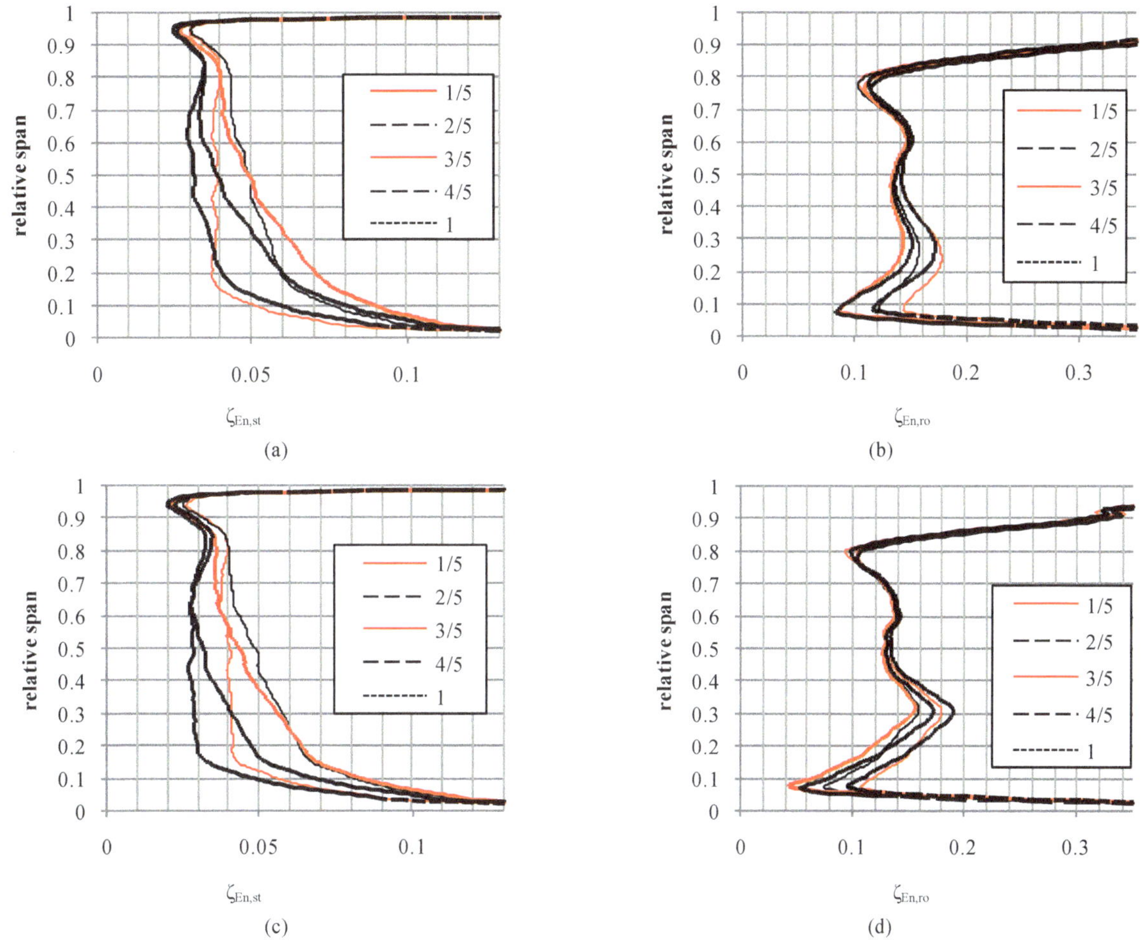

Figure 7. Distributions of the energy loss coefficient for reference analyses (low temperatures): (a) stator, SST; (b) rotor, SST; (c) stator, SAS; (d) rotor, SAS.

of the rotor blade ring. The biggest difference between the values of $z_{En,ro}$ for both turbulence models is visible in the bottom part of the channel. At ~0.1 of the blade relative height it exceeds 0.03. In this area there is also a difference between the values of fluctuations in $z_{En,ro}$. These fluctuations feature slightly bigger amplitudes in the case of results obtained using the SST turbulence model.

At 0.3 of the channel relative height (where previously an especially big increment in the entropy of the medium was observed in the trailing edge region) the SAS turbulence model gave a much larger increase in the energy loss coefficient than that modelled by the SST model. This may indicate a significant impact of the turbulence phenomena in this region, which are better mapped by the SAS model, on the energy loss coefficient value.

In the figures presenting distributions of the energy loss coefficient values for the rotor channel no intersections of curves are observed for time steps subsequent to each other.

In this situation the fluctuations in average values of $z_{En,ro}$ calculated for the entire blade can be seen quite clearly.

Moreover, it can be noticed that low values of $z_{En,st}$ correspond to high values of $z_{En,ro}$. The changes in values of the energy loss coefficient for individual blade rings will in a way balance each other, reducing fluctuations in the average value of the loss coefficient calculated for the entire stage.

4.3. Pressure Loss Coefficient Analysis

Figures 8(a) and **(c)** present distributions of the pressure loss coefficient for the channel of the stator blade ring. Their shape is similar to the shape of $z_{En,st}$ curves because both coefficients are functions of the same parameters of the medium. As it was the case for the energy and entropy loss coefficients, the values of $z_{P,st}$ are smaller for the calculation results obtained using the SAS turbulence model. The differences between the results of the two models reach 0.02 (time step "1" and 0.2 of the channel relative height).

The range of changes in the values of the pressure loss

coefficient for the stator channel decreases with its height. At 0.2 of the channel relative height fluctuations in the values of $z_{P,st}$ amount to as much as 0.055. At the same time it can be seen that the changes in values of the pressure loss coefficient along the channel entire length do not run in the same direction. This is proved by several intersections of curves for time steps 1/5, 2/5 and 3/5.

Figures 8(b) and **(d)** present distributions of the pressure loss coefficient for the rotor blade ring. The most significant difference between the values of this coefficient for the SST and SAS turbulence models occurs in the bottom part of the flow channel. Here, for the SAS model, the pressure loss coefficient assumes values smaller even by 0.07. At the same time, fluctuations in the values of $z_{P,ro}$ for this particular turbulence model (SAS) feature bigger amplitudes. The maximum range of the fluctuations is achieved at 0.1 of the channel relative height, where it assumes the value of 0.11. Fluctuations in $z_{P,ro}$ in the upper part of the blade are very slight. In principle, a noticeable amplitude occurs only at the point of deflection of the curves, at 0.8 of the channel relative height. A change in the value of $z_{P,ro}$ in this region may result from unsteady operation of the seal.

Considering the impact of the mutual stator/rotor positioning on the $z_{P,ro}$ value, a drop in value of the pressure loss coefficient can be observed for the first three time steps (mutual stator/rotor positions). In the next steps the value rises. This change takes place smoothly along the channel entire height, which is proved by the fact that there is no clear intersection of curves formed for two time steps subsequent to each other.

5. Conclusions

Based on the results of the numerical analyses performed for the gas turbine stage, it can be stated that the SST turbulence model obtained higher values of loss coefficients compared to the SAS model. This statement is correct for the part of the channel where there are no intense phenomena resulting in energy dissipation. In the upper part of the rotor channel, where the impact of the seal is significant, and at 0.3 of the channel relative

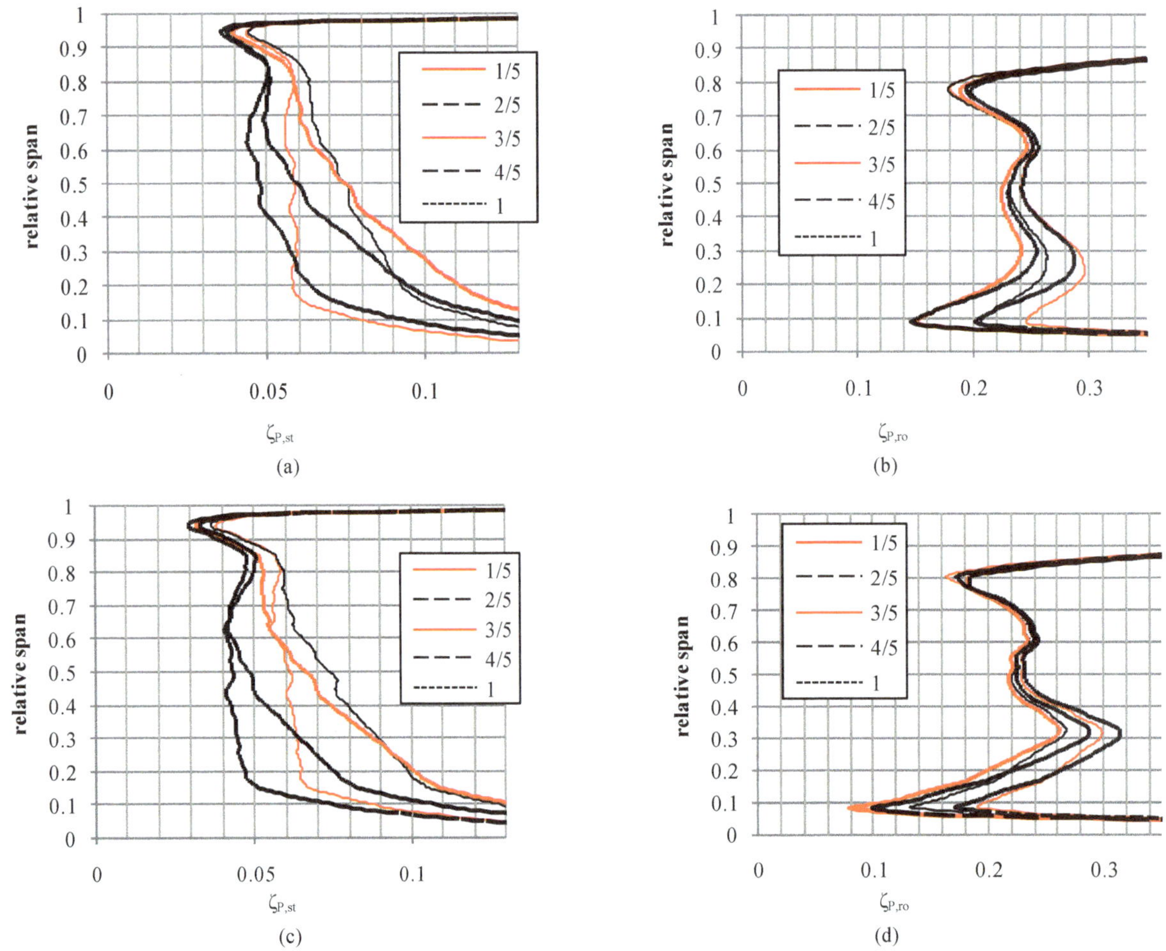

Figure 8. Distributions of the pressure loss coefficient for reference analyses (low temperatures): (a) stator, SST; (b) rotor, SST; (c) stator, SAS; (d) rotor, SAS.

height, where an especially big increment in entropy occurs, the values of the loss coefficients obtained for the SAS turbulence model are higher. This probably results from the fact that the turbulent processes taking place in these areas are mapped better by the hybrid Scale-Adaptive Simulation (SAS) turbulence model.

Comparing the two turbulence models, attention should also be drawn to the shape of the deflection of the curves in the area at ~0.3 of the blade relative height. Here, the curves representing changes in loss coefficients vary mildly for the SST model, whereas for the SAS turbulence model a considerable refraction of the curves occurs. Thus the area of high values of the loss coefficient for the SAS model comprises a much smaller part of the channel.

Intersections of curves for time steps subsequent to each other were observed for the energy loss coefficient and the pressure loss coefficient calculated for the stator channel. This proves that there is a phase shift in the direction of the changes in the coefficient value on both sides of the intersection of the curves, which leads to a reduced fluctuation in the average value of the loss coefficient calculated for the entire channel.

For all loss coefficients under analysis, the low values in the rotor region correspond to their high values in the region of the stator blade channel. This levels out the fluctuations in the values of individual coefficients within the entire stage.

6. Acknowledgements

The authors would like to thank the Polish Ministry of Science and Higher Education for the financial support for the research project UMO-2011/01/B/ST8/03488.

REFERENCES

[1] J. D. Denton, "Loss Mechanisms in Turbomachines," *Journal of Turbomachinery*, Vol. 115, No. 4, 1993, pp. 621-656.

[2] N. Wei, "Significance of Loss Models in Aerothermodynamic Simulation for Axial Turbines," Doctoral Thesis, Royal Institute of Technology.

[3] S. Dykas, W. Wróblewski and H. Łukowicz, "Prediction of Losses in the Flow through the Last Stage of Low-Pressure Steam Turbine," *International Journal for Numerical Methods in Fluids*, Vol. 53, No. 6, 2007, pp. 933-945.

[4] P. Lampart, "Investigation of Endwall Flows and Losses in Axial Turbines, Part I. Formation of Endwall Flows and Losses," *Journal of Theoretical and Applied Mechanics*, Vol. 47, No. 2, 2009, pp. 321-342.

[5] W. Wróblewski, A. Gardzilewicz, A. Dykas and P. Kolovratnik, "Numerical and Experimental Investigation of Steam Condensation in LP Part of Large Power Turbine," *Journal of Fluids Engineering*, *Trans*, *ASME*, Vol. 131, No. 4, 2009, pp. 1-11.

[6] T. P. Moffit, E. M. Szanca and W. J. Whitney, "Design and Cold-Air Test of Single-Stage Uncooled Core Turbine With High Work Output," NASA Report TP-1680, 1980.

[7] F. R. Menter, "Two-Equation Eddy-Viscosity Turbulence Models for Engineering Applications," *AIAA Journal*, Vol. 32, No. 8, 1994, pp. 1598-1605.

Symbols

ζ: loss coefficient
h: specific enthalpy, $J \cdot kg^{-1}$
p: pressure, Pa
s: specific entropy, $J/kg^{-1} \cdot K^{-1}$
$\varkappa$: adiabatic exponent

Subscripts

0: stator inlet
1: stator/rotor interface
2: rotor outlet
En: energy
rel: rotating reference frame
ro: rotor
s: entropy
st: stator
t: total

10

MHD Mass Transfer Flow past a Vertical Porous Plate Embedded in a Porous Medium in a Slip Flow Regime with Thermal Radiation and Chemical Reaction

Nazibuddin Ahmed, Kishor Kumar Das
Department of Mathematics, Gauhati University, Guwahati, India

ABSTRACT

This problem presents the effects of thermal radiation and chemical reaction on MHD unsteady mass transfer flow past a semi-infinite vertical porous plate embedded in a porous medium in a slip flow regime with variable suction. A magnetic field of uniform strength is assumed to be applied transversely to the direction of the main flow. Perturbation technique is applied to transform the non-linear coupled governing partial differential equations in dimensionless form into a system of ordinary differential equations. The resulting equations are solved analytically and the solutions for the velocity, temperature and concentration fields are obtained. The effects of various flow parameters on velocity, temperature and concentration fields are presented graphically. For different values of the flow parameters involved in the problem, the numerical calculations for the Nusselt number, Sherwood number and skin-friction co-efficient at the plate are performed in tabulated form. It is seen that chemical reaction causes the velocity field and concentration field to decrease and the chemical reaction decreases the rate of viscous drag at the plate.

Keywords: Porous Medium; Slip Flow; Rarefaction; Viscous Drag; Free Convection

1. Introduction

In recent years, free convective flow and heat transfer problems in the presence of magnetic field through a porous medium have attracted the attention of a number of scholars because of their possible application in many branches of science and technology such as fiber and granular insulation, geothermal system, etc. In engineering science, it finds its application in MHD pumps, MHD bearing, MHD power generators, etc. The phenomena of heat and mass transfer are also very common in theory of stellar structure and observable effects are detectable on the solar structure.

Analytical solutions of the problem of convective flows, which arise in the fluids due to interaction of the force of gravity and density differences caused by simultaneous diffusion of thermal energy and chemical species, have been presented by many authors due to application of such problems in Geophysics and Engineering. Some of them are Bejan and Khair [1], Trevisan and Bejan [2], Acharya *et al.* [3], Rapits and Kafousias [4], and Das *et al.* [5]. The study through porous medium has got its importance because of its occurrence in movement of water. Investigations of such problems also have importance in purification process, petroleum technology and in the field of agricultural engineering. Study of flow problems through porous medium is heavily based on Darcy's experimental law [6]. Wooding [7] and Brinkman [8,9] have modified Darcy's law, which are used by many authors on study of convective flow in porous media. Recently Chaudhary and Jain [10] have studied the combined heat and mass transfer effects on MHD free convective flow through porous medium.

The study of the effect of chemical reaction on heat and mass transfer in a flow is of great practical importance to the engineers and scientists because of its universal occurrence of many branches of science and technology. In processes such as drying, distribution of temperature and moisture over agricultural fields, energy transfer in a wet cooling tower and flow in a cooler heat and mass transfer occur simultaneously. Possible applications of this type of flow can be found in many Industries. Many investigations have studied the effect of chemical reaction in different convective heat and mass transfer flows. Chambre and Young [11] have presented a first order chemical reaction in the neighbourhood of horizontal plate. Muthucumaraswamy [12] has presented heat and mass transfer effects on a continuously moving isothermal vertical surface with uniform suction by taking in to account the homogeneous chemical reaction of first order. Muthucumaraswamy and Meenakshisundaram [13] investigated theoretical study of chemical reaction effects on vertical oscillating plate with variable temperature and mass diffusion.

Radiation is a process of heat transfer through electromagnetic waves. Radiative convective flows are encountered in countless industrial and environmental process. For example, heating and cooling chambers, fossil fuel combustion energy processes, evaporation from large open water reservoirs, astrophysical flows etc. Radiative heat and mass transfer play an important role in space related technology. The effect of radiation on various convective flows under different conditions have been studied by many researchers including Hussain and Thakar [14], Ahmed and Sarmah [15], Rajesh and Varma [16], Pal and Mondal [17], Samad and Rahman [18], S. Karthikeyan *et al.* [19], Das *et al.* [20] and Pal *et al.* [21].

At the macroscopic level, it is accepted that the boundary condition for a viscous fluid at a solid wall is one of "no slip". While no-slip boundary condition has been proven experimentally to be accurate for a number of macroscopic flows, it remains an assumption that is based on physical principles. In fact nearly two hundred years ago, Navier's [22] proposes a general boundary condition that incorporates the possibility of fluid slip at a solid boundary. Navier's proposed condition assumes that the fluid slip velocity at a solid surface is proportional to the shear stress at the surface. The mathematical form of the Navier's proposed condition on slip velocity as emphasized by Goldstein [23] is $v=\gamma\left(\frac{\mathrm{d}v}{\mathrm{d}y}\right)$ where γ being the slip coefficient, v the slip velocity and y the normal coordinate. The fluid slippage phenomena at solid boundaries appear in many applications such as in micro-channels or nano channels and in applications where a thin film of light oil is attached to the moving plates or when the surface is coated with special coating such as thick monolayer of hydrophobic octadecyltrichorosilane. Due to practical applications of the fluid slippage phenomemenon at solid boundaries, several scholar have carried out their research work in that literature, the names of whom (Yu & Amed [24], Waltanebe *et al.* [25], Jain and Sharma [26], Khaled and Vafai [27] and Poonia and Chaudhary [28]) are worth meaning.

The present work is concerned with the effect of thermal radiation and chemical reaction on magneto-hydrodynamic convective mass transfer flow of an unsteady viscous incompressible eclectically conducting fluid past a semi-infinite vertical permeable plate embedded in a porous medium in slip flow regime. In this paper, we have generalized the work done by Karthikeyan *et al.* [19] by considering mass transfer with chemical reaction effect. The classical model for radiation effect introduced by Cogley *et al.* [29] is used. Perturbation technique is applied to convert the governing non-linear partial differential equations in to a system of ordinary differential equations which are solved analytically.

2. Mathematical Analysis

We consider a two-dimensional unsteady flow of a laminar, incompressible, electrically conducting and heat absorbing fluid past a semi-infinite vertical porous plate embedded in a uniform porous medium. We introduce the coordinate system $(\bar{x},\bar{y},\bar{z})$ with X axis is chosen along the plate , Y axis perpendicular to it and directed in the fluid region and Z axis along the width of the plate as shown in the **Figure 1**. A uniform magnetic field of strength B_0 in the presence of radiation is imposed transversely in the direction of Y axis. The induced magnetic field is neglected under the assumption that the magnetic Reynolds number is small. It is assumed that there is no applied voltage which implies the absence of any electrical field. The radiative heat flux in the X direction is considered negligible in comparison to that in Y direction. The governing equations for this study are based on the conservation of mass, linear momentum, energy and species concentration. Taking in to consideration the assumptions made above, these equations in Cartesian frame of reference are given by Equation of continuity:

$$\frac{\partial\bar{v}}{\partial\bar{y}}=0 \tag{1}$$

Momentum equation:

$$\begin{aligned}&\bar{v}\frac{\partial\bar{u}}{\partial\bar{y}}+\bar{v}\frac{\partial\bar{u}}{\partial\bar{y}}\\&=-\frac{1}{\rho}\frac{\partial\bar{p}}{\partial\bar{x}}+g\beta\left(\bar{T}-\bar{T}_\infty\right)+g\bar{\beta}\left(\bar{C}-\bar{C}_\infty\right)\\&+\nu\frac{\partial^2\bar{u}}{\partial\bar{y}^2}-\frac{\sigma B_0^2}{\rho}\bar{u}-\frac{\nu\bar{u}}{K^*}\end{aligned} \tag{2}$$

Energy equation

$$\begin{aligned}&\bar{v}\frac{\partial\bar{T}}{\partial\bar{y}}+\frac{\partial\bar{T}}{\partial\bar{t}}\\&=\frac{k}{\rho C_p}\frac{\partial^2\bar{T}}{\partial\bar{y}^2}-\frac{1}{\rho C_p}\frac{\partial q_r^*}{\partial\bar{y}}+\frac{Q_0\left(\bar{T}_\infty-\bar{T}\right)}{\rho C_p}\end{aligned} \tag{3}$$

Species continuity equation:

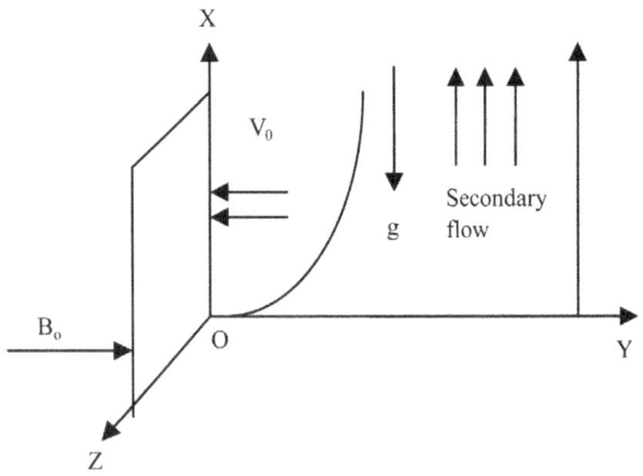

Figure 1. Physical model of the problem.

$$\bar{v}\frac{\partial \bar{C}}{\partial \bar{y}}+\frac{\partial \bar{C}}{\partial \bar{t}}=D_M\frac{\partial^2 \bar{C}}{\partial \bar{y}^2}+\bar{K}\left(\bar{C}_\infty-\bar{C}\right) \tag{4}$$

where $\bar{x}$ and $\bar{y}$ are the dimensional distances along and perpendicular to the plate respectively. $\bar{u}$ and $\bar{v}$ are the components of the dimensional velocities along $\bar{x}$ and $\bar{y}$ respectively. ρ is the density of the medium, g is the acceleration due to gravity, ν is the kinematic viscosity, σ is the fluid electrical conductivity, K^* is the permeability of the porous medium, $\bar{K}$ is first order chemical reaction, β is the coefficient of thermal expansion, $\bar{\beta}$ is the coefficient of mass expansion. $\bar{T}$ is the dimensional temperature of the fluid near the plate, $\bar{T}_\infty$ is the dimensional free stream temperature, $\bar{C}$ is the dimensional concentration of the fluid near the plate, $\bar{C}_\infty$ is the dimensional free stream concentration, k is constant pressure, q_r^* is the radiative heat flux and Q_0 is the dimensional heat absorption coefficient.

Cogley *et al.* [29] showed that, in the optically thin limit for a non-gray gas near equilibrium, the radiative heat flux is represented by the following form:

$$\frac{\partial q_r^*}{\partial \bar{y}}=4\left(\bar{T}-\bar{T}_\infty\right)I^* \tag{5}$$

where $I^*=\int K_{\lambda w}\frac{\partial e_{b\lambda}}{\partial \bar{T}}\mathrm{d}\lambda$, $K_{\lambda w}$ is the absorption coefficient at the wall and $e_{b\lambda}$ is the Planck's function.

Under the assumption, the appropriate boundary conditions for velocity involving slip flow, temperature and concentration fields are given by

$$\begin{aligned}&\bar{u}=\bar{u}_{\text{slip}}=\bar{h}\frac{\partial \bar{u}}{\partial \bar{y}},\ \bar{T}=\bar{T}_w+\varepsilon\left(\bar{T}_w-\bar{T}_\infty\right)\mathrm{e}^{n^*t^*},\\&\bar{C}=\bar{C}_w+\varepsilon\left(\bar{C}_w-\bar{C}_\infty\right)\mathrm{e}^{n^*t^*}\quad \text{at } \bar{y}=0\end{aligned} \tag{6}$$

$$\begin{aligned}&\bar{u}\to\bar{U}_\infty=U_0\left(1+\varepsilon \mathrm{e}^{n^*t^*}\right),\ \bar{T}\to\bar{T}_\infty,\\&\bar{C}\to\bar{C}_\infty\quad \text{as } \bar{y}\to\infty\end{aligned} \tag{7}$$

where $\bar{T}_w$ and $\bar{C}_w$ are the dimensional temperature and species concentration at the wall respectively and $\bar{h}$ is the characteristic dimension of the flow fluid. Science the suction velocity normal to the plate is a function of time only, it can be taken in the exponential form as

$$\bar{v}=-V_0\left(1+\varepsilon A\mathrm{e}^{n^*t^*}\right) \tag{8}$$

where A is a real positive constant, ε and εA are small quantities less than unity and V_0 is a scale of suction velocity which is a non-zero positive constant.

Outside the boundary layer, Equation (2) gives

$$-\frac{1}{\rho}\frac{\mathrm{d}\bar{p}}{\mathrm{d}\bar{x}}=\frac{\mathrm{d}\bar{U}_\infty}{\mathrm{d}\bar{t}}+\frac{\sigma B_0^2}{\rho}\bar{U}_\infty+\frac{\nu}{K^*}\bar{U}_\infty \tag{9}$$

Now we introduce the dimensionless variables as follows

$$\left.\begin{aligned}&u=\frac{\bar{u}}{U_0},v=\frac{\bar{v}}{V_0},y=\frac{\bar{y}V_0}{\nu},U_\infty=\frac{\bar{U}_\infty}{U_0},t=\frac{\bar{t}V_0^2}{\nu},\\&\theta=\frac{\bar{T}-\bar{T}_\infty}{\bar{T}_w-\bar{T}_\infty},\phi=\frac{\bar{C}-\bar{C}_\infty}{\bar{C}_w-\bar{C}_\infty},n=\frac{n^*\nu}{V_0^2},\\&\alpha=\frac{K^*V_0^2}{\nu^2},\mathrm{Pr}=\frac{\mu C_p}{k},M=\frac{\sigma B_0^2\nu}{\rho V_0^2},\\&Gr=\frac{\nu\beta g\left(\bar{T}_w-\bar{T}_\infty\right)}{U_0V_0^2},Gm=\frac{\nu\bar{\beta}g\left(\bar{C}_w-\bar{C}_\infty\right)}{U_0V_0^2},\\&Q=\frac{Q_0\nu}{\rho C_pV_0^2},R=\frac{4\nu I^*}{\rho C_pV_0^2},K=\frac{\bar{K}\nu}{V_0^2},h=\frac{V_0\bar{h}}{\nu}\end{aligned}\right\} \tag{10}$$

where *Pr* is the Prandtl number, *M* is the magnetic field parameter, *Gr* is Grashof number for heat transfer, *Gm* is the Grashof number for mass transfer, *Q* is the heat source parameter, α is the permeability parameter, θ is the non dimensional temperature, ϕ is the non dimensional concentration, *R* is radiation parameter, *h* is the rarefaction parameter and *K* is the chemical reaction parameter.

In view of Equations (8) to (10) the governing Equations (2), (3) and (4) reduce the following non-dimensional form:

$$\begin{aligned}&\frac{\partial u}{\partial t}-\left(1+\varepsilon A\mathrm{e}^{nt}\right)\frac{\partial u}{\partial y}\\&=\frac{\mathrm{d}U_\infty}{\mathrm{d}t}+\frac{\partial^2 u}{\partial y^2}+Gr\theta+Gm\phi+N\left(U_\infty-u\right)\end{aligned} \tag{11}$$

$$\frac{\partial \theta}{\partial t}-\left(1+\varepsilon A\mathrm{e}^{nt}\right)\frac{\partial \theta}{\partial y}=\frac{1}{\mathrm{Pr}}\frac{\partial^2\theta}{\partial y^2}-R\theta-Q\theta \tag{12}$$

$$\frac{\partial \phi}{\partial t}-\left(1+\varepsilon A\mathrm{e}^{nt}\right)\frac{\partial \phi}{\partial y}=\frac{1}{Sc}\frac{\partial^2\phi}{\partial y^2}-K\phi \tag{13}$$

where $N=M+\dfrac{1}{\alpha}$

The boundary conditions (6) and (7) in the dimensionless form can be written as

$$u = u_{\text{slip}} = h\frac{\partial u}{\partial y}, \theta = 1 + \varepsilon e^{nt}, \phi = 1 + \varepsilon e^{nt} \text{ at } y = 0 \quad (14)$$

$$u \to U_\infty = 1 + \varepsilon e^{nt}, \theta \to 0, \phi \to 0 \quad \text{as } y \to \infty \quad (15)$$

3. Solution of the Problem

Equations (11) to (13) are coupled non-linear partial differential equations and these can be solved in closed form. However, these equations can be reduced to a set of ordinary differential equations, which can be solved analytically. These can be done by representing the velocity, temperature and concentration of the fluid in the neighbourhood of the plate as

$$u = u_0(y) + \varepsilon e^{nt} u_1(y) + O(\varepsilon^2) \quad (16)$$

$$\theta = \theta_0(y) + \varepsilon e^{nt} \theta_1(y) + O(\varepsilon^2) \quad (17)$$

$$\phi = \phi_0(y) + \varepsilon e^{nt} \phi_1(y) + O(\varepsilon^2) \quad (18)$$

Substituting (16) to (18) in Equations (11) to (13) and equating the harmonic and non harmonic terms and neglecting the coefficient of $O(\varepsilon^2)$ we get the following pairs of equations for (u_0, θ_0, ϕ_0) and (u_1, θ_1, ϕ_1).

$$u_0'' + u_0' - Nu_0 = -N - Gr\theta_0 - Gm\phi_0 \quad (19)$$

$$\begin{aligned} &u_1'' + u_1' - (N+n)u_1 \\ &= -Au_0' - Gr\theta_1 - Gm\phi_1 - (N+n) \end{aligned} \quad (20)$$

$$\theta_0'' + \Pr\theta_0' - \Pr(R+Q)\theta_0 = 0 \quad (21)$$

$$\theta_1'' + \Pr\theta_1' - \Pr(R+Q+n)\theta_1 = -A\Pr\theta_0' \quad (22)$$

$$\phi_0'' + Sc\phi_0' - KSc\theta_0 = 0 \quad (23)$$

$$\phi_1'' + Sc\phi_1' - Sc(K+n)\theta_1 = -ASc\phi_0' \quad (24)$$

where the primes denote the differentiation with respect to y

The corresponding boundary conditions can be written as

$$\begin{aligned} &u_0 = hu_0',\ u_1 = hu_1',\ \theta_0 = 1, \\ &\theta_1 = 1,\ \phi_0 = 1, \phi_1 = 1 \ \text{ at } y = 0 \end{aligned} \quad (25)$$

$$\begin{aligned} &u_0 = 1,\ u_1 = 1,\ \theta_0 \to 0,\ \theta_1 \to 0, \\ &\phi_0 \to 0, \phi_1 \to 0 \ \text{ at } y \to \infty \end{aligned} \quad (26)$$

The solutions of Equations (18) to (23) which satisfy the boundary conditions (24) and (25) are given by

$$u_0(y) = 1 + D_5 e^{-m_5 y} + D_4 e^{-m_3 y} + D_3 e^{-m_1 y} \quad (27)$$

$$\begin{aligned} u_1(y) = {} & 1 + D_{11} e^{-m_6 y} + D_6 e^{-m_1 y} - D_7 e^{-m_2 y} \\ & + D_8 e^{-m_3 y} - D_9 e^{-m_4 y} + D_{10} e^{-m_5 y} \end{aligned} \quad (28)$$

$$\theta_0(y) = e^{-m_1 y} \quad (29)$$

$$\theta_1(y) = (1 - D_1) e^{-m_2 y} + D_1 e^{-m_1 y} \quad (30)$$

$$\phi_0(y) = D_2 e^{-m_3 y} \quad (31)$$

$$\phi_1(y) = (1 - D_2) e^{-m_4 y} + D_2 e^{-m_3 y} \quad (32)$$

where,

$$m_1 = \frac{\Pr + \sqrt{\Pr^2 + 4\Pr(R+Q)}}{2},$$

$$m_2 = \frac{\Pr + \sqrt{\Pr^2 + 4\Pr(R+Q+n)}}{2},$$

$$m_3 = \frac{Sc + \sqrt{Sc^2 + 4KSc}}{2},$$

$$m_4 = \frac{Sc + \sqrt{Sc^2 + 4Sc(K+n)}}{2},$$

$$m_5 = \frac{1 + \sqrt{1 + 4N}}{2},$$

$$m_6 = \frac{1 + \sqrt{1 + 4(N+n)}}{2}$$

$$D_1 = \frac{A\Pr m_1}{m_1^2 - \Pr m_1 - \Pr(\Pr + Q + n)},$$

$$D_2 = \frac{AScm_3}{m_3^2 - Scm_3 - Sc(K+n)}, \quad D_3 = \frac{-Gr}{m_1^2 - m_1 - N},$$

$$D_4 = \frac{-Gm}{m_2^2 - m_2 - N},$$

$$D_5 = \frac{-(1 + D_3 + D_3 hm_1 + D_4 + D_4 hm_3)}{1 + m_5 h},$$

$$D_6 = \frac{AD_3 m_1 - GrD_1}{m_1^2 - m_1 - (N+n)},$$

$$D_7 = \frac{-Gr(1 - D_1)}{m_2^2 - m_2 - (N+n)},$$

$$D_8 = \frac{AD_4 m_3 - GmD_2}{m_3^2 - m_3 - (N+n)}, \quad D_9 = \frac{-Gm(1 - D_2)}{m_4^2 - m_4 - (N+n)},$$

$$D_{10} = \frac{AD_5 m_5}{m_5^2 - m_5 - (N+n)},$$

$$D_{11} = \frac{\left(-(1 + D_6 - D_7 + D_8 - D_9 + D_{10}) + h(-D_6 m_1 + D_7 m_2 - D_8 m_3 + D_9 m_4 - D_{10} m_5)\right)}{1 + hm_6}$$

Substituting equations (27)-(32) in Equations (16)-(18), we obtain the velocity, temperature and concentration distributions in the boundary layer as follows:

$$u(y,t)=1+D_5e^{-m_5y}+D_4e^{-m_3y}+D_3e^{-m_1y} \\ +\varepsilon e^{nt}\left(1+D_{11}e^{-m_6y}+D_6e^{-m_1y}-D_7e^{-m_2y}\right. \\ \left.+D_8e^{-m_3y}-D_9e^{-m_4y}+D_{10}e^{-m_5y}\right) \quad (33)$$

$$\theta(y,t)=e^{-m_1y}+\varepsilon e^{nt}\left((1-D_1)e^{-m_2y}+D_1e^{-m_1y}\right) \quad (34)$$

$$\phi(y,t)=e^{-m_3y}+\varepsilon e^{nt}\left((1-D_2)e^{-m_4y}+D_1e^{-m_3y}\right) \quad (35)$$

4. Skin Friction

The non dimensional skin friction at the plate is given by:

$$C_f=\left(\frac{\partial u}{\partial y}\right)_{y=0}=\left(\frac{\partial u_0}{\partial y}+\varepsilon e^{nt}\frac{\partial u_1}{\partial y}\right)_{y=0} \\ =-\left(m_5D_5+m_3D_4+m_1D_3\right) \\ +\varepsilon e^{nt}\left(-m_6D_{11}-m_1D_6+m_2D_7\right. \\ \left.-m_3D_8+m_4D_9-m_5D_{10}\right) \quad (36)$$

5. Nusselt Number

The non-dimensional form of the rate of heat transfer in terms of Nusselt number at the plate is given by:

$$N_u=-\left(\frac{\partial\theta}{\partial y}\right)_{y=0}=-\left(\frac{\partial\theta_0}{\partial y}+\varepsilon e^{nt}\frac{\partial\theta_1}{\partial y}\right)_{y=0} \\ =m_1+\varepsilon e^{nt}\left(m_2(1-D_1)+m_1D_1\right) \quad (37)$$

6. Sherwood Number

The non-dimensional form of the rate of mass transfer in terms of Sherwood number at the plate is given by:

$$S_h=-\left(\frac{\partial\phi}{\partial y}\right)_{y=0}=-\left(\frac{\partial\phi_0}{\partial y}+\varepsilon e^{nt}\frac{\partial\phi_1}{\partial y}\right)_{y=0} \\ =m_3+\varepsilon e^{nt}\left(m_4(1-D_2)+m_3D_2\right) \quad (38)$$

7. Results and Discussion

In order to get physical insight in to the problem, we have carried out numerical calculations for non-dimensional velocity field, temperature field, concentration field, co-efficient of skin friction C_f at the plate, the rate of heat transfer Nu and the rate of mass transfer in terms of Sherwood number Sh by assigning specific values to the different values to the parameters involved in the problem, viz., Magnetic parameter *M*, Chemical reaction parameter *K*, Radiation parameter *R*, Grashof number for heat transfer Gr, Grashof number for mass transfer Gm, permeability parameter α, heat source parameter *Q* and rarefaction parameter *h*. Throughout our investigation the value of Prandtl number *Pr* is kept constant at 0.71 which corresponding to air at 20°C. The Schmidt number *Sc* are taken in such a way that they represent the diffusing chemical species of common interest in air (for example $Sc=0.30$ for He, Sc = 0.30 for H_2O and $Sc=0.78$ for NH_3), time t = 1, n = 0.1, A = 1 and the values of other parameters are chosen arbitrarily. The numerical results are demonstrated through different graphs and table and their results are interpreted physically.

Figure 2 plots the velocity profiles against the spanwise coordinate y for different magnetic parameters. This illustrates that velocity decreases as the existence of magnetic field becomes stronger. This conclusion agrees with the fact that the magnetic field exerts retarding force on the free-convection flow.

Figure 3 illustrates the effect of radiation on the velocity. It is seen from this figure that there is a steady increase in the velocity with the increase in radiation parameter *R*. The increase in this parameter *R* leads to increase the boundary layer thickness and to reduce the

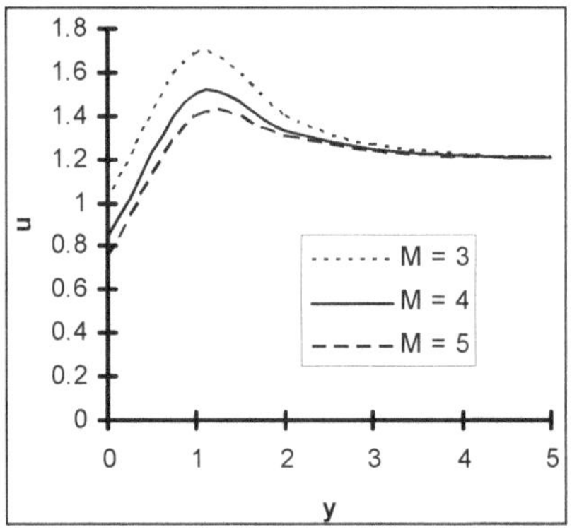

Figure 2. Velocity *u* versus *y*, under the effect of *M*, for *R* = 2, *K* = 1, *Q* = 1, *h* = 0.3, ε = 0.2, *Sc* = 0.6, *Gr* = 6, *Gm* = 4, *α* = 1.

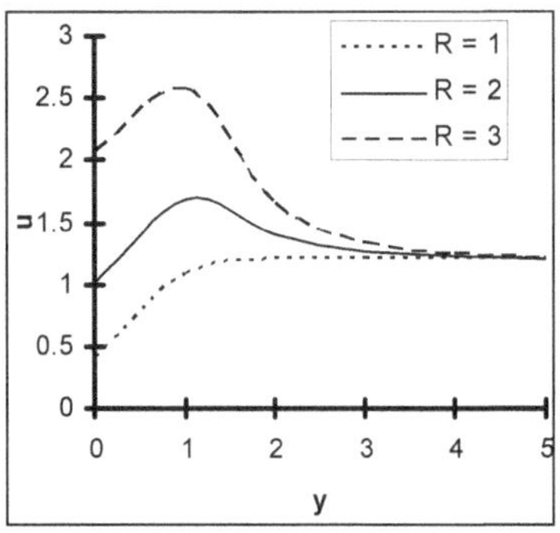

Figure 3. Velocity *u* versus *y*, under the effect of *R*, for *M* = 3, *K* = 1, *Q* = 1, *h* = 0.3, ε = 0.2, *Sc* = 0.6, *Gr* = 6, *Gm* = 4, *α* = 1.

heat transfer rate in the presence of thermal buoyancy force.

The change of velocity profile due to different chemical reaction parameters is plotted in **Figure 4**. This figure shows that the fluid motion is retarded on account of chemical reaction. This shows that the consumption of chemical species leads to fall in the concentration field which in turn diminishes the buoyancy effects due to concentration gradients. Consequently, the flow field is decelerated.

Figure 5 depicted the effect of heat source parameter on velocity field. It is seen from this figure that the heat source parameter Q leads the fluid motion to retard.

Figure 6 indicates the fact that an increase in Schmidt number Sc decelerates the fluid flow. In other words, mass diffusivity causes the fluid velocity to increase.

It is observed from **Figure 7** that an increase in Grashof number for heat transfer leads to a rise in the values of velocity u due to enhancement in buoyancy force.

The plot of velocity profile for different values of Grashof number for mass transfer is given in **Figure 8**. It is observed that velocity increases for the increasing values of Grashof number for mass transfer.

The change in velocity profile due to different permeability of porous medium is plotted in **Figure 9**. Here it is seen that due to increase of porosity of the medium fluid motion is accelerated. Moreover **Figures 10** and **11** displays that the velocity u increases as ε and rarefaction parameter h are increased indicating the fact that slips at the surface accelerates the fluid motion.

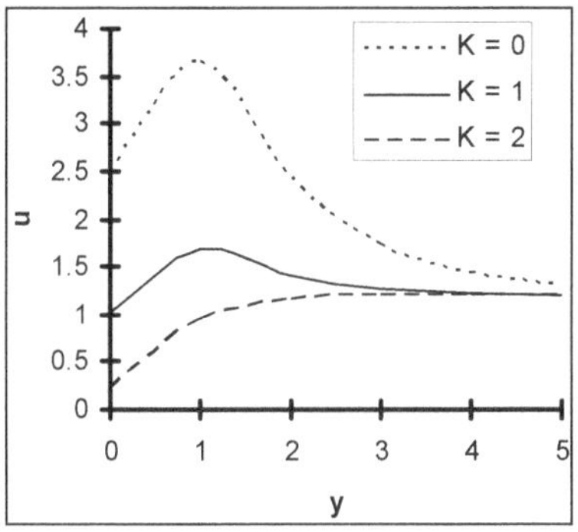

Figure 4. Velocity u versus y, under the effect K, for $R = 2$, $M = 3$, $Q = 1$, $h = 0.3$, $\varepsilon = 0.2$, $Sc = 0.6$, $Gr = 6$, $Gm = 4$, $a = 1$.

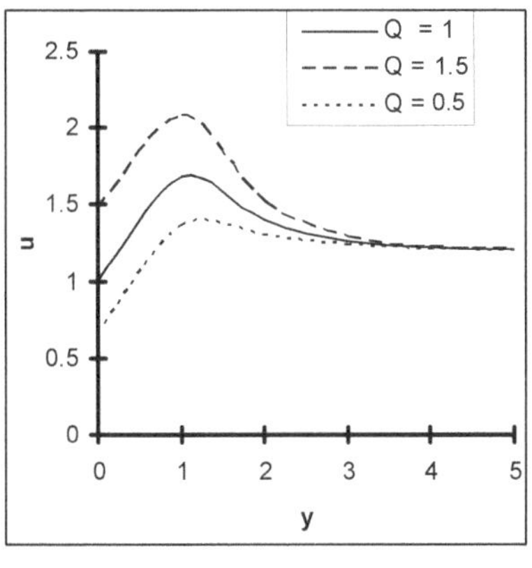

Figure 5. Velocity u versus y, under the effect Q, for $R = 2$, $M = 3$, $K = 1$, $h = 0.3$, $\varepsilon = 0.2$, $Sc = 0.6$, $Gr = 6$, $Gm = 4$, $a = 1$.

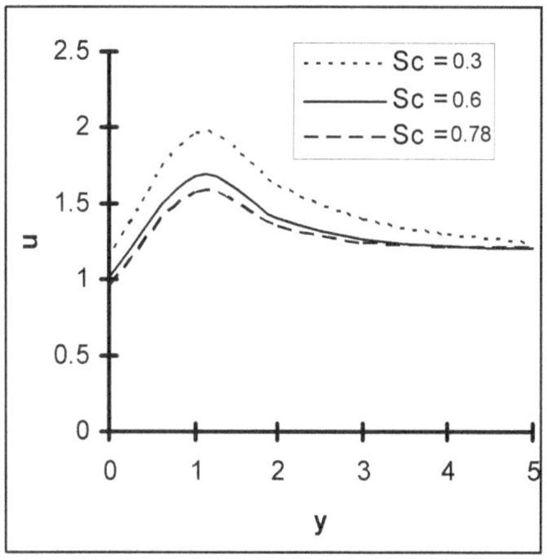

Figure 6. Velocity u versus y, under the effect Sc, for $R = 2$, $M = 3$, $Q = 1$, $h = 0.3$, $\varepsilon = 0.2$, $K = 1$, $Gr = 6$, $Gm = 4$, $a = 1$.

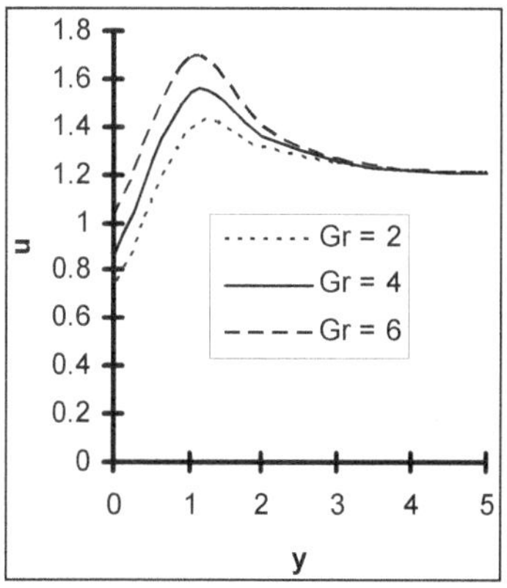

Figure 7. Velocity u versus y, under the effect of Gr, for $R = 2$, $M = 3$, $K = 1$, $Q = 1$, $h = 0.3$, $\varepsilon = 0.2$, $Gm = 4$, $a = 1$.

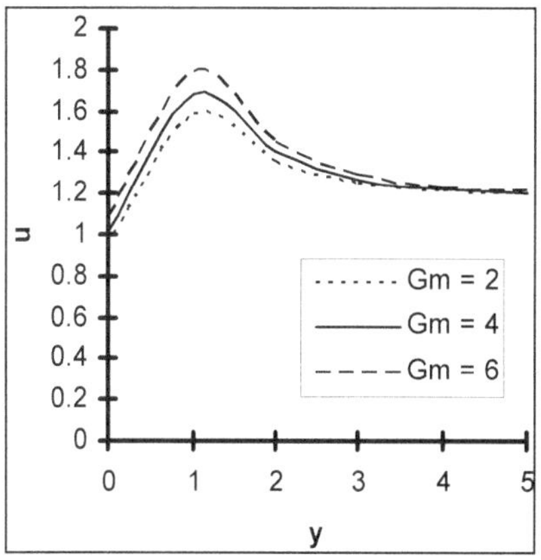

Figure 8. Velocity u versus y, under the effect of Gm, for $R = 2$, $M = 3$, $K = 1$, $Q = 1$, $h = 0.3$, $\varepsilon = 0.2$, $Gr = 6$, $a = 1$.

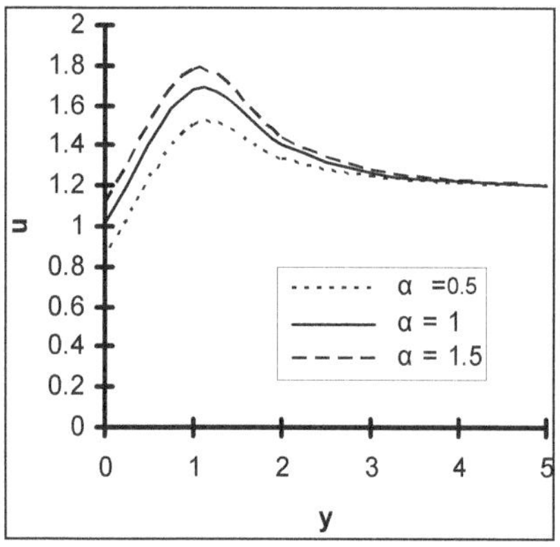

Figure 9. Velocity u versus y, under the effect of α, for $R = 2$, $M = 3$, $K = 1$, $Q = 1$, $h = 0.3$, $\varepsilon = 0.2$, $Gr = 6$, $Gm = 4$.

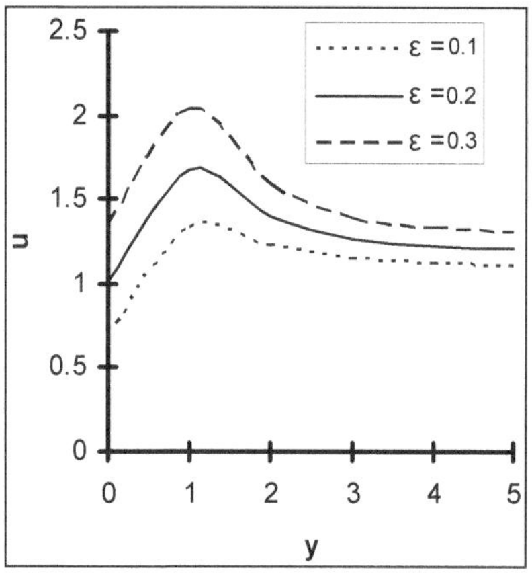

Figure 10. Velocity u versus y, under the effect of ε, for $R = 2$, $M = 3$, $K = 1$, $Q = 1$, $h = 0.3$, $Gm = 4$, $Gr = 6$, $\alpha = 1$.

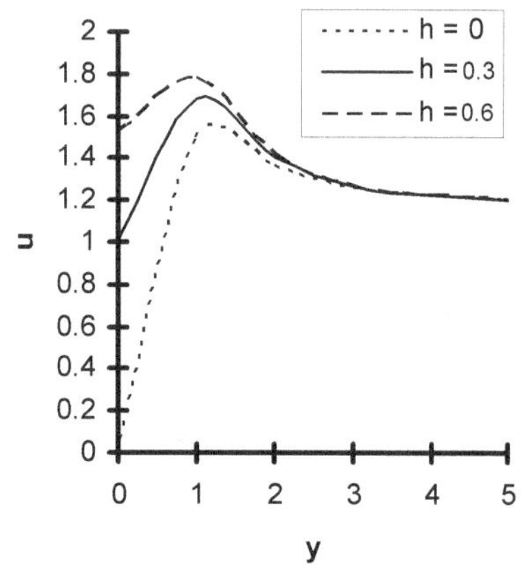

Figure 11. Velocity u versus y, under the effect of h, for $R = 2$, $M = 3$, $K = 1$, $Q = 1$, $\varepsilon = 0.2$, $Gm = 4$, $Sc = 0.6$, $Gr = 6$, $\alpha = 1$.

The effects of Radiation parameter R, heat source parameter Q and ε on temperature field against y are displayed in **Figures 12-14**.

It is observed from **Figure 12** that the temperature θ decreases as the radiation parameter R increases. This result qualitatively agrees with expectation, since the

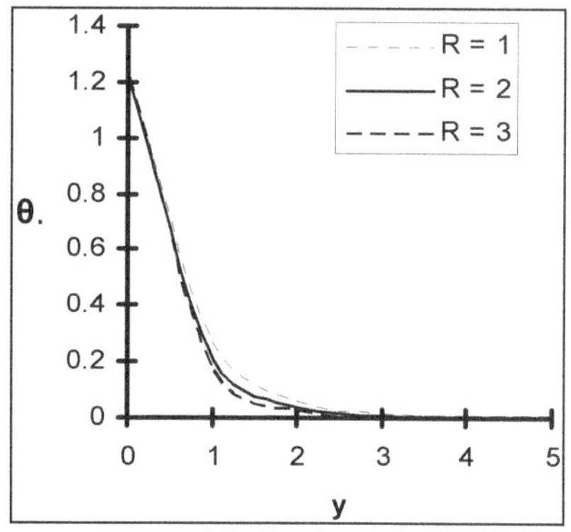

Figure 12. Temperature for different radiation parameter R with $Q = 1$, $\varepsilon = 0.2$.

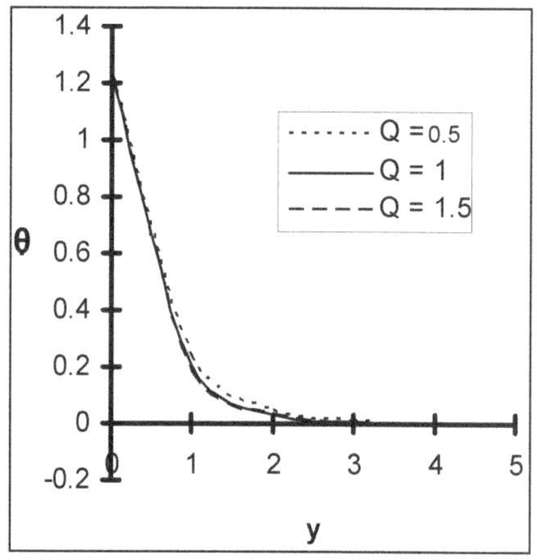

Figure 13. Temperature for different heat source parameters Q with $R = 2$, $\varepsilon = 0.2$.

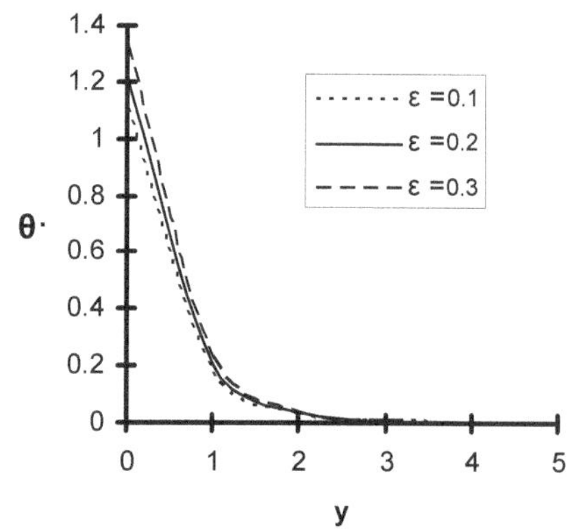

Figure 14. Temperature for different radiation parameter ε with $Q = 1$, $R = 2$.

effect of radiation is to decrease the rate of energy transport to the fluid, thereby decreasing the temperature of the fluid.

It is seen from **Figure 13** that increases in the heat source parameter decreases the temperature profile.

Moreover from **Figure 14** it reveals that the temperature increases as ε increases.

Figures 15-17 exhibit the variation of species concen-

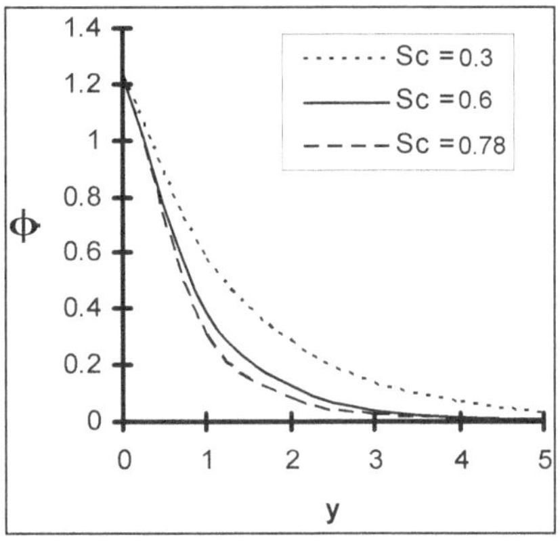

Figure 15. Concentration profile for different *Sc* with $Q = 1$, $\varepsilon = 0.2$, $K = 1$.

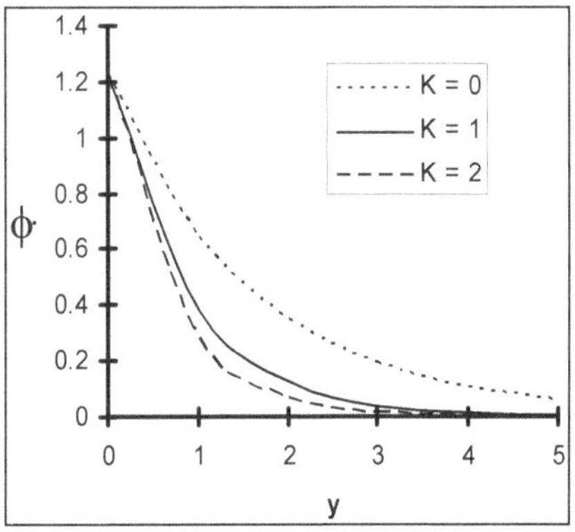

Figure 16. Concentration profile for different *K* with $Q = 1$, $\varepsilon = 0.2$, $Sc = 0.6$.

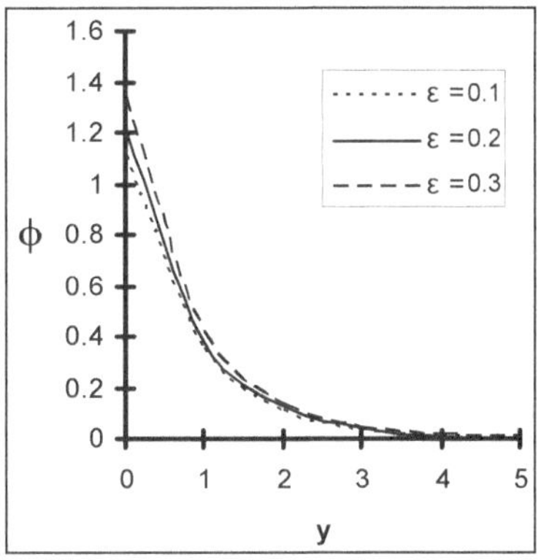

Figure 17. Concentration profile for different with ε, $Q = 1$, $K = 1$, $Sc = 0.6$.

tration against spanwise co-ordinate y under the influence of chemical reaction parameter K, Schmidt number Sc and ε. It is seen from **Figure 15** that the concentration level of the fluid drops due to increasing Schmidt number indicating the fact that the mass diffusivity raises the concentration level steadily. Further it is observed from **Figures 15** and **16** that concentration falls under the effect of chemical reaction parameter K whereas it rises due to the effect of ε.

The numerical values for skin-friction, Nusselt number and Sherwood number are computed for various values of the parameters M, Sc, K, α, Q, h, Gr, R and Gm. These results are presented in table 1. It is seen from this table that the effect of increasing values of M, Sc and K is to decrease skin-friction coefficient whereas increasing values α, Q, Gr, Gm, h and R increases skin-friction coefficient.

It is observed from the **Table 1** that there is no effect of M, Sc, K, α, h, Gr and Gm is seen on Nusselt number. But Nusselt number decreases with increase in Q and R.

Similarly, no effect of M, R, α, h, Gr and Gm is seen on Sherwood number whereas it is decrease with the increasing values of Sc and K respectively.

8. Conclusions

Our investigation of the problem setup leads to the following conclusions:

- The fluid velocity decreases as the existence of the magnetic field parameter becomes stronger.
- The fluid velocity is decelerated in the region adjacent to the plate, due to the effects of Schmidt number as well as chemical reaction.
- The fluid velocity is accelerated under the effects of thermal radiation, Grashof number for heat and mass transfer, heat source parameter, ε, rear fraction parameter h and porosity of the medium.
- There is a steady drop in temperature for high radiation and chemical reaction.
- The mass diffusivity raises the concentration level steadily, *i.e.*, the concentration level of the fluid, falls due to increasing Schmidt number.
- Increase in chemical reaction decreases the temperature whereas temperature increases as ε increases.
- The viscous drag at the plate in the direction of the buoyancy force may be successfully inhibited on application of strong magnetic field in operation.
- An increase in Grashof number for heat and mass transfer, thermal radiation, heat source parameter, rear fraction, ε and porosity of the medium results in a growth in the drag force and it falls under the effects of chemical reaction and Schmidt number.
- The rate of heat transfer (from the plate to the fluid) decreases due to the effects of thermal radiation and heat source parameter.

The mass flux from the plate to the fluid is reduced under the influence of Schmidt number and chemical reaction.

9. Acknowledgements

The author is highly thankful to CSIR-HRDG for funding this research work under Research Grant-in-aid No.

Table 1. Skin friction, Nusselt number and Sherwood number for various values of *M, R, Gr, Gm, Sc, K, α* with Pr = 0.7, *n* = 0.1, *A* = 1, ε = 0.2.

Sc	*M*	*K*	*α*	*Q*	*h*	*Gr*	*R*	*Gm*	Cf	Nu	Sh
	3								3.194506	2.358945	1.394154
0.6	5	1	1	1	0.3	6	2	4	2.679543	2.358945	1.394154
	7								2.389400	2.358945	1.394154
0.3									3.6053425	2.358945	0.885951
0.6	3	1	1	1	0.3	6	2	4	3.1945066	2.358945	1.394154
0.78									2.9820064	2.358945	1.670695
		0							7.9651191	2.358945	0.755311
0.6	3	1		1	0.3	6	2	4	3.1945066	2.358945	1.394154
		2							0.5904764	2.358945	1.766132
			0.5						2.6795643	2.358945	1.394154
0.6	3	1	1	1	0.3	6	2	4	3.1945066	2.358945	1.394154
			10.5						3.4687398	2.358945	1.394154
				0.5					2.1014135	2.2108416	1.394154
0.6	3	1	1	1	0.3	6	2	4	3.1945066	2.3589459	1.394154
				10.5					4.6123157	2.4960342	1.394154
						2			2.2527005	2.358945	1.394154
0.6	3	1	1	1	0.3	4	2	4	2.7236035	2.358945	1.394154
						6			3.1945066	2.358945	1.394154
								2	2.9747134	2.358945	1.394154
0.6	3	1	1	1	0.3	6	2	4	3.1945066	2.358945	1.394154
								6	3.4142988	2.358945	1.394154
					0				4.772427	2.358945	−1.394154
0.6	3	1	1	1	0.3	6	2	4	3.194506	2.358945	−1.394154
					0.6				2.400774	2.358945	−1.394154

25(0209)/12/EMR-II.

REFERENCES

[1] A. Bejan and K. R. Khair, "Heat and Mass Transfer in Po- rous Medium," *International Journal of Heat and Mass Transfer*, Vol. 28, No. 5, 1985. 902-918.

[2] O. V. Trevisan and A. Bejan, "Natural Convection with Combined Heat and Mass Transfer Buoyancy Effects in a Porous Medium," *Heat and Mass Transfer*, Vol. 28, No. 8, 1985, pp. 1597-1611.

[3] M. Acharya, G. C. Das and L. P. Singh, "Magnetic Field Effects on the Free Convection and Mass Transfer Flow through Porous Medium with Constant Suction and Constant Heat Flux," *Indian Journal Pure Applied Mathematics*, Vol. 31, No. 1, 2000, pp. 1-18.

[4] A. Rapits and N. Kafousias, "Magneto Hydrodynamics Free Convective Flow and Mass Transfer through a Porous Medium Bounded by an Infinite Vertical Porous Plate with constant Heat Flux," *Canadian Journal of Physics*, Vol. 60, No. 12, 1982, pp. 1724-1729.

[5] U. N. Das, N. Ahmed and D. Sharma, "Three-Dimensional Free Convective MHD Flow and Heat Transfer through Porous Medium," *FJAM*, Vol. 4, No. 3, 2000, pp. 357-369.

[6] H. P. G. Darcy, "Lee Fort Airs Publication La Villede Dijan Victor Dalonat," Paris, Vol. 1, 1857, p. 41.

[7] R. Wooding, "Steady State Free Thermal Convection of Liquid in a Saturated Permeable Medium," *Journal of Fluid Mechanics*, Vol. 2, No. 3, 1957, pp. 273-285.

[8] H. C. Brickmam, "A Calculation of the Viscous Force Exerted by a Flowing Fluid on a Dense Swarm of Particles," *Applied Sciences Research*, Vol. A1, 1947, pp. 27-34.

[9] H. C. Brickman, "On the Permeability of Media Consisting of Closely Packed Porous Particles," *Applied Sciences Research*, Vol. A1, 1947, pp. 81-86.

[10] R. C. Chaudhary and A. Jain, "Combined Heat and Mass Transfer Effects on MHD Free Convection Flow Past an Oscillating Plate Embedded in Porous Medium," *Romanian Journal of Physics*, Vol. 52, No. 5-7, 2007, pp. 505-524.

[11] P. L. Chambre and J. D. Young, "On the Diffusion of a Chemically Reactive Species in a Laminar Boundary Layer Flow," *Physics of Fluids*, Vol. 1, No. 1, 1958, pp. 48-54.

[12] R. Muthucumaraswamy, "Effects of a Chemical on a Moving Isothermal Surface with Suction," *Acta Mechanica*, Vol. 155, No. 1-2, 2002, pp. 65-70.

[13] R. Muthucumaraswamy and S. Meenakshisundaram, "Theroretical Study of Chemical Reaction Effects on Vertical Oscillating Plate with Variable Temperature," *Theoretical and Applied Mechanics*, Vol. 33, No. 3, 2006, pp. 245-257.

[14] M. A. Hussain and H. S. Takhar, "Radiation Effect on Mixed Convection along a Vertical Plate in Presence of Heat Generation or Absorption," *Journal of Heat Transfer*, Vol. 31, No. 4, 1996, pp. 243-248.

[15] N. Ahmed and H. K. Sarmah, "The Radiation Effect on a Transient MHD Flow Mass Transfer Past an Impulsively Fixed Infinite Vertical Plate," *International Journal of Applied Mathematics and Mechanics*, Vol. 5, No. 5, 2009, pp. 87-98.

[16] V. Rajesh and S. V. K. Varma, "Radiation Effects on MHD Flow through a Porous Medium with Variable Temperature and Mass Diffusion," *International Journal of Applied Mathematics and Mechanics*, Vol. 6, No. 11, 2010, pp. 39-57.

[17] D. Pal and H. Mondal, "Radiation Effects on Combined Convection over a Vertical Flat Plate Embedded in a Porous Medium of Variable Porosity," *Meccanica*, Vol. 44, No. 2, 2009, pp. 133-144.

[18] A. M. D. Samad and M. M. Rahman, "Thermal Radiation Interaction with Unsteady MHD Flow past a Vertical Porous plate immersed in a porous medium, *Journal of Naval Architecture and Marine Engineering*, Vol. 3, No. 1, 2006, pp. 7-14.

[19] S. Karthikeyan, S. Sivasankaran and S. Rajan, "Thermal Radiation Effects on MHD Convective Flow over a Vertical Porous Plate Embedded in a Porous Medium by Perturbation Technique," *Proceedings of International Conference on Fluid Dynamics and Its Applications*, Bangalore, 20-22 July 2010, pp. 484-491.

[20] S. Das, M. Jana and R. N. Jana, "Radiation Effect on Natural Convection near a Vertical Plate Embedded in Porous Medium with Ramped Wall Temperature," *Open Journal of Fluid Dynamics*, Vol. 1, No. 1, 2011, pp. 1-11.

[21] D. Pal and B. Talukdar, "Perturbation Analysis of Unsteady Magnetohydrodynamic Convective Heat and Mass Transfer in a Boundary Layer Slip Flow past a Vertical Permeable Plate with Thermal Radiation Chemical Reaction," *Communications in Nonlinear Science and Numerical Simulation*, Vol. 15, No. 7, 2010, pp. 1813-1830.

[22] C. L. M. H. Navier, "Memoire Surles du Movement des," *Mem Acad. Sci. Inst. France*, Vol. 1, No. 6, 1823, pp. 414-416.

[23] S. Yu and T. A. Ahem, "Slip Flow Heat Transfer in Rectangular Micro Channel," *International Journal of Heat and Mass Transfer*, Vol. 44, No. 22, 2002, pp. 4225-4234.

[24] S. Goldstein, "Modern Development in Fluid Dynamics," Vol. 2, Dover, New York, 1965, p. 676.

[25] K. Watannebe and Y. H. Mizunuma, "Slip of Newtonian Fluids at Solid Boundary," *JSME International Journal Series B*, Vol. 41, No. 3, 1998, p. 525.

[26] S. R. Jain and P. K. Sharma, "Effect of Viscous Heating on Flow past a Vertical Plate in Slip-Flow Regime without Periodic Temperature Variations," *Journal of Rajasthan Academy of Physical Sciences*, Vol. 5, No. 4, 2006, pp. 383-398.

[27] A. R. A. Khaled and K. Vafai, "The Effect of the Slip Condition on Stokes and Couette Flows Due to an Oscillating Wall, Exact Solutions," *International Journal of Non-Linear Mechanics*, Vol. 39, No. 5, 2004, pp. 759-809.

[28] H. Poonia and R. C. Chaudhary, "The Influence of Radiative Heat Transfer on MHD Osciallating Flow in a Planner Channel with Slip Condition," *International Journal of Energy & Technology*, Vol. 4, No. 2, 2012, pp. 1-7.

[29] A. C. L. Cogley, W. G. Vincenti and E. S. Giles, "Differential Approximation for Radiative Heat Transfer in a Non-Grey Gas near Equilibrium," *American Institute of Aeronautics and Astronautics*, Vol. 6, No. 3, 1968, pp. 551-553.

11

A Study of the Factors Affecting Transporting Solid—Liquid Suspension through Pipelines

A. I. Abd Al Aziz, H. I. Mohamed
Public Authority of Applied Education & Training, Kuwait, Kuwait

ABSTRACT

An experimental investigation was carried out on the transport of solid liquid mixture through pipelines. The principal aim of this was to study how to transport slurries through pipeline systems. The experimental tests include measurements of main parameters affecting transport of solid liquid mixture, like sand slurry and mud slurry. These parameters are deduced by applying non-dimensional approach, which includes Reynolds number, Froude number, concentration, specific gravity, and ratio of particle to pipe diameter. Preliminary results include the following general trends: 1) Increasing input concentration increases the pressure gradient, whereas decreases the efficiency of solid transport; 2) As specific gravity of solid material increases, the pressure gradient increases and the efficiency of transport decreases; 3) As mixture velocity increases, the efficiency of transport increases; 4) Solids with fine grain size are preferred than with coarse grain size from the view points of pressure gradient and efficiency of transport. Also, the present experimental data has been compared with the correlations developed before by different authors. Such correlations relate the pressure gradient to flow velocity, specific gravity, and efficiency of transport to grain size of solid material, and input solid concentration.

Keywords: Sand Slurry; Solid Concentration; Pressure Gradient; Efficiency of Transport; Solid Diameter

1. Introduction

The present study is originated in order to investigate the problem of blocking the pipelines that transport water and sold mixture. The slurry is the product of many industrial process like oxygen converter in iron and steel industries, and drilling industries.

These lines have a considerably low flow velocity, which leads to enhanced rate of deposition. As a result of this operation, pipeline systems suffer from fast blocking.

The present study is also of interest because it is applicable to problems related to the handling and transporting of solid raw materials and products, e.g. coal.

The experimental study for determining the pressure gradient and efficiency of transport is burdened by some different constraints arising mainly from the following three characteristics of the problem.

1) The abundance of variables which govern the flow.

2) The vast range over which most of the variables may vary.

3) The inherent limitation on accuracy and reproducibility of the data that can be obtained with heterogeneous settling slurries.

The basic problem in hydraulic transport of solids is the determination of the fluid forces acting upon the solid particles and the effects of these forces upon the behavior of such particles and upon the resultant energy losses.

The experimental results were compared with the correlations (13-15) developed later by other authors.

2. Literature Survey

2.1. Introduction

The transport of solid particles by a liquid flowing in a pipeline is governed by a number of parameters that can be classified as follows: pipeline parameters, liquid parameters, solid particle parameters and system parameters e.g. mixture flow velocity.

It would be useful to study the effects of these parameters upon the performance characteristics such as: the pressure gradient ($\Delta P/L$) and efficiency of transport (η).

During the last few years, there have been several investigations concerning the connections between these parameters with the pressure drops in addition to the effect of these parameters upon the characteristics of centrifugal pumps. The findings of the previous investigations for the flow of slurries are summarized in the following sections:

2.2. Flow Regimes and Critical Velocity

2.2.1. Flow Regimes

There are four distinct regimes of particle conveyance, (see Newit *et al.* [1]) who carried out experiments with solids of sizes from 0.062 to 4.67 mm. and specific gravity ranging between 1.18 and 4.6. The experiments were performed in a 2.54 cm. pipe and resulted in the following regimes.

1) Conveyance as a homogenous suspension in which mean mixture velocity is high to prevent sedimentation during transport process.

2) Conveyance as a heterogeneous suspension, which is maintained by turbulence heterogeneous flow.

3) Conveyance by saltation, in which the particles are alternatively picked up by the liquid and deposited further along the pipe.

4) Conveyance as a sliding bed, that is a regime in which the principal mechanism is sliding motion of solids along the bottom of the pipe.

Another classification of these regimes is derived by Durand [2]. This classification depends on particle size ranges as follows:

1) Particles of a size less than 40 μm, are transported as a homogenous suspension.

2) Particles of a size between 40 μm and 0.15 mm are transported as suspension that is maintained by turbulence.

3) Particles of a size among 0.15 and 1.5 mm are transported by a suspension and saltation.

4) Particles of a size greater than 1.5 mm are transported by saltation.

The flow regimes given in this classification are related having a specific gravity of 2.65.

2.2.2. Critical Velocity

The critical velocity conditions have been the study of extensive research effort and yet can still be confusing, as many definitions have been used for this term.

The most important critical velocities are those which another phase of flow is produced as laminar/turbulent transport velocity in case of homogenous slurry flow or the deposit velocity in case of settling slurry flow.

2.2.3. Critical Deposit Velocity

A further result of Durand's work is an empirical equation for predicting the critical deposit velocity of slurry, *i.e.* the velocity below which a stationary deposit of solids forms in the pipe. The critical deposit velocity correlation is given by:

$$v_c = F_L\sqrt{2gD(s-1)} \tag{1}$$

where F_L is a dimensionless function of particle diameter given by Durand and can be obtained using **Figure 1**, which requires only the mean particle size and concentration distribution of solids to find the value of F_L. The result in equation is of considerable practical value for the critical velocity, if sufficiently high in a given case, may be the limiting parameter.

2.3. Pressure Drop through Slurry Pipeline

The need to predict pressure drop for flowing slurries is an eminently practical one.

2.3.1. Durand's Equation for Pressure Losses in Horizontal Pipes

The empirical correlation published by DURAND [2] is related to the pressure drops associated with the flow of sand-water and gravel-water mixtures with particles of sizes ranging from 0.2 to 25 mm. and pipe diameters from 3.8 to 58 cm. with solids concentrations up to 60% by volume.

A principal result of the work by DURAND [2] is the correlations:

$$\frac{i-i_w}{i_w C_o} = k\left[\frac{v^2}{gD}\sqrt{C_D}\right]^{-1.5} \tag{2}$$

where i and i_w are the pressure drop of slurry and of water respectively, k is a constant and C_D is the drag coefficient for the free failing particle at its terminal velocity in the stagnant unbounded liquid, assumed spherical and of diameter d_s given by:

$$C_D = \frac{4gd_s[s-1]}{3v_\infty^2} \tag{3}$$

The constant k is found to be 3.18 for sand slurry ac-

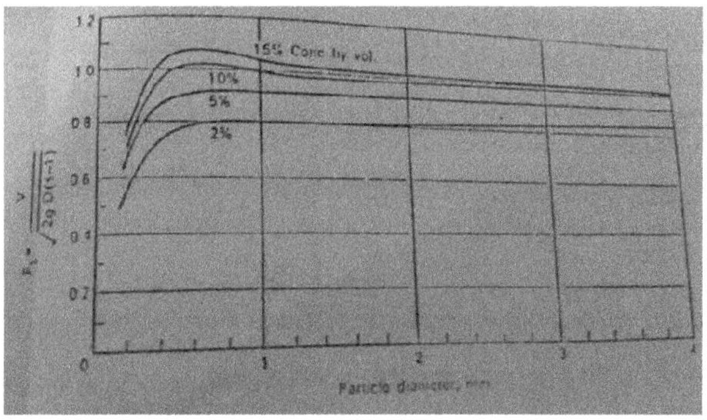

Figure 1. Values of F_L (Durand *et al.* (1953)).

cording to WORSTER [3]. To account for the effect of particle density, moreover, a modification of Equation (1) is given by

$$\frac{i-i_w}{i_w C_o} = k\left[\frac{v^2}{gD(S-1)}\sqrt{C_D}\right]^{-1.5} \quad (4)$$

where S is the specific gravity of solid material.

In this equation Durand's group used $(S-1) = 1.65$, and, by comparing Equation (2) and Equation (4) $k = 150$.

Durand's correlation, Equation (2) gave 18.0% absolute average deviation when compared with data by Turain and Yuan [6].

2.3.2. Worster's Equation for Pressure Losses in Horizontal Pipes

Worster [3] studied flow phenomena associated with large particles in particular coal particles, and produced a similar correlation to that of Durand and his coworkers except for the inclusion of a term in take account of specific gravity of the solid material and the absence of any parameter of particle size.

$$\frac{i-i_w}{i_w C_o} = 120\left[\frac{\sqrt{gD}}{v}\sqrt{\frac{\rho_{s-p}}{\rho}}\right]^{3} \quad (5)$$

2.3.3. Improved Durand—Equation for Multiple Applications

The applicability of Durand's equation could be improved for general use by applying suitable parameters representing the grain size distribution. Thus, the Durand's equation can not only describe polydisperse (pseudo)-homogenous or heterogeneous transportation, but also solid- liquid mixtures containing a certain amount of fine particles. Even non Newtonian influences can be taken into account.

The applicability of the extended Durand's equation for polydisperse mixtures was demonstrated by measurement data. With respect to this, the transition between pseudo-homogeneous and heterogeneous transport has been considered on the basis of measured concentration profiles.

WAGNER [4] empirically could show with special polydisperse mixtures the grain size distribution can be taken into account more efficiently by using d_{s90} and d_{s10} in the following correlation factor, *m*; viz.

$$m = 2 - \left(\frac{d_{s90}}{d_{s10}}\right)^{-0.4} \quad (6)$$

where d_{s90} and d_{s10} are particle diameter at 90% and 10% passing sieve, respectively.

Thus the following extended version of Durand's equation can be written for polydisperse-heterogenous Newtonian mixture:

$$\Delta P = \left[83^{\frac{1}{m}}\left[\frac{gD}{\upsilon^2}(\rho_s-\rho_f)\rho_f\sqrt{C_D}\right]^{\frac{1.5}{m^3}} C_o + 1\right] \cdot \lambda_f \rho_f 2\upsilon^2 \frac{L}{D} \quad (7)$$

where ρ_s, ρ_f are solid and fluid densities, respectively. C_o is the delivered total concentration, λ_f is the pipe friction factor. Adapting this equation also to homogeneous mixtures by taking account of grain size distribution, the heterogeneous part of the equation can be partly or totally eliminated. This can be done by defining a homogeneous part '*F*' of the grain size distribution, which is namely fine grain and a heterogeneous part $(1 - F)$.

The delivered concentration of the fine material could be represented as follows.

$$C_F = F \cdot C_o \quad (8)$$

The delivered concentration of the coarse material could be represented as follows.

$$C_c = (1-F)C_o \quad (9)$$

The density of the enriched carrier fluid is given by:

$$\rho_{fs} = FC_o\rho_s + (1-FC_o)\rho_f \quad (10)$$

Substituting ρ_f in Equation (7) by ρ_{fs} and adding $(1 - F)$ as a factor of C_o.

The generally valid equation for polydisperse heterogeneous and homogeneous Newtonian mixtures that was improved by M. WEBER, [5] can be stated as follows:

$$[C\_o+1]\lambda_f \rho_{fs} 2v_m^2 \frac{L}{D} \quad (11)$$

$$\Delta P = \left[83^{\frac{1}{m}}\left[\frac{gD}{v_m^2}(\rho_s-\rho_{fs})\rho_f\sqrt{C_D}\right]^{\frac{1.5}{m^3}} (1-F)C_o + 1\right] \cdot \lambda_f \rho_{fs} 2v_m^2 \frac{L}{D} \quad (12)$$

2.3.4. Turain's Equation for Pressure Drop in Slurry Pipeline

The slurry flow correlation established in Turain and Yuan [6] is represented by relations:

$$\lambda = \lambda_w + 0.3202\xi \quad \text{for } \xi \geq 1 \quad (13)$$

$$\lambda = \lambda_w + 0.3202\xi^{1.338} \quad \text{for } \xi \leq 1 \quad (14)$$

where

$$\xi = C_o^{0.4925}\lambda_w^{0.5586}C_D^{-0.1095}\left[\frac{gD(S-1)}{v^2}\right]^{0.9982} \quad (15)$$

C_o is the discharge concentration of solids, volume %.

$C_D = \dfrac{4gd(\rho_s - \rho)}{3\rho v_\infty^2}$, Drag coefficient for free falling sphere, dimensionless, where S is the specific gravity of solids, D is the inside diameter of pipe, g is the gravitational constant, $\lambda = (\Delta PD/2\rho v^2 L)$, friction factor for pipe flow of slurry, dimensionless, and

$\lambda_w = \Delta P_w D/2\rho v^2 L$, friction factor for pipe flow in absence of solids, dimensionless.

This empirical correlation was based upon a collection of 1511 data points pertained to flow in 5.04, 2.54, and 1.27 on pipelines of water suspensions of twelve different sizes of solid particles ranging from 31 μm. to 4380 μm. and covering, in round figures, suspended solid densities of 2300, 3000, 4400, 7500, and 11300 kg/m^3.

The correlation represented by Equation (8) and (9) predicts the slurry friction factor with absolute average deviations of 9.43% for all 1511 data points, with 76 points having deviations exceeding 30%, and only 9 points having deviations exceeding 50%.

2.4. Optimal Grain Distribution for Transport with High Concentration

Hydraulic conveying at high solid's concentration is very attractive and a large amount of experimental data has accumulated to date, but many fundamental problems still need to be answered. H.C. HOU [7] carried out theoretical analysis and answered some questions dealing with concentration transport.

2.4.1. Two Forms of Flow

There are two forms of transport. The first case is the case of dilute concentration with fine and/or coarse materials, in which the solid particles are suspended individually in the flowing water. In this case the energy expenditure for suspension of particle is compensated by the turbulence energy, which is taken from the kinetic flow field with the result that the von Karman constant for flow with sediment suspension should be less than that for pure water. For a given particle—composition it was found that the higher the value of concentration, the lower the value of von Karman constant. The reduction of von Karman's constant with the presence of sediment means that the increase of dissipated energy that is directly withdrawn from the mean flow field needs a corresponding increase in the energy-supply from outside the system. Therefore, the hydraulic conveying of solids by means of Newtonian fluids is not economic.

Another form of hydrotransport is the transport with hyper concentration including some of fine particles in the range $(d_s \leq 0.02 - 0.04\ \text{mm})$. This forms a non-Newtonian system which is complex and requires thorough analysis.

2.4.2. Critical Volume Concentration

There are critical volume concentrations $(C_i)_{CR} = 6\%$, by which the system—transfers from Newtonian fluid (dilute concentration) to hyper concentration *i.e.* non-Newtonian fluid. This limit was determined by H.C. HOU [7] after carrying some current experiments concerning solid-liquid mixture transport.

2.4.3. Specific Energy Consumption in the Slurry Pipelines

An experimental approach had been carried out by M.S. Lee, V. Matousek, C.K. Chung and Y.N. Lee [8] the results indicate that the specific energy consumption at velocities near the deposition limit is not very sensitive to the pipe size.

It also indicate that the low concentrated slurries of about 7% exhibit higher specific energy consumption values for all three pipes of diameters (155, 204, and 305 mm).

2.4.4. Closure

The above review has shown a lot of experimental data that have been accumulated during the last period. However most of these data concentrates on general flow-types rather than on specific types. The present work, on the other hand concentrates on the special problem of transport slurry of the by-product solids of oxygen converter shop.

3. The Present Work

The present work divided into two branches; the first one was:

Theoretical and dimensional analysis, the second one was the experimental tests.

3.1. Theoretical and Dimensional Analysis

In this section the dimensional analysis is used to plan experiments and present data compactly. It is noteworthy to mention that many workers also used it in theoretical studies, as well.

Basically, dimensional analysis is a method for reducing the number and complexity of experimental variables that affect a given physical phenomenon, using a sort compacting technique. By applying this techniques for the problem of transporting solid-liquid mixture we proceed as follows:

Under isothermal conditions, at least 12 variables are needed to describe suspension flow behavior, provided that pipeline roughness and solids density distribution are excluded from consideration.

The input variables can be listed as follows:

Pipeline parameters:

1. Pipe diameter, assuming section to be circular D

2. Inclination to horizontal g
3. Pipe length. L
Liquid parameters:
1. Density. ρ
2. Viscosity. μ
Solid particles parameters:
1. Density. ρ_s
2. Size distribution (dimensionless) ψ
3. Shape factor ε
System parameters:
1. Velocity of flow. v
2. Solid-liquid ratio of the flowing mixture at inlet. C_i

The output variables or the performance characteristic variables are:

1. The pressure drop for pipeline flow of solid – liquid suspension (ΔP).
2. The efficiency of transport (η), defined as the ratio between output to input concentrations per unit volume, and can be represented as follows: $\eta = \frac{C_o}{C_i} \times 100\%$.

The total number of 12 variables can be reduced to 9 independent dimensionless groups by applying the conventional Π – theorem. One way of expressing this, is through a relation of the type.

$$\left(\eta, \frac{\Delta P}{2\rho\upsilon^2} \frac{D}{L}\right) = \phi\left(\frac{D\rho\upsilon}{\mu}, \frac{d_s}{D}, \frac{L}{D}, \frac{\rho_s}{\rho}, \frac{v^2}{gD}, C_o, \psi, \varepsilon\right) \tag{16}$$

Under some conditions, certain simplifications can be made in Equation (16).

For example, if the pipe length is large the dependence on the $\left(\frac{L}{D}\right)$ ratio becomes insignificant, and this term can be dropped, based on the measurement of the axial variations in pressure drop. Few other simplifications drop both ψ and ε.

Accordingly, for this case Equation (16) takes the form:

$$\left(\eta, \frac{\Delta P}{2\rho\upsilon^2} \frac{D}{L}\right) = \phi\left(\frac{D\rho\upsilon}{\mu}, \frac{d_s}{D}, \frac{\rho_s}{\rho}, \frac{v^2}{gD}, C_o\right) \tag{17}$$

Despite the substantial reduction in the number of variables in going from Equation (16) to Equation (17), the functional dependence is still unknown.

3.2. Experimental Test Rig

To establish the relationship stated in Equation (17) a purely experimental approach designed. The model used for test is shown in **Figure 2**.

The model was chosen in such a way to permit the observation of effects of dimensionless groups that control the rate of deposition of solids and pressure gradient through pipeline system. These dimensionless groups are, Reynolds number, Froude number, relative density, particle grain size, and input concentration. The model consists of two cast steel reservoirs 6 cubic meter capacity each of conical-shaped bottom, and connected to a slurry-type centrifugal pump with diameters of 5.08 cm suction and 2.54 cm discharge pipes. The pumps are connected to 7.5 Kw. A/C induced motor with variable speeds. the pipeline is 285 meter long, and the solid concentration by volume was varied between 0.027% and 10%. Average grain size range between 0.25 mm and 1 mm and specific gravities ranging from 2.65 to 3.540 the Reynolds number was varied from 10907 to 63699.

The measured velocity was varied from 0.5 m/s to 2.5 m/s.

In the tests five variables were determined namely:

1. Pressure gradient along the pipeline system.
2. Flow rate measurements.
3. Input concentration to pipeline (C_i).
4. Output concentration from pipeline (C_o).
5. Input grain size.

3.2.1. Pressure Measurement

Pressure measurements were determined by means of bourdon tube manometers that were calibrated by means of dead weight testers. The bourdon gauges are fitted to the pipeline at an equal distance of 16 meter between any two successive manometers.

To avoid wrong readings of manometers as a result of existing solid material through the flow field, special connections were made by mounting a helical-shaped pipe full of oil that is lighter than water.

3.2.2. Flow Measurements

Flow measurements were determined by measuring the flow rate during one hour operation from suction tank to discharge tank.

During the experimental work, the flow rate controlled by means of adjustment of delivery valves only while keeping the suction valves fully opened.

3.2.3. Concentration Measurement

In this experimental work, the following steps have been carried out to determine the input and output concentrations,

1. Prepare a clean, empty bottle and weigh it.
2. Fill the empty bottle with 50 cm^3 sample from the homogenous mixture of the suction tank.
3. Put the bottle on a heater until water is completely evaporated and hence the solids are completely dried.
4. Put the solid sample into cold dryer to cool it.
5. Weigh the solid sample.
6. Subtract the two weights in steps 5 and 1 respectively.

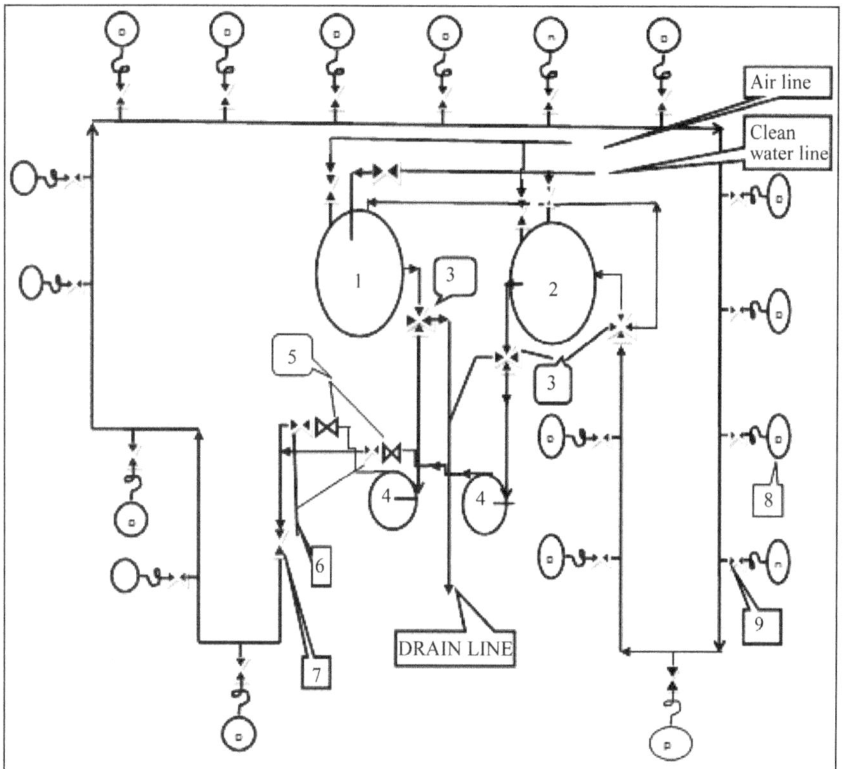

No	Description	No	Description
1	Tank 1	6	Delivery valves
2	Tank 2	7	Control valve
3	Three way suction valves	8	Pressure gauges
4	Centrifugal pumps	9	Pressure gauge valve
5	Non return valves		

Figure 2. Schematic outline of experimental test rig.

The difference is the total solid existed in the sample (suspension solid + soluble solid).

7. Take another 50 cm^3 sample from the same mixture of the suction tank and filter it, so as the water constitutes the soluble solids only.

8. Repeat the steps from (3 to 4) listed before to determine the weight of soluble solids.

9. Subtract the result of step number 8 from that of number 6 to obtain the T.S.M (Total Suspended Material) which represents the transported material through the pipeline system.

3.2.4. Input Grain Size Measurements

The input grain size was determined using a set of standard screen meshes that are available in the laboratories of the factory.

3.3. Experimental Procedure

Preliminary measurements were made to ensure that the system produces a stable (no pulsation), with no leakage from test devices. The Reynolds number (Re_D) for the tests is varied from (10907 to 63699).

The model was provided with 18 pressure gauges, fitted at locations of equal distance of 16 meter apart to ensure precise plotting of the pressure drop variation with length and velocity.

4. Results and Discussions

In the present section the results obtained from the experimental model are presented and discussed.

4.1. Performance Characteristics of Water in Absence of Solids

From section (3) it was noted that the governing equation of slurry flow system may be written in the form:

$$\left(\eta, \frac{\Delta P}{2\rho\upsilon^2}\frac{D}{L}\right) = \phi\left(\frac{D\rho\upsilon}{\mu}, \frac{d_s}{D}, \frac{\rho_s}{\rho}, \frac{v^2}{gD}, C_o\right) \quad (17)$$

This equation can be reduced to the following equation for the clean water

$$:\left(\frac{\Delta P}{2\rho \upsilon^{2}}\frac{D}{L}\right)=\phi\left(\frac{D\rho\upsilon}{\mu}\right) \tag{18}$$

From this equation, the parameter describing the flow performance is represented by the pressure gradient, whereas the Reynolds number of clean water Re_w is the only affecting variable.

4.2. Effect of Varying Reynolds Number Re and Froude Number upon the Performance Characteristics

In the present experimental work, it was found that the simplest way to vary the value of Reynolds number would be through adjustment of the flow velocity. Therefore, Reynolds number as well as Froude number vary together.

1) Effect of varying Re, Fr, upon pressure gradients

The Reynolds number has been varied from (10907 – 63699) and Froude number from (0.5 - 17).

It has been noted from **Figures 3** to **8** that as Reynolds

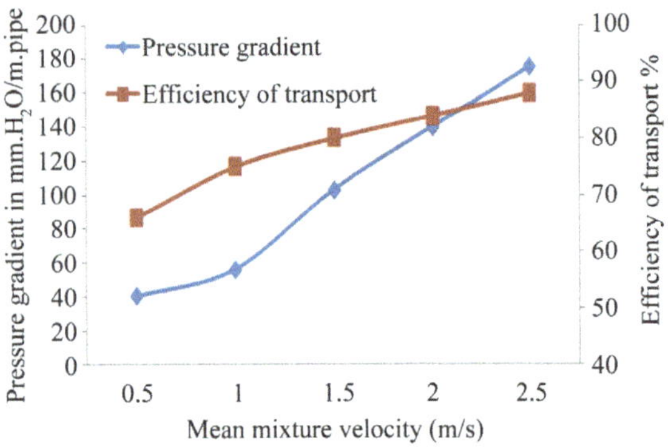

Figure 3. Variation of slurry performance characteristics for sand slurry with solid concentration 1% by volume and particle diameter d_s = 0.25 mm. versus mean mixture velocity.

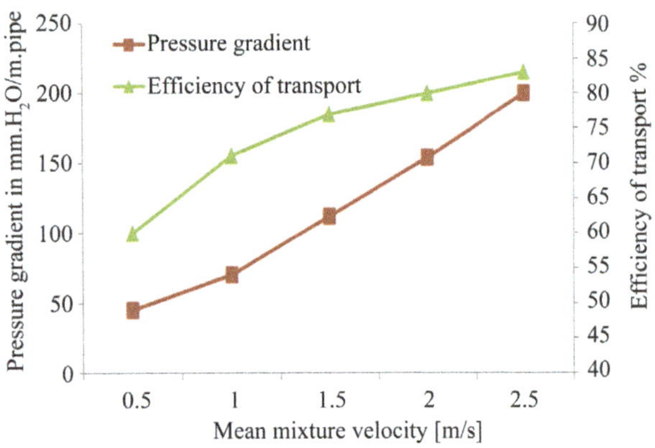

Figure 4. Variation of slurry performance characteristics for sand slurry with solid concentration 1% by volume and particle diameter d_s = 0.5 mm. versus mean mixture velocity.

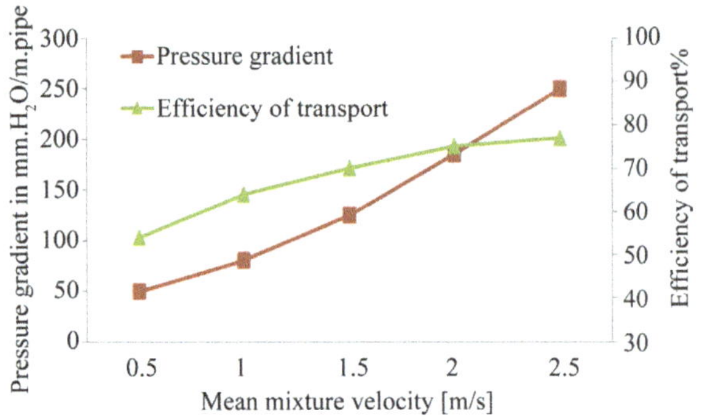

Figure 5. variation of slurry performance characteristics for sand slurry with solid concentration 1% by volume and particle diameter d_s = 1 mm. versus mean mixture velocity.

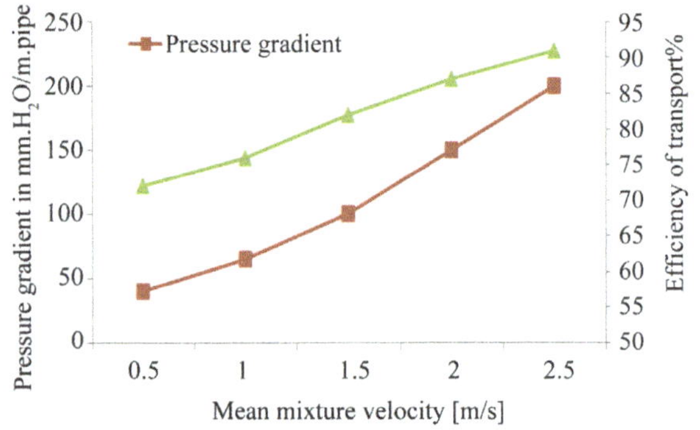

Figure 6. Variation of slurry performance characteristics for slurry of iron oxide with solid concentration 1% by volume and particle diameter d_s = 0.25 mm. versus mean mixture velocity.

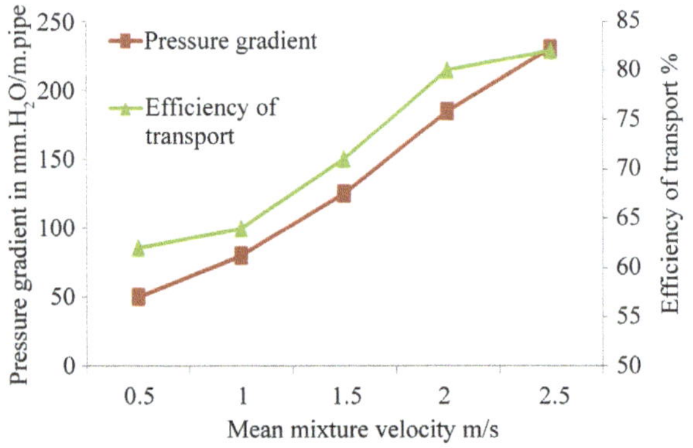

Figure 7. Variation of slurry performance characteristics for slurry of iron oxide with solid concentration 1% by volume and particle diameter d_s = 0.5. mm. versus mean mixture velocity.

and Froude numbers increase the pressure gradient $\left(\frac{\Delta P}{L}\right)$, which is a waste of power available increases.

Also these experimental results concerning pressure gradient have close results in comparison to Turian and Yuan [6] correlations. this comparison is shown in **Figures 9** and **10**.

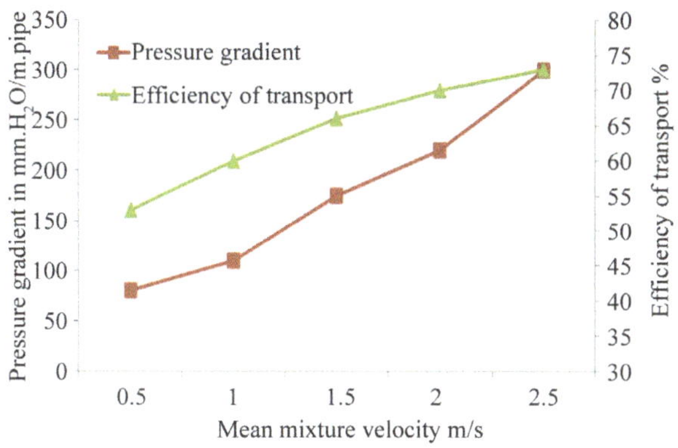

Figure 8. Variation of slurry performance characteristics for slurry of iron oxide with solid concentration 1% by volume and particle diameter d_s = 1 mm. versus mean mixture velocity.

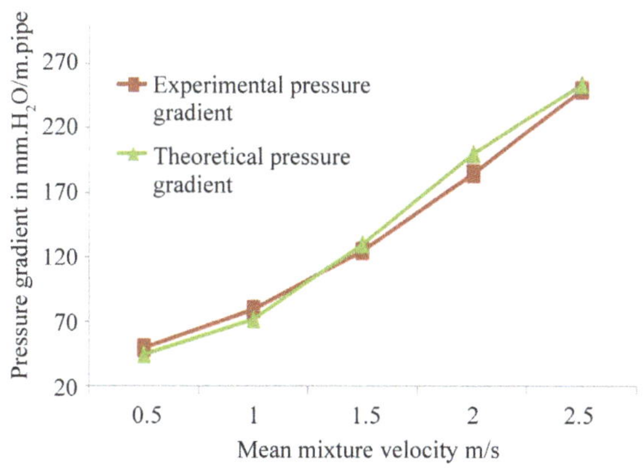

Figure 9. Comparison between experimental and theoretical pressure gradient for sand slurry with solid concentration 1% by volume with grain size diameter d_s = 1 mm.

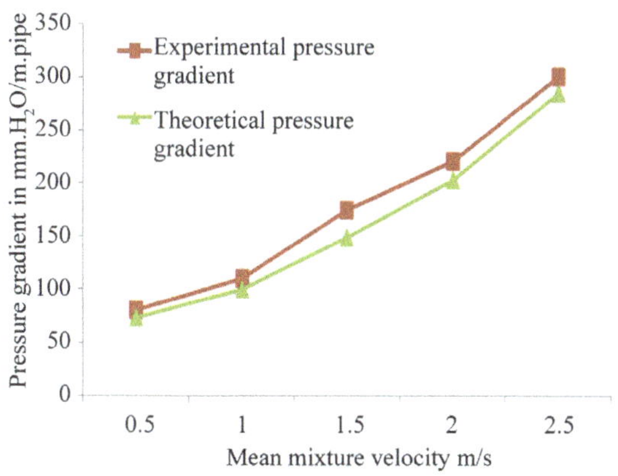

Figure 10. Comparison between experimental and theoretical pressure gradient for slurry of iron oxide with solid concentration 1% by volume with grain size diameter d_s = 1 mm.

2) Effects of varying Re, Fr, upon efficiency of transport

The effect of varying Reynolds and Froude numbers upon the efficiency of slurry transport (η) is shown in **Figures** from **3** to **8**. It can be noticed that as Reynolds and Froude numbers increases the efficiency of transport (η) is increased. This remark also encourages the use of high velocities to avoid blocking slurry pipelines.

4.3. Effect of Varying Particle Diameter (d_s) upon Performance Characterstics of Slurry Pipeline

1) Effect of varying particle diameter (d_s) upon pressure gradient $\left(\frac{\Delta P}{L}\right)$:

Experimental tests have been carried out using samples of solid materials of arithmetic mean diameter from 0.25 mm to 1 mms. at the same input concentration by volume. The effects of changing particle diameter on pressure gradient are shown in **Figures** from **3** to **8** From these figures it was noticed that as the particle diameter increases the pressure drop increases slightly.

2) Effects of varying solid particle diameter (d_s) upon efficiency of transport (η):

Figures from **3** to **8**, indicates also the effects of varying particle grain diameter on efficiency of transport (η). As particle diameter increases the efficiency of transport (η) decreases.

4.4. Effect of Varying Specific Gravity of Particles upon Performance Characteristics

1) Effects of varying specific gravity upon pressure gradient:

The specific gravity was varied from 2.65 for sand and 3.540 for iron oxide.

Figure 11 indicates the effects of varying specific gravity upon the pressure drop $\left(\frac{\Delta P}{L}\right)$. as specific gravity increases, the pressure drop increases; indicating the relative

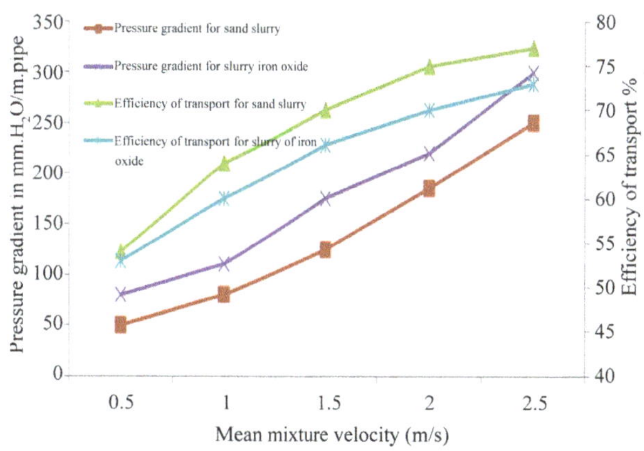

Figure 11. Effect of varying specific gravity upon characteristics of transporting solid-liquid mixtures with solid concentration 1% and particle diameter 1 mm.

merit of transporting light materials other than heavy materials through pipeline system

2) Effects of varying specific gravity upon efficiency of transport (η):

From **Figure 12** as specific gravity increases the efficiency of transport decreases which encourages using light materials through hydraulic transport of solid liquid mixtures in pipelines.

4.5. Effect of Varying input Concentration upon Performance Characterstics

1) Effects of varying input concentration on pressure gradient:

The input concentration of slag of iron oxide and sand are varied from (1% - 10%) by volume. The effect of varying input concentration on pressure gradient is shown in **Figures** from **13** to **17**. From these figures, as input concentration increases, the pressure gradient increases. This increase is directly proportional to input concentra-

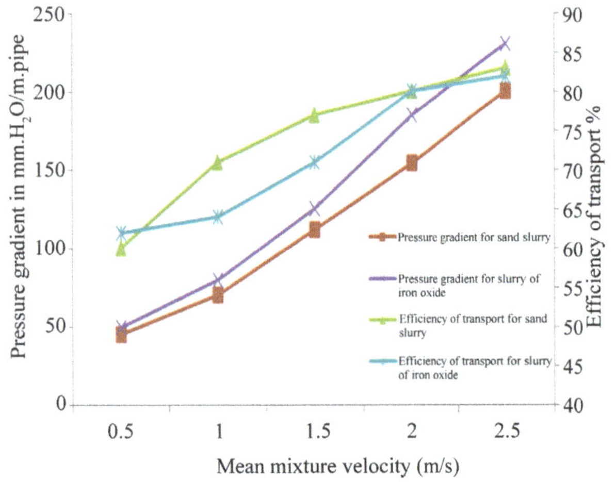

Figure 12. Effect of varying specific gravity upon characteristics of transporting solid-liquid mixtures with solid concentration 1% and particle diameter = 0.5 mm.

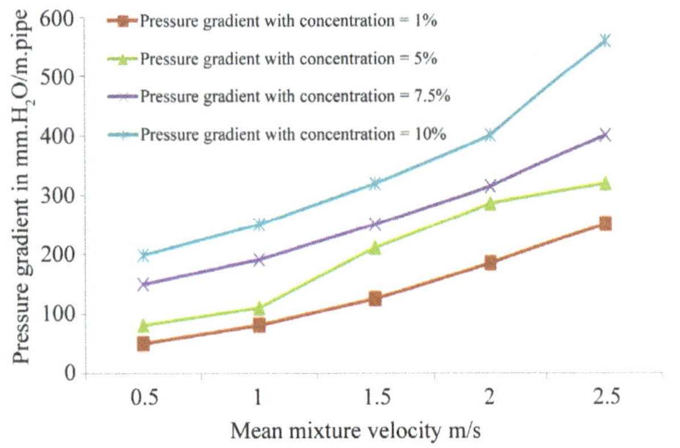

Figure 13. Effect of varying input solid concentration on pressure gradient for sand slurry of grain size diameter d_s = 1 mm. versus mean mixture velocity.

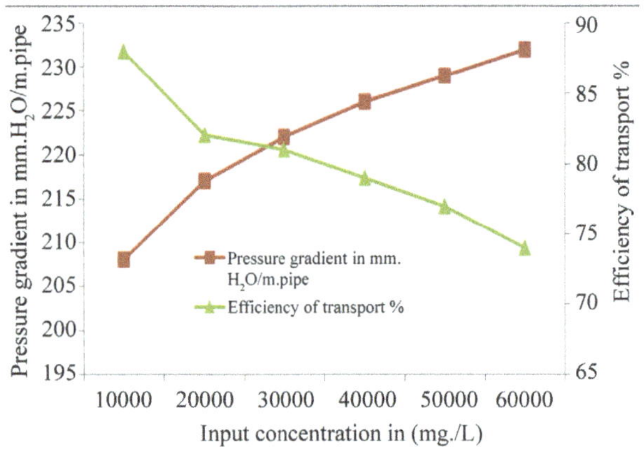

Figure 14. Variation of slurry performance characteristics for slurry of iron oxide versus input solid concentration at Reynolds number = 63699.

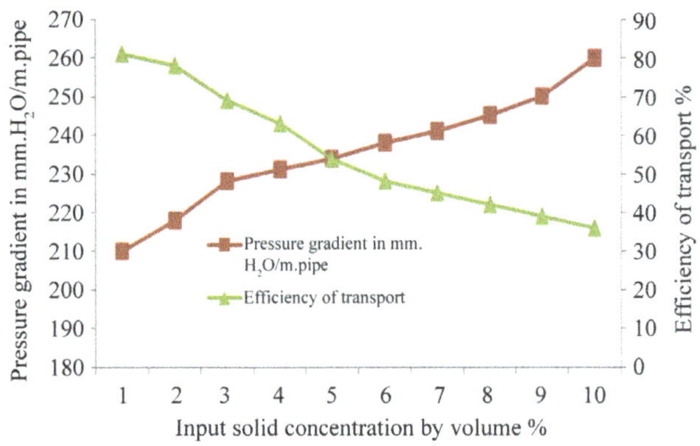

Figure 15. Variation of slurry performance characteristics for sand versus input solid concentration at Re = 63699.

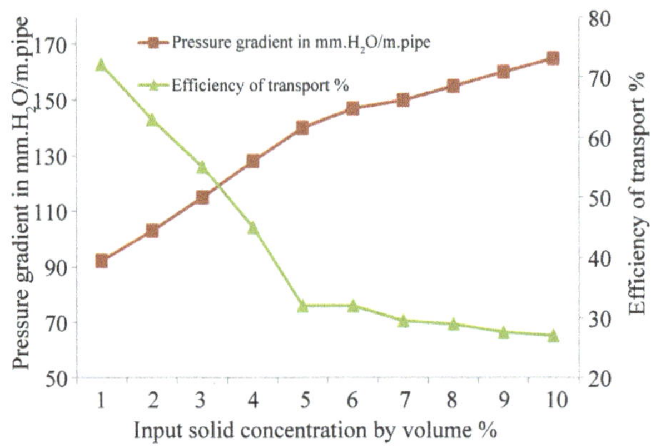

Figure 16. Variation of slurry performance characteristics for sand versus input solid concentration at Re = 37396.

tion by volume and this increase may be from probability of separating a part from input solid material.

2) Effects of varying input solid concentration upon efficiency of transport:

Figures from **14** to **18** indicate the effects of varying input concentration on efficiency of transport. It has been noticed that as input concentration increases the efficiency

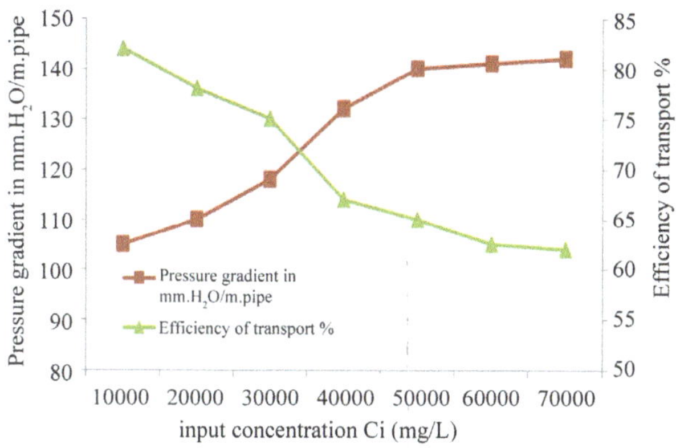

Figure 17. Variation of slurry performance characteristics for iron oxide versus input solid concentration at Re = 37396.

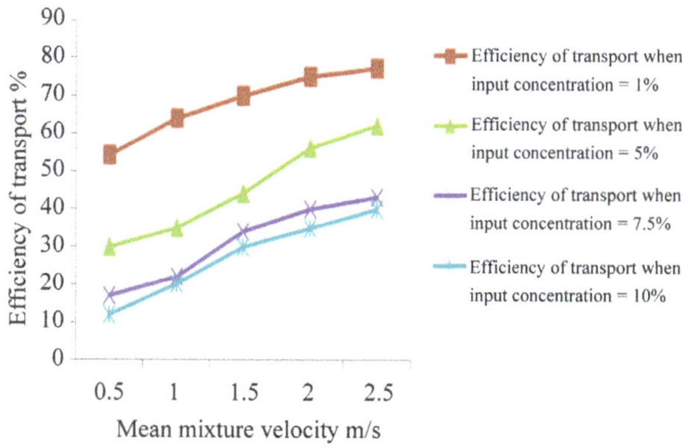

Figure 18. Effect of varying input solid concentration on the efficiency of transport of sand slurry of grain size diameter d_s = 1 mm. versus mean mixture velocity.

of transport decreases sharply. This remark encourages using low input concentrations to avoid sharp decrease in efficiency; hence avoid fast blocking of the pipe.

4.6. Effect of Reynolds Number on Grain Size Distribution

Figure 19 shows the effect of Reynolds number on grain size distribution at pipe outlet for concentration of 0.119% by volume.

From this figure we notice that:

1) High values of Reynolds number increases the ability to transport large grain sizes.

2) Transporting fine grain size does not need high Reynolds number.

5. Conclusions

The main conclusions that can be extracted from the results and observations of this work are mentioned below:

1) The input concentration plays an important role in the slurry transportation process. Increasing input concentration causes an increase in the pressure drop and

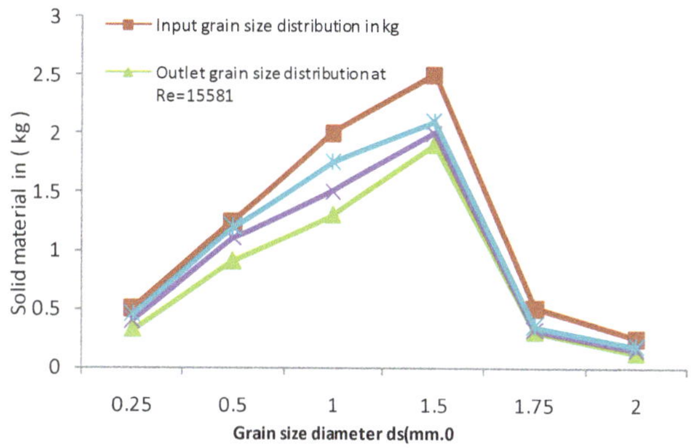

Figure 19. Effect of Reynolds number's on grain size distribution at pipe outlet at total input concentration of 0.119% by volume of iron oxide.

a decrease in efficiency of transport, probably due to agglomeration and flocculation of solid particles.

2) Increasing the grain size increases the pressure loss and decreases efficiency of transport.

3) As the Reynolds number increases, transport efficiency increases with an increase in pressure drop, and subsequently input power increases as well.

4) The process of transportation can be improved by using fine materials rather than course materials.

5) The specific gravity of solid materials plays an important role in efficiency of transport and also in pressure gradient. As specific gravity increases, the tendency to settle down increases, which can lead to decrease in efficiency of transport and increase in pressure drop.

6. Recommendations

1) The present work was carried out to investigate and deduce the dimensionless groups that affected the transporting process, but the functional dependence was still unknown. So it's recommended to use the data summed up here and define the governing equations.

2) It is also recommended to carry out experiments to investigate how the slurry transport affects the character istics of the pumps.

REFERENCES

[1] D. M. Newitt, J. F. Richardson, M. Abbot and R. B. Turtle, "Hydraulic Conveying of Solids in Horizontal Pipes," *Transactions of the Institution of Chemical Engineers*, Vol. 33, 1955, pp. 93-113.

[2] R. Durand, "The Hydraulic Transport of Coal and Solid Materials in Pipes," *Proceedings of Colloquium on the Hydraulic Transport of Coal*, National Coal Board, London, 1952, pp. 39-55.

[3] R. C. Worster, "Proceedings of Colloquium on Hydraulic Transport of coal, National Coal Board, London, 1952.

[4] K. Wagner, "Investigations on the Influence of the Grain

Size Distribution for Horizontal Hydraulic Conveying of Solids in the Heterogeneous Regime," Thesis, Karlsruhe, 1982.

[5] M. Weber, "Improved Durand–Equation for Multiple Application," *International Symposium on Slurry Flows*, 1986, pp. 61-66

[6] R. M. Turain, T. F. Yuan, *et al.*, "Pressure Drop Correlation Pipeline Flow of Solid Liquid Suspensions," *AICHE Journal*, Vol. 17, No. 4, 1971, pp. 809-817.

[7] H. C. Hou, "Investigation of Optimal Grain Distribution for Transport with High Concentration," *Hydraulic Transport of Solids in Pipes*, city, 29-31 October 1986, pp. 177-183.

[8] M. S. Lee, V. Matousek, C. K. Chung and Y. N. Lee, "Pipe Size Effect on Hydraulic Transport of Jumoonjin Sand: Experiments in a Dredging Test Loop," Terra et Aqua-number 99, Hamburg, June 2005, pp. 1-10.

Nomenclatures

C_c: Delivered concentration of coarse material, volume %

C_D: Drag coefficient for free falling particle $= \dfrac{4gd(\rho_s - \rho)}{3\rho\vartheta_\infty^2}$ dimensionless.

C_f: Delivered concentration of fine material, volume %.

C_i: Input concentration of solids, volume %.

C_o: Output discharge concentration of solids, volume %.

D: inside pipe diameter in meter [m]

D_s: Diameter of solid particle assumed spherical, in meter. [m]

F: Fraction of fine material.

$(1 - F)$: Fraction of coarse material.

F_L: Coefficient for deposit velocity.

$\lambda = \dfrac{\Delta PD}{2\rho\upsilon^2 L}$: Friction factor for pipe flow of slurry, dimensionless.

$\lambda_w = \dfrac{\Delta PD}{2\rho\upsilon^2 L}$: Friction factor for pipe flow in absence of solids, dimensionless.

G: Gravitational constant, m/s^2.

i: Head loss for mixture.

i_w: Head loss for water.

k: Constant dimensionless.

L: Pipe length in meter.

S: Ratio of solid to liquid densities dimensionless.

v: Mean mixture velocity in m/s.

υ_∞ : Terminal deposit velocity for particles m/s.

$\left[\dfrac{\Delta P}{L}\right]$: Pressure gradient measured along pipeline in mm, water column per meter pipe length.

$\mathrm{Re} = \dfrac{\rho v D}{\mu}$: Reynolds number for pipe flow, dimensionless.

$Fr = \dfrac{\upsilon}{\sqrt{gD}}$: Froude number for pipe flow, dimensionless.

Greek Letters

ε : Shape factor for suspended particles, dimensionless.

$\eta = \dfrac{C_o}{C_i} \times 100\%$: Efficiency of transport.

ψ : Particle size distribution function, dimensionless.

12

Structure and Oscillation of Underexpanded Jet

Hiromasa Suzuki[1], Masaki Endo[2], Yoko Sakakibara[2]
[1]Graduate School of Advanced Science and Technology, Tokyo Denki University, Saitama, Japan
[2]Division of Electronic and Mechanical Engineering, Tokyo Denki University, Saitama, Japan

ABSTRACT

In this study, flow characteristics are experimentally and numerically discussed, especially concerning shock-cell structure and oscillatory phenomena of supersonic jet. The jet becomes underexpanded when the nozzle pressure ratio of convergent nozzle exceeds the critical value. The underexpanded jet is not uniform because of the presence of the expansion wave, the compression wave and the shock wave formed in it. Many vortices are induced in jet boundary region by shearing stress generated between supersonic jet flow and atmospheric air. They interact with shock waves composing cell node of jet, which is closely related to a noise radiating from the jet. In experiment, flow field is visualized and the shape of cell structure is examined. Furthermore, frequency of vortex generation is measured. The jet is simulated using TVD scheme and the results are compared with the experimental ones. Distortion of cell structure caused by three dimensional oscillation of jet and behavior of vortices are discussed.

Keywords: Underexpanded Jet; Plate Shock; Incident Shock; Visualization; TVD Scheme

1. Introduction

Underexpanded jet is one of the supersonic jets; when the nozzle pressure ratio of the convergent nozzle is higher than the critical value, the jet is underexpanded. In case of air, the critical pressure ratio is about 1.893. The expansion waves, which are generated at the nozzle lip, propagate in the jet and reflected at jet boundary as the compression waves. They converge to form a shock wave, and the expansion waves are generated again when the shock wave reaches the jet boundary. The jet is not uniform because these phenomena repeat and typical shock cell structure of the underexpanded jet is formed.

When the flat plate is placed perpendicularly to the jet, the strong wave called "plate shock" appears in front of the plate. The air passes through the plate shock, and its velocity becomes subsonic. The flow changes its direction and the wall jet is formed along the plate.

The structure of underexpanded impinging jet is shown in **Figure 1**. Such a flow field is often observed in the assist gas of laser cutting [1], the cooling jet of glass tempering process [2,3] and the exhaust jet of VSTOL aircraft or rocket. The flow field causes several issues, for example, behavior of jet disturbed by vortex of the jet boundary and high frequency noise generation called "screech tone" [4-7].

In this study, shadowgraph method and Schlieren method were used to visualize the flow field. Furthermore, a period of generation cycle of vortex near the jet boundary was measured using a device comprising a laser and a photoelectric sensor.

The behavior of jet and vortices near the jet boundary were examined. In addition, under the same conditions, the numerical study was carried out using TVD scheme.

2. Experimental Apparatus and Numerical Technique

2.1. Experimental Considerations

The piping system of the experimental apparatus is shown in **Figure 2**. The air is compressed by screw

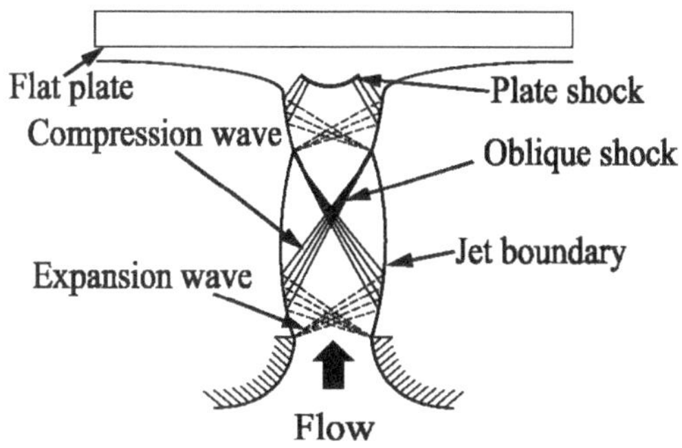

Figure 1. Schematic of underexpanded impinging jet.

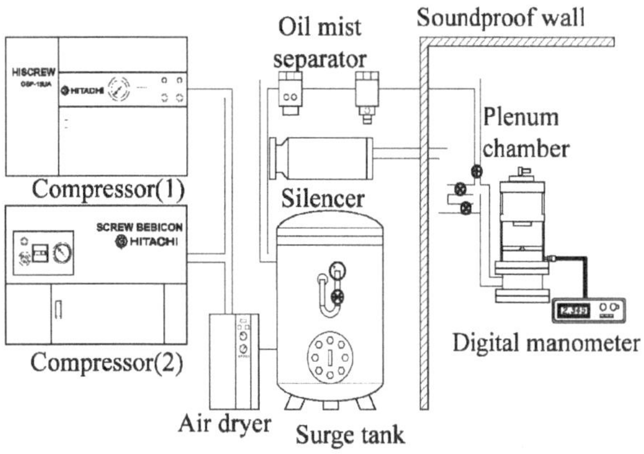

Figure 2. Piping system.

compressors and passes through the air dryer, surge tank, and oil mist separator. The compressed air is supplied into the plenum chamber to which convergent nozzle is attached in the soundproof room. The stagnation pressure of the air in the plenum chamber can be manually regulated by two gate valves and is measured by a digital manometer. An acoustic absorbent sheet is glued to the exit plane of the convergent nozzle. In this study, the convergent nozzle is adopted. The diameter of the convergent nozzle exit is D = 10 mm and the radius of the internal surface of the nozzle R = 20 mm, which is well finished. And a flat plate is placed to be perpendicular to the jet axis, downstream of the nozzle.

The experiment is carried out under the condition of the nozzle pressure ratio p_0/p_a = 3.0 to 3.8 with 0.1 steps and the nozzle plate distance l/D = 1.5 to 2.5 with 0.1 steps, where p_0 is the pressure in the plenum chamber and p_a the atmospheric pressure. The behavior of jet, vortex moving along jet boundary and plate shock above plate was examined under the conditions as mentioned above.

The flow fields were visualized using Schlieren photography and shadowgraphy. The spark bulb is employed as source of light, which emits flushing light duration a period of 180 ns.

Furthermore, the advective motion of vortex near the jet boundary is measured using a device composed of a laser and a photoelectric sensor, *i.e.* a simple Schlieren system. A semi-conductor type laser is used the beam width being l_w = 3 mm. The vortex measurement system is shown in **Figure 3**. The knife edge is installed to cause it to block out the light. Going through the vortex, the laser beam is refracted in a direction away from the vortex core, and then, the illumination of beam reachable the sensor increases. Thus, the sensor detects the moving vortex. The laser is set at the position of antinode of second cell where the vortex starts to be organized. The antinode position is measured by means of the Schlieren pictures.

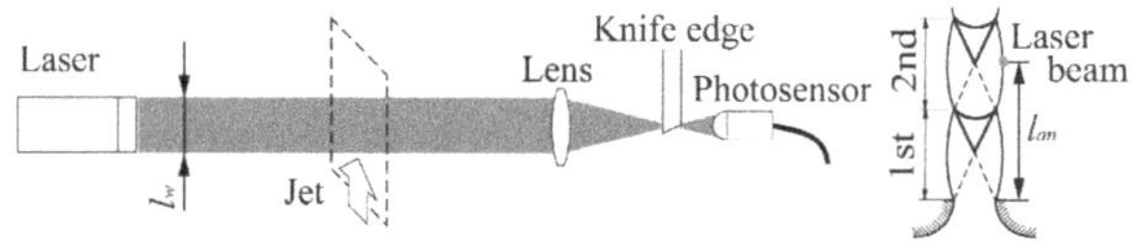

Figure 3. Vortex measurement system.

2.2. Numerical Scheme

The underexpanded jet impinging on the plate without an elastic deformation is numerically simulated. The computational flow field is assumed to be axisymmetric and Euler equations are employed as the governing equations. They are solved using 2nd-order TVD scheme proposed by Yee [8], with minmod function as a stable limiter function [9].

The computational region used in this study is covered with a structured mesh. The radial distance of the region is 7.5 diameters of the converging nozzle, where 150 computational grids are clustered towards the jet axis. Along the jet axis there are 100 grids at uniform intervals. The critical condition estimated from the stagnant properties in the plenum chamber is specified at the exit plane of the nozzle. The boundary condition is given to be reflective at the jet axis. The nozzle wall and the plate are slip wall boundary and, at the other boundaries, the Riemann variables are kept at constant.

3. Results and Discussion

3.1. Behavior of Free Jet

Figure 4 shows the flow field visualized using Schlieren method. These pictures were taken at random under the same condition. In these figures, the underexpanded free jet issues from the nozzle in the lower side of the picture at the pressure ratio p_0/p_a = 3.5. The knife edge is placed in the direction normal to jet axis so that the expansive region is dark and compressive region bright. As can be seen in these pictures, the dark region and the bright region are deformed and the deformation becomes stronger as further away from the nozzle. Thus, the jet is found to be unsteady and oscillate laterally or helically.

In order to quantitatively evaluate the unsteady behavior of the jet, the statistical analysis is conducted through the visualization of the flow field. 50 pictures were taken at random at each pressure ratio, and the instantaneous shape of the jet was measured, e.g. the cell length l_c and the inclination of cell node α_n as shown in **Figure 5.** The cell node between the 1st and 2nd cell, *i.e.* end node of cell, is defined as 1st cell node, and so on.

3.2. Cell Length

Figure 6 shows deviations of cell length of 1st, 2nd and 3rd cell at each pressure ratio. The empty plots show the

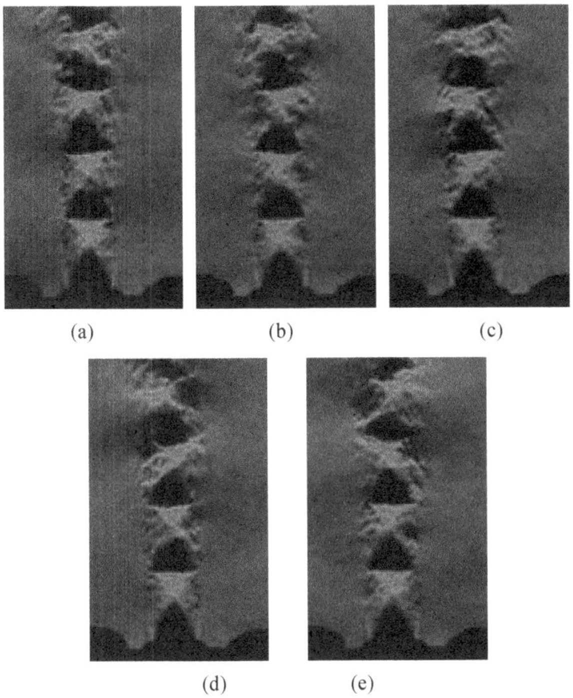

Figure 4. Schlieren photograph ($p_0/p_a = 3.5$).

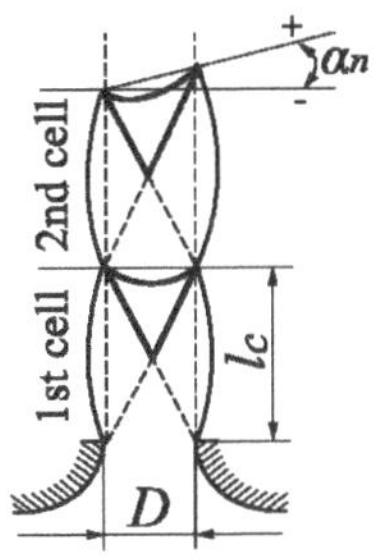

Figure 5. Measurement points.

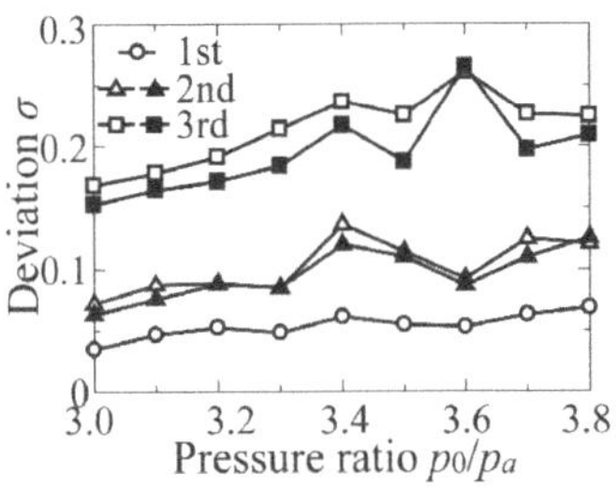

Figure 6. Deviation of cell length.

lengths of individual cell, and the solid ones total lengths between the nozzle exit and the end node of corresponding cell. Overall, these plots gradually rise as the pressure ratio increases and these take extreme values at p_0/p_a = 3.4 and 3.6. Concerning the individual cell length, the deviation for downstream cell is seen to swell two-fold. Thus, the amplitude of the oscillation in length becomes larger downstream.

Comparing the deviations between the individual and the total lengths, the deviation of total length is smaller than that of individual length regardless of the pressure ratio. This means that the 3rd cell shrinks if the 2nd cell stretches, that is, the changes of length caused by oscillation of the neighboring cells are cancelled as a whole. Observing the difference between the deviations at each pressure ratio carefully, the oscillation at $p_0/p_a = 3.6$ may have the same phase.

3.3. Cell Node Angle

Figure 7 shows deviation of cell node angles α_n of 1st, 2nd and 3rd cell. This deviation is related to three-dimensional oscillation because α_n means the angle of node against jet axis as shown in **Figure 5**. The value of the deviation almost linearly rises with the increase in the pressure ratio and becomes larger downstream, e.g., σ_1 = 0.9, σ_2 = 2.2 and σ_3 = 4.3 at p_0/p_a = 3.4. Hence, this means that the hydrodynamic instability along the jet boundary grows downstream.

3.4. Behavior of Impinging Jet

The experiment is carried out under the same condition of p_0/p_a as the free jet and the pictures are statistically analyzed. 50 pictures were taken at random in each nozzle plate distance. The geometric characteristics of the jet structure were measured, e.g. the positions of vertex and base of the plate shock (l_{vp} and l_{bp}), the plate shock diameter d_p and the inclination of plate shock α_p as shown in **Figure 8**.

Figure 9 shows the deviation against the pressure ratio. The deviation σ in the ordinate is normalized by the deviation σ_s at l/D = 1.5. It can be seen that the deviation of α_p undergoes abrupt change from l/D = 1.6 to 1.8 and that the fluctuation of l_{vp} stands out at l/D = 1.8. These comparatively high values represent that the jet strongly oscillates under this condition. Especially at l/D = 1.8, deviation of l_{vp} stands out and this means that the center part of plate shock strongly oscillates in axially-symmetric pattern.

Furthermore, the deviation of α_p also has high value at l/D = 1.7, and this means that lateral or helical motion occurs simultaneously. Thus, the plate shock intensely

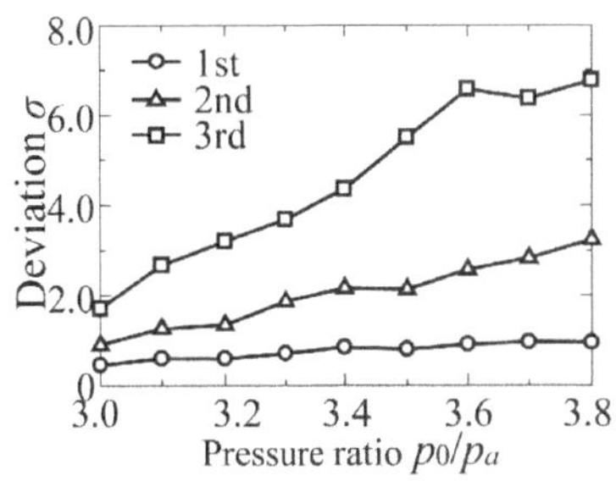

Figure 7. Deviation of angle of cell node.

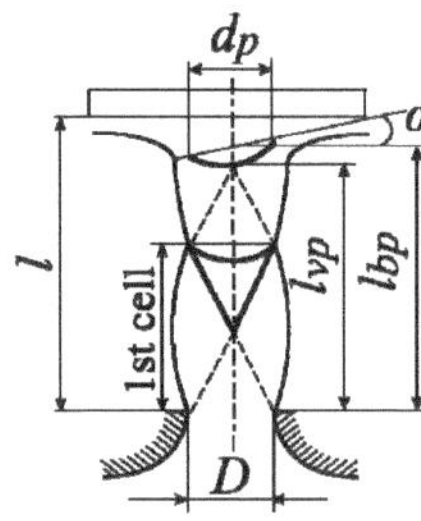

Figure 8. Measurement points.

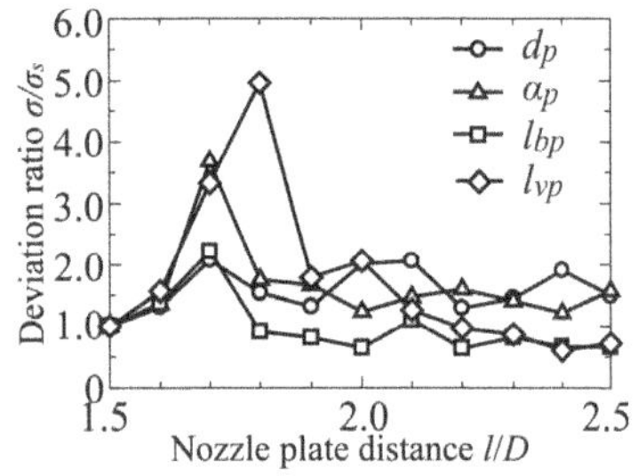

Figure 9. Deviation of plate shock.

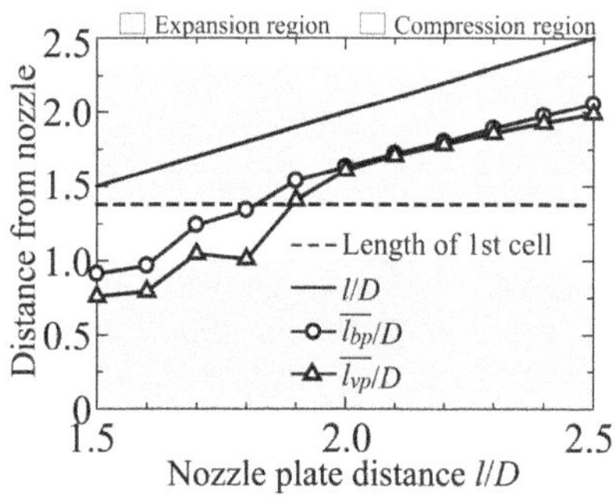

Figure 10. Distance of plate shock from nozzle.

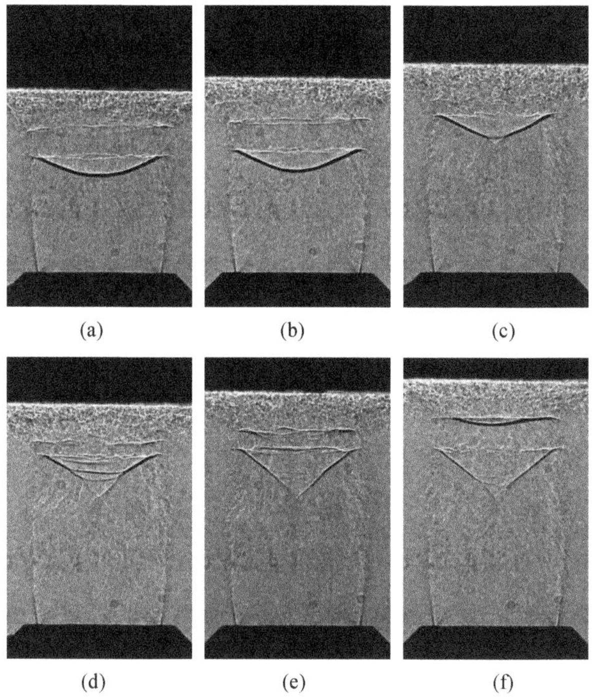

Figure 11. Shadowgraph photograph (p_0/p_a = 3.4). (a) *l/D* = 1.5; (b) *l/D* = 1.6; (c) *l/D* = 1.7; (d) *l/D* = 1.8; (e) *l/D* = 1.9; (f) *l/D* = 2.0.

oscillates at a certain nozzle plate distance.

Figure 10 shows averages of l_{vp} and l_{bp} at p_0/p_a = 3.4. The dashed line shows the position of 1st cell node obtained from pictures of free jet. Position of plate shock moves rapidly downstream to approach the plate at between *l/D* = 1.6 and *l/D* = 1.7. Distance between the vertex and base of the plate shock increase in the range from *l/D* = 1.7 to *l/D* = 1.8. These phenomena seem to be affected by location of the vertex of plate shock; whether the vertex is in the compression region or not. In the compression region, the plate shock crosses the incident shocks from the nozzle lip before they reach the jet axis, while in the expansion region, the incident shocks intersect each other on jet axis as shown in the next.

Figure 11 shows shadowgraph pictures of impinging jet at various nozzle plate distances from *l/D* = 1.5 to 2.5. The jet issues from the nozzle exit on the bottom of each picture. Plate shock crosses the incident shocks at *l/D* = 1.5 and 1.6, and at *l/D* > 1.7, the intersection of incident shock appears upstream of plate shock. In addition, a weak shock wave can be seen in the region between the plate shock and the plate. This weak shock appears up to *l/D* = 1.6 and disappears at *l/D* = 1.7. At *l/D* > 1.8, the plate shock is in the expansion region of second cell.

Details of shock structure can be seen in the results obtained by the numerical simulation. **Figures 12** show the density contours (left-side half of each figure) and the pressure contours (right-side half of each figure) at different nozzle plate distance between *l/D* = 1.4 and 2.0. The flow in the region between the plate shock and the plate is seen to be complex with *l/D* smaller than *l/D* = 1.6. Under such conditions, because the intersection of the incident shocks in the jet is located downstream of the plate shock, the strength of plate shock near the jet axis is stronger than that in the surrounding region. Therefore, contact surface is generated downstream of the plate shock and the iteration of reflection of waves between jet boundary and the contact surface occurs. Weak shock wave seen in **Figures 11(a)** and **(b)** is considered to be the shock wave which consists of converged compression wave during this iteration. The weak shock wave disappears at *l/D* = 1.7, where the intersection of the incident shocks is upstream of plate shock (see **Figure 11(c)** and **Figure 12(e)**). In **Figure 12(f)**, plate shock forms close to the incident shock and they are partly piled up. This can be seen also in **Figure 11(d)**.

The position of plate shock changes into second cell at larger *l/D*.

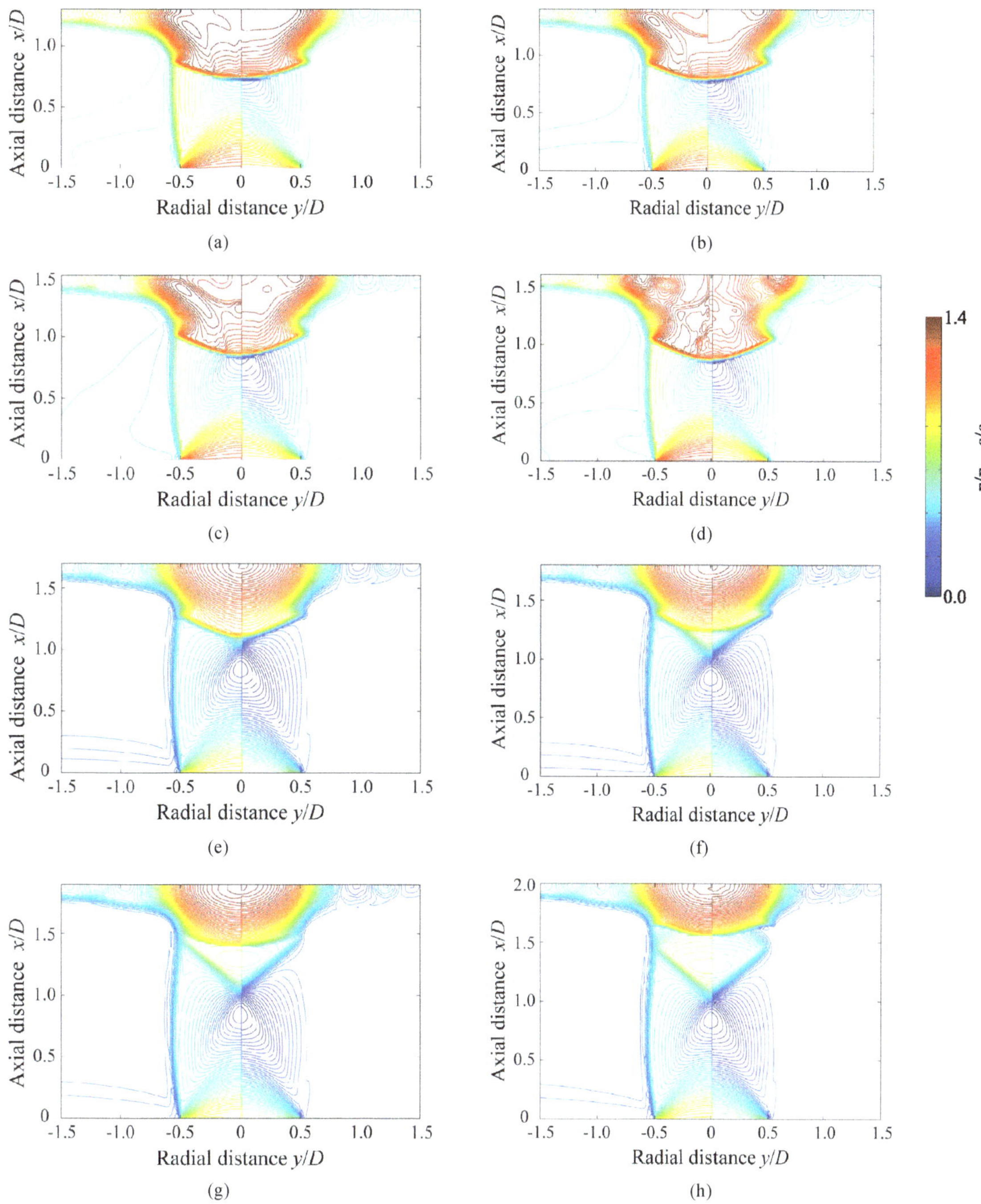

Figure 12. Numerical result (p_0/p_a = 3.4): (a) l/D = 1.3; (b) l/D = 1.4; (c) l/D = 1.5; (d) l/D = 1.6; (e) l/D = 1.7; (f) l/D = 1.8; (g) l/D = 1.9; (h) l/D = 2.0.

3.5. Frequency of the Vortex Near the Free Jet Boundary

Figure 13 shows the output signal history measured by the photoelectric sensor at p_0/p_a = 3.0. The sharp drop of the voltage means that a vortex passed through a laser. The vortices are seen to intermittently go through the measuring point. To search the periodicity of the vortex movement, the output signal is analyzed using FFT. The change of the dominant frequencies against the pressure rise is shown in **Figure 14**. The empty plot denotes the dominant frequency obtained though the FFT analysis of output data as in **Figure 13** at each pressure ratio. The solid plots are obtained from the experimental results of the sound pressure wave, which comes from the paper by S. Tamura and J. Iwamoto [10]. The sound pressure wave has two components of the frequencies. Those dominant frequencies lie around 15 kHz and 25 kHz,

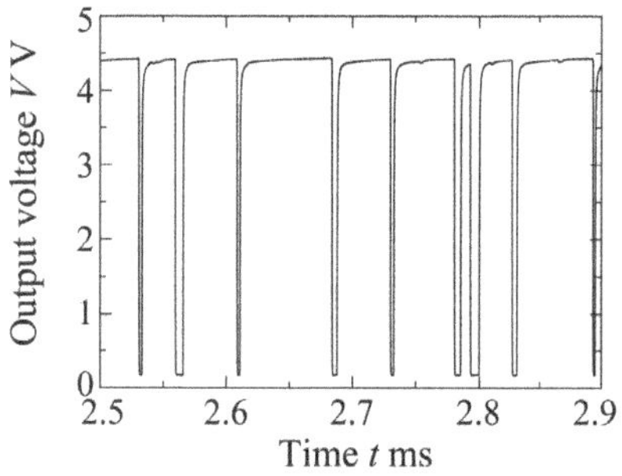

Figure 13. Output signal wave (p_0/p_a = 3.0).

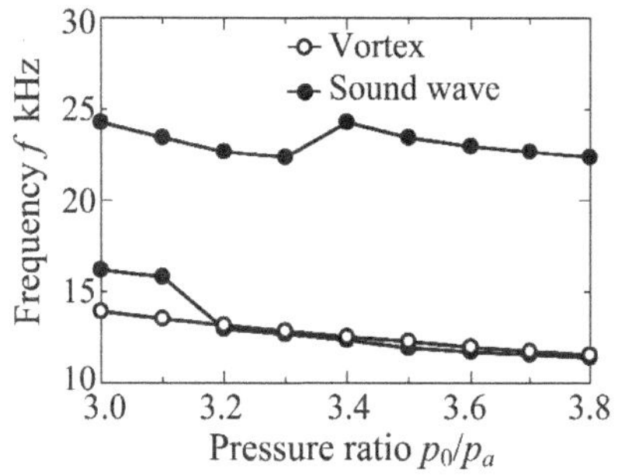

Figure 14. Frequency of vortex generation and sound wave.

slightly decreasing with the rise of pressure ratio. The higher frequency jumps at p_0/p_a = 3.3 and lower one drops at 3.1. These discontinuous changes are known to be caused by the structure of the underexpanded jet. As can be seen in **Figure 14**, the change of lower frequency quantitatively agrees well with that of the vortex. It is evident that the sound emitted form the jet is cause by the behavior of the vortex moving along the jet boundary.

4. Conclusions

To investigate the oscillatory phenomena of underexpanded jet, the flow field was visualized using the schlieren photography, and shadowgraphy and the behavior of vortex near the jet boundary was observed. As a result, the following conclusions were drawn:

- The deviation of cell node angle linearly increases with rise of pressure ratio.
- The deviation of cell length takes maxima and minima at some pressure ratios.
- The weak shock is formed in the downstream region of plate shock, when the vertex of plate shock locates near the end of expansion region.
- Good agreement is found between frequency of vortex generation and that of sound wave.

REFERENCES

[1] J. Fieret, M. J. Terry and B. A. Ward, "Overview of Flow Dynamics in Gas-Assisted Laser Cutting," *SPIE High Power Lasers*, Vol. 801, 1987, pp. 243-250.

[2] S. Aratani and N. Ojima, "Effect of Shock Waves on Fracture and Quenching in Tempered Glass," *Proceedings of the 18th International Symposium of Shock Waves*, Sendai, 21-26 July 1991, pp. 1277-1282.

[3] J. M. Barsom, "Fracture of Tempered Glass," *Journal of the American Ceramics Society*, Vol. 51, No. 2, 1968, pp. 75-78.

[4] A. Powell, Y. Umeda and R. Ishii, "Observation of the Oscillation Modes of Choked Circular Jets," *Journal of the Acoustical Society of America*, Vol. 92, No. 5, 1992, pp. 2823-2839.

[5] A. Powell, "The Sound-Producing Oscillations of Round Underexpanded Jets Impinging on Normal Plates," *Journal of the Acoustical Society of America*, Vol. 83, No. 2, 1988, pp. 515-533.

[6] C. K. W. Tam, "Broadband Shock Associated Noise from Supersonic Jets Measured by a Ground Observer," *AIAA Journal*, Vol. 30, No. 10, 1992, pp. 2395-2401.

[7] J. Panda, "Shock Oscillation in Underexpanded Screeching Jets," *Journal of Fluid Mechanics*, Vol. 363, 1998, pp. 173-198.

[8] H. C. Yee, "Upwind and Symmetric Shock-Capturing Scheme," National Aeronautics and Space Administration (NASA), Washington DC, 1987.

[9] Y. Sakakibara and J. Iwamoto, "Numerical Study of Oscillation Mechanism in Underexpanded Jet Impinging on Plate," *Journal of Fluids Engineering*, Vol. 120, No. 3, 1998, pp. 477-481.

[10] S. Tamura and J. Iwamoto, "Study on the Noise Generated from Underexpanded Free and Impinging Jet," *Transactions of the Japan Society of Mechanical Engineers, Series B*, Vol. 60, No. 579, 1994, pp. 283-289.

Nomenclature

D: Diameter of nozzle exit
d_p: Diameter of plate shock
f: Frequency
l: Distance between nozzle exit and flat plate
l_c: Length of cell
l_{an}: Position of antinode of second cell
l_{bp}: Position of base of plate shock
$\overline{l_{bp}}$: Average of position of base of plate shock
l_{vp}: Position of vertex of plate shock
$\overline{l_{vp}}$: Average of position of vertex of plate shock
l_w: Width of laser beam
p_a: Atmospheric pressure
p_0: Stagnation pressure in a plenum chamber
R: Radius of convergent nozzle
α_n: Angle of cell node
α_p: Angle of plate shock
σ: Deviation
σ_1: Deviation of 1st cell of free jet
σ_2: Deviation of 2nd cell of free jet
σ_3: Deviation of 3rd cell of free jet
σ_s: Deviation at $l/D = 1.5$

Study on the Wake Shape behind a Wing in Ground Effect Using an Unsteady Discrete Vortex Panel Method

Cheolheui Han[1*], Spyros A. Kinnas[2]
[1]Department of Aeronautical and Mechanical Design, Korea National University of Transportation, Chungju, Republic of Korea
[2]Department of Civil, Architectural and Environmental Engineering, University of Texas at Austin, Austin, USA

ABSTRACT

The unsteady evolution of trailing vortex sheets behind a wing in ground effect is simulated using an unsteady discrete vortex panel method. The ground effect is included by image method. The present method is validated by comparing the simulated wake roll-up shapes to published numerical results. When a wing is flying in a very close proximity to the ground, the optimal wing loading is parabolic rather than elliptic. Thus, a theoretical model of wing load distributions is suggested, and unsteady vortex evolutions behind lifting lines with both elliptic and parabolic load distributions are simulated for several ground heights. For a lifting line with elliptic and parabolic loading, the ground has the effect of moving the wingtip vortices laterally outward and suppressing the development of the vortex. When the wing is in a very close proximity to the ground, the types of wing load distributions does not affect much on the overall wake shapes, but parabolic load distributions make the wingtip vortices move more laterally outward than the elliptic load distributions.

Keywords: Wake Shape; Wing in Ground Effect; Unsteady Discrete Vortex Method

1. Introduction

When a wing is flying near the ground, the aerodynamic characteristics of the wing is changed due to the interaction of the wing and the ground. This phenomenon is called as wing-in-ground (WIG) effect. New conceptual vehicles utilizing the WIG effect has been suggested focusing on fuel efficiency and stability [1].

Panel methods have been frequently applied to the conceptual design of wing-in-ground effect vehicles [2]. Many free-wake analysis techniques have been developed in order to represent the complicated wake vortex behavior behind a wing (Hoeijmakers and Vaatstra [3], Ribeiro and Kroo [4], Zhu and Takami [5], for example). Lamarre and Paraschivoiu [6] showed that the large-scale behavior of the roll-up of the aircraft trailing vortices can be simulated using the shape of 2-dimensional vortex sheet at any pseudo time because it resembles the shape of the cross section of the 3-dimensional sheet at a distance behind the wing (see [7] for a comprehensive review). Pullin [8], Krasny [9,10] investigated the roll-up of an initially plane semi-infinite vortex sheet. Morky [11] developed a 2-dimensional algorithm (based on the constant vortex sheet method) for modeling the evolution and propagation of aircraft trailing vortices in a sheared environment near the ground. Han and Cho [12] investigated the unsteady evolution of trailing vortex sheets behind a wing in ground effect using a discrete vortex method.

Han and Cho [12] assumed that the wings in close formation flight near the ground have the elliptic load distributions. Windall and Barrow [13] showed, based on a linear approach, that semi-elliptic wing planform with parabolic wing loading is optimal for a wing in ground effect for all aspect ratios. When a large aspect ratio wing is flying very close to the ground (extreme ground effect: EGE), based on a nonlinear inviscid theory, the optimal spanwise distribution of loading is parabolic rather than elliptic [2].

The objective of the present work is to investigate the effect of wing load distributions on the unsteady wake vortex evolution behind a wing in ground effect (IGE) and extreme ground effect (EGE). A new load distribu-

*Corresponding author.

tion is suggested in order to match the change of the load distributions due to the change of the ground height. A lifting line with an initial load distribution is discretized with discrete vortex elements. The trailing wake vortices from each vortex element are represented by free vortices that deform freely by the assumption of a force-free position during the simulation.

2. Numerical Method

2.1. Wing Loading

For a lifting line solution of a symmetrical thin rectangular wing in ground effect, Tan and Plotkin [14] confirmed that the lifting line solution agrees well with vortex-lattice numerical method. Formal mathematical development of Lifting Line Theory is well described in many text books on aerodynamics [15,16]. Following the notations in [17], solving an integro-differential equation for spanwise circulation distribution in terms of geometrical characteristics of the wing with boundary conditions at both wingtips provides the aerodynamic characteristics of the wing.

$$\alpha(y_0)-\alpha_0(y_0) = \frac{2\Gamma(y_0)}{m_0(y_0)Uc(y_0)} + \frac{1}{4\pi U}\int_{y=-\frac{b}{2}}^{y=\frac{b}{2}} \frac{\frac{\mathrm{d}\Gamma}{\mathrm{d}y}(y)}{y_0-y}\mathrm{d}y \quad (1)$$

where $m_0(y_0)$ is the local lift curve slope, $c(y_0)$ is the local wing chord, and $\alpha(y_0)$ and $\alpha_0(y_0)$ are respectively geometric and zero-lift angles of attack at the spanwise location y_0. It can be easily verified that the solution of Equation (1) for an elliptic planform, untwisted wing out of ground effect (OGE) has an elliptic spanwise distribution of bound circulation and the elliptic wing loading is the optimal value for minimum induced drag within the limitation of flat rigid wake assumption. When a wing is flying in close proximity to the ground, the optimal spanwise distribution of bound circulation is parabolic [13]. Equation (2) shows the equations for bound circulation distributions (for a wing in and out of ground effect) along the spanwise direction, where bound circulations are divided by their maximum values for each case. Thus, it can be said that the nondimensional bound circulation distribution of a wing in ground effect is confined in the envelope with its upper one as an elliptic distribution Equation (2a), and its lower one as a parabolic distribution Equation (2c) (See **Figure 1**).

Out of Ground Effect: Elliptic Loading,

$$\Gamma(y)=\left(1-y^2\right)^{\frac{1}{2}}, |y|\le 1, t=0 \quad (2a)$$

In Ground Effect:

$$\Gamma(y)=\left(1-y^2\right)^{p}, |y|<1, \frac{1}{2}<p<1\ t=0 \quad (2b)$$

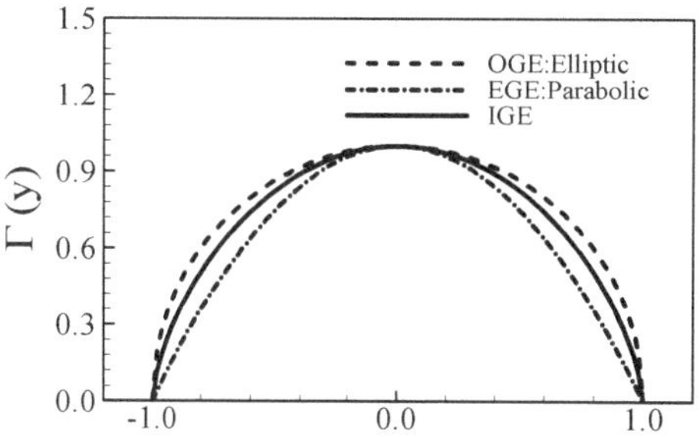

Figure 1. Comparison of spanwise circulation distributions.

Extreme Ground Effect: Parabolic Loading:

$$\Gamma(y)=\left(1-y^2\right), |y|\le 1, t=0 \quad (2c)$$

When applying Equation 2 to the calculation of bound circulation distributions, it is also required to understand the relationship between bound circulation and wing height.

Shigemi *et al.* [18] performed a wind tunnel test in order to investigate the ground effect on an 8.9% model of HOPE ALFLEX (Automatic Landing Flight Experiment) vehicle. They found an equation for the change of lift coefficient due to the change of the wing height as follows. After modifying the equation in order to match with Equation (2),

$$C_L = C_{L\infty} + B\exp(-Ch), B=\left(C_{LE}-C_{L\infty}\right) \quad (3)$$

where $C_{L\infty}$ is the lift coefficient of a wing in free flight (OGE), C_{LE} is the lift coefficient of a wing in extreme ground effect (EGE) and C is the coefficient that makes Equation (3) matching with the experimental data and should have a positive value. h represents the distance between the mid chord to the ground.

As shown in Equation (3), the lift coefficient is a nonlinear function of wing height, which also results in the change of circulation distribution. Thus, in the present paper, it is assumed that p in Equation (2b) changes as follows.

$$p = p_0 + \frac{10}{18}(1-h) \quad (4)$$

where p_0 is set to 1/2. p has the values of 0 and 1 respectively for $h = 1.0$ and 0.1.

2.2. Unsteady Discrete Vortex Panel Method

Present method is well described and validated in Ref. where the effect of smoothing schemes or vortex models on the numerical accuracy is shown by comparing computed results with the published data of Krasny [10]. In this approach, a lifting line with an initial load distribution is discretized with discrete vortex panel elements (See **Figure 2**). Each panel segment is assumed to have a constant-strength vortex. This constant-strength vortex panel element is replaced by a single point vortex with

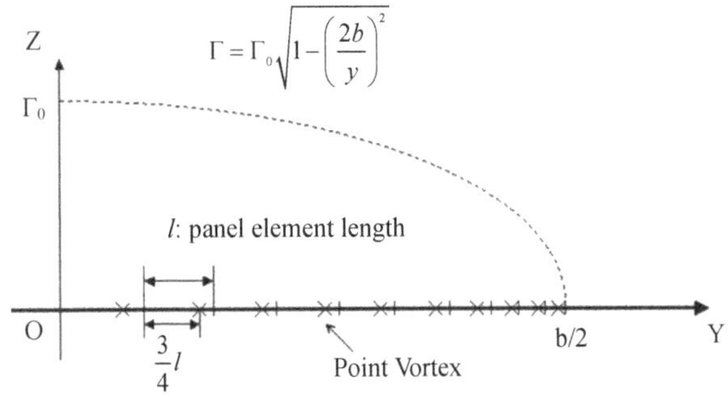

Figure 2. Representation of a lifting line with elliptic loading with discrete vortex panel elements.

the same vorticity. On each line segment, the point vortex is located at the 3/4 point. The trailing wake vortices from each vortex panel element are represented by free point vortices that can deform freely with an assumption of a force-free position.

The distributions of the dimensionless tangential velocities induced by a point vortex are represented using the Lamb (Oseen) model [19]. The Lamb model has a Gaussian distribution of vorticity. Ling *et al.* [20] proposed a model that is based on the solution for the actual velocity in a viscous fluid. The core radius, r_c, is approximately equal to the radial distance of the point where the maximum velocity is induced. This is expressed

$$r_{cj} = r_{cj0} + 3.17\sqrt{\frac{\Delta t_j}{R_e}} \tag{5}$$

The second term on the right-hand side of Equation (5) provides an estimate for the growth in the core radius of the vortex. The vortex Reynolds number is related to the time scale over which the laminar diffusion processes in the vortex core region become turbulent [21]. In the present work, an infinite Reynolds number is assumed. The tangential velocity induced by a point vortex with its circulation, Γ_j, is as follows.

$$v_{\theta j} = \frac{\Gamma_j}{2\pi r}\left\{1-\exp\left[-1.25643\left(\frac{r}{r_{cj}}\right)^2\right]\right\} \tag{6}$$

Several smoothing schemes are used in order to obtain accurate solutions by circumventing the singularity behavior when the distance is very small. Krasny [10] avoids the singularity in the induced velocity of a point vortex through a smoothing factor, δ. The induced velocities because of other vortices can then be written as follows

$$\vec{w}_j = \frac{G_i}{2\pi}\sum_{i=1}^{N}\left(\frac{(z_j - z_i)}{r_{ji}^2+\delta^2}\vec{i} - \frac{(y_j - y_i)}{r_{ji}^2+\delta^2}\vec{j}\right) \tag{7}$$

where $G(y)$ is represented as follows depending on the wing distance to the ground.

$$\text{OGE}: y\left(1-y^2\right)^{-\frac{1}{2}}dy \quad h \ge 1.0 \tag{8a}$$

$$\text{IGE}: 2py\left(1-y^2\right)^{p-1}dy,\ p = p_0 + \frac{1}{2}\exp(Kh),\quad 0.1 \le h \le 1.0 \tag{8b}$$

$$\text{EGE}: 2ydy \quad h \le 0.1 \tag{8c}$$

In order to enforce the no penetration condition at the ground, an image method is used. Since the wake is force-free, the evolution of each vortex is investigated by moving the positions of point vortices using an Euler convection scheme.

$$(\Delta y, \Delta z)_j = (u, v)_j\, dt \tag{9}$$

3. Results and Discussion

Figure 3 shows the validation of the present method by comparing the wake shapes behind a lifting line having an elliptic loading with the published result of Krasny [10]. It is assumed that the wing is placed at $h = 0$ and is out of ground effect. In both calculations, the in-flow velocity and the maximum circulation at the midspan are assumed to be unity. The smoothing factor for the present calculation is set to 0.05 [the same as Krasny's]. As shown in **Figure 4**, an unrealistic vortex sheet crossing is not observed in the present results. A time step size of 0.04 sec is used in the present calculation whereas Krasny [10] used a time step size of 0.01 sec. Thus, it can be said that the present results are in agreement with Krasny's.

Figure 4 shows the comparison of wake shapes behind wings in and out of ground effect. Both wings are placed at the same ground height of h = 1.0. As shown in the figure, wake vortices behind a wing in ground effect cannot fully develop due to the insufficient distance between the wing and the ground. The positions of wingtip vortices of a wing in ground effect are moved laterally

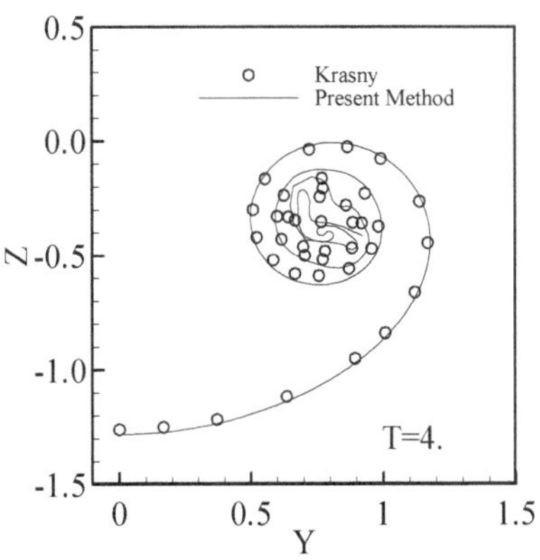

Figure 3. Comparison of wake shapes between present method and Krasny's.

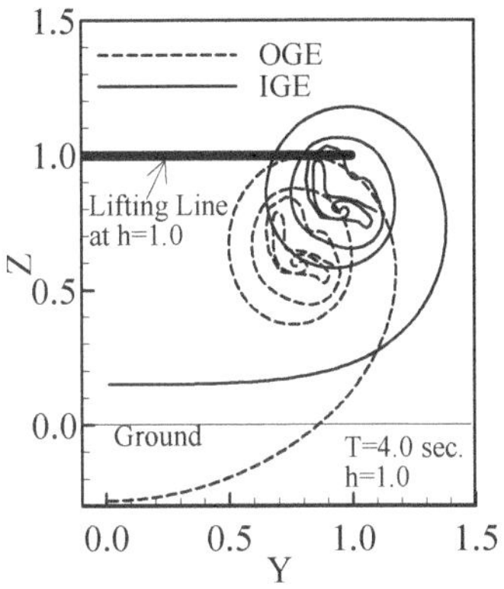

Figure 4. Wake shapes behind a wing in and out of ground effect.

more outward and vertically more upward than those of a wing out of ground effect. When the wing is in ground effect, the desingularized point vortices become confined to a small region and coalesce. Present method does not include the viscous diffusion model of interacting point vortices, but it can be said that the interaction of point vortices while merging will decrease further the viscous diffusion. The total circulation of wingtip vortices behind a wing in ground effect will be decreased.

Figure 5 shows the wake shapes behind a wing in close proximity to the ground. When the wing is placed at $h = 0.2$, from Equation (4), $p = 17/18$. As shown in Equation (2), $p = 0.5$ means the elliptic circulation distribution whereas $p = 1.0$ represents the parabolic circulation distribution. When $h = 0.2$, the value of exponent p approaches to $p = 1.0$. **Figure 5(a)** shows the wake shapes behind a wing at $h = 0.2$ with three cases of circulation distributions. **Figure 5(b)** shows the wake shapes behind a wing at $h = 0.1$ for elliptic and parabolic circulation distributions. As shown in both **Figures 5(a)** and **(b)**, overall wingtip vortex shapes are similar to each other and does not dependent on the circulation distributions, however, the position of the wingtip vortices for elliptic loading is moved laterally more outward than the other cases. Thus, it can be said that, for a single wing in ground effect, the assumption of elliptic load distributions does not affect much the wake shapes but negligibly changes the position of wingtip vortices.

Figure 6 shows the changes of wake shapes behind a wing in ground effect. As the wing is placed close to the ground, the wingtip vortices departed from the wing is moving vertically downward and laterally outward. As shown in **Figure 6(b)**, when the wing is in close proximity to the ground ($h = 0.1$), the area of confined vorticity is decreased compared to the case of the wing at $h = 0.5$.

4. Conclusions

In the present paper, a new formula that matches elliptic loading for OGE and parabolic loading for EGE is suggested.

From the computed results on the unsteady wake evolution using the suggested model for wing loading, it was found that the overall wingtip vortex shapes were not affected by the wing load distributions. But, as the wing approaches very close to the ground, the parabolic wing load distribution calculation produced more laterally inward wingtip positions than the elliptic load distribution case.

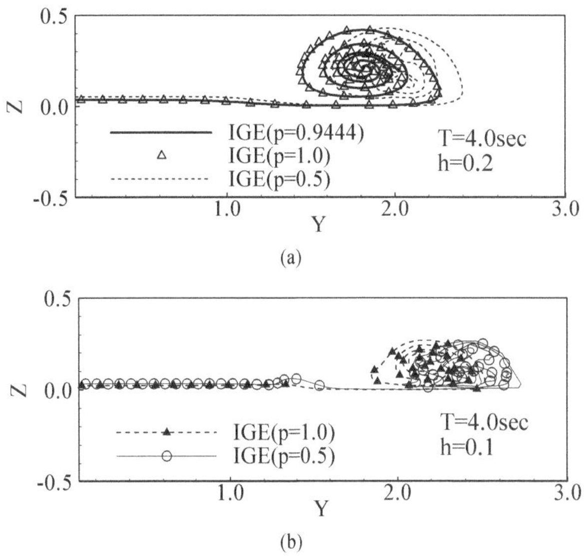

Figure 5. Wake shapes behind a wing in close proximity to the ground with various wing loadings.

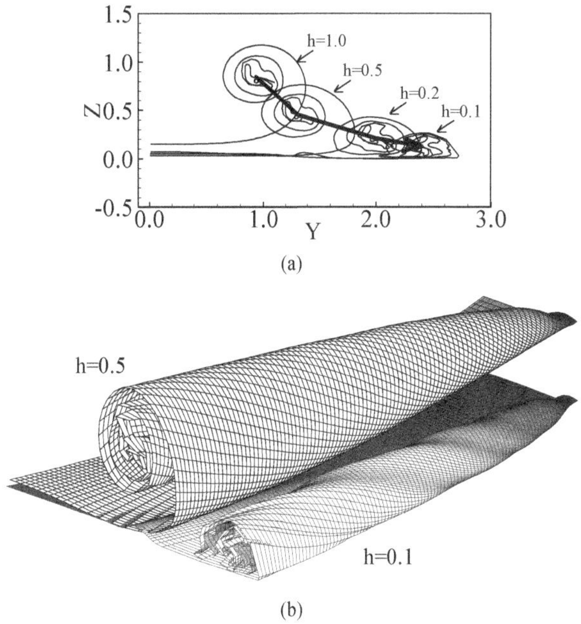

Figure 6. Wake shapes for different wing heights. (a) Wake shape in the Trefftz plane, T = 4.0; (b) Three dimensional view of unsteady wake evolutions.

In future, the present model will be applied to the unsteady wake evolution behind wings in formation where the significant interaction between wingtip vortices is expected.

5. Acknowledgements

The research was supported by a grant from the 2011 program for visiting professors overseas in Korea National University of Transportation.

REFERENCES

[1] K. V. Rozhdestvensky, "Wing-in-Ground Effect Vehicles," *Progress in Aerospace Sciences*, Vol. 42, No. 3, 2006, pp. 183-211.

[2] K. V. Rozhdestvensky, "Aerodynamics of a Lifting System in Extreme Ground Effect," Springer, Heidelberg, 2000.

[3] H. W. M. Hoeijmakers and W. Vaatstra, "A Higher Order Panel Method Applied to Vortex Sheet Roll-up," *AIAA Journal*, Vol. 21, No. 4, 1983, pp. 516-523.

[4] R. S. Ribeiro and I. Kroo, "Vortex-in-Cell Analysis of Wing Wake Roll-Up," *Journal of Aircraft*, Vol. 32, No. 5, 1995, pp. 962-969.

[5] K. Zhu and H. Takami, "Effect of Ground on Wake Roll-Up behind a Lifting Surface," *Proceedings of the 37th Japan National Congress for Applied Mechanics*, Tokyo, 1987, pp. 115-123.

[6] F. Lamarre and I. Paraschivoiu, "Efficient Panel Method for Vortex Sheet Roll-Up," *Journal of Aircraft*, Vol. 29, No. 1, 1992, pp. 28-33.

[7] T. Sarpkaya, "Computational Methods with Vortices—The 1988 Freeman Scholar Lecture," *Journal of Fluids Engineering*, Vol. 111, No. 1, 1989, pp. 5-52.

[8] D. I. Pullin, "The Large-scale Structure of Unsteady Self-Similar Rolled-Up Vortex Sheets," *Journal of Fluid Mechanics*, Vol. 88, No. 3, 1978, pp. 401-430.

[9] R. Krasny, "A Study of Singularity Formation in a Vortex Sheet by the Point-Vortex Approximation," *Journal of Fluid Mechanics*, Vol. 167, 1986, pp. 65-93.

[10] R. Krasny, "Computation of Vortex Sheet Roll-Up in the Trefftz Plane," *Journal of Fluid Mechanics*, Vol. 184, 1987, pp. 123-155.

[11] M. Morky, "Numerical Simulation of Aircraft Trailing Vortices Interacting with Ambient Shear or Ground," *Journal of Aircraft*, Vol. 38, No. 4, 2001, pp. 636-643.

[12] C. Han and J. Cho, "Unsteady Trailing Vortex Evolution Behind a Wing in Ground Effect," *Journal of Aircraft*, Vol. 42, No. 2, 2005, pp. 429-434.

[13] S. E. Windall, and T. M. Barrows, "An Analytic Solution for Two- and Three-Dimensional Wings in Ground Effect," *Journal of Fluid Mechanics*, Vol. 41, No. 4, 1970, pp. 769-792.

[14] A. Plotkin and C. H. Tan, "Lifting-Line Solution for a Symmetrical Thin Wing in Ground Effect," *AIAA Journal*, Vol. 24, No. 7, 1986, pp. 1193-1194.

[15] J. Anderson, "Fundamentals of Aerodynamics," 5th Edition, McGraw Hill, New York, 2010.

[16] J. Katz and A. Plotkin, "Low Speed Aerodynamics," 2nd Edition, Cambridge University Press, Cambridge, 2001.

[17] I. G. Sheldon, "Wing Tip Vortices," In: I. G. Sheldon, Ed., *Fluid Vortices*, Kluwer Academic Publishers, Berlin, 1995, Chapter X.

[18] M. Shigemi, T. Fujita, A. Iwasaki, T. Ohnuki, K. Rinoie, H. Nakayasu and M. Sagisaka, "Experimental Investigation of Static and Dynamic Ground Effect on HOPE ALFLEX Vehicle," Technical Report of National Aerospace Laboratory TR-1236, National Aerospace Laboratory, Tokyo, 1994.

[19] A. Ogawa, "Vortex Flow-CRC Series on Fine Particle Science and Technology," CRC Press, Boca Raton, 1940.

[20] G. C. Ling, P. W. Bearman and J. M. R. Graham, "A Further Simulation of Starting Flow around a Flat Plate by a Discrete Vortex Model," *Internal Seminar on Engineering Applications of the Surface and Cloud Vorticity Methods*, Vol. 51, No. 14, 1986, pp. 118-138.

[21] J. G. Leishman, "Principles of Helicopter Aerodynamics," 2nd Edition, Cambridge University Press, Cambridge, 2000.

14

Effect of Laryngopharyngeal Anatomy on Expiratory Airflow and Submicrometer Particle Deposition in Human Extrathoracic Airways

Xiuhua Si[1], Jinxiang Xi[2*], Jongwon Kim[2]
[1]Department of Engineering, Calvin College, Grand Rapids, USA
[2]Department of Mechanical and Biomedical Engineering, Central Michigan University, Mount Pleasant, USA

ABSTRACT

The objective of this study is to systematically assess the influences of the larynopharyneal anatomical details on airflow and particle behaviors during exhalation by means of image-based modeling. A physiologically realistic nose-throat airway was developed with medical images. Individual airway anatomy such as uvula, pharynx, and larynx were then isolated for examination by progressively simplifying this image-based model geometry. Low Reynolds number (LRN) k - ω model and Langrangian tracking model were used to simulate the dynamics of airflow and particle transport for a wide range of exhalation conditions (4 - 45 L/min) and particle sizes (1 nm - 1 μm). Results showed that pharyngeal anatomical details exerted a significant impact on breathing resistance and particle profiles. Abrupt pressure drop resulting from the uvula-related airway obstruction was observed. Even though the total deposition rate in the nasal airway is largely unaffected by the upstream effect, the local deposition patterns vary notably. Results of this study also indicate that the pressure drop appears to be an appropriate parameter to characterize the geometric variations for diffusive depositions. Inclusion of pressure drop ($D^{0.5}Q^{-0.62}\mathrm{dp}^{0.07}$) gives an improved correlation than using the conventional diffusion factor ($D^{0.5}Q^{-0.28}$).

Keywords: Medical Image-Based Modeling; Nasal Airflow; Uvula; Breathing Resistance; Nanoparticle Deposition

1. Introduction

A significant issue in evaluating nasal airflow and aerosol deposition during exhalation is determining the extent to which the larynx and pharynx affect their behaviors before entering the nasal cavity. The larynx consists of vocal folds that form an elliptical or triangular cross-sectional area of flow constriction (*i.e.*, glottal aperture). It is approximately 6 cm long and its cross-sectional area varies with mean flow rate [1] and oscillates during a breathing cycle [2]. The pharynx is a region that connects the nose, mouth, larynx, and esophagus, and is highly variable in its morphology. Especially, the uvula, which is a projection of the tissue suspended from the soft palate and resides between the naso- and oro-pharynx, can remarkably alter the pharyngeal airway structure. Air exhaled from the lungs enters the larynx, travels through the pharynx and nasal passages, and exits the nostrils. In addition to the geometric curvature from the larynx through the nasopharynx, the airflow experiences two dramatic geometric constrictions (*i.e.*, the glottal aperture and the uvula) before it enters the nasal cavity, yielding recirculating zones downstream of the glottis and within the nasopharynx. As a result, the airflow and particle profiles entering the nasal cavity are far from uniform.

The nasal deposition of submicrometer aerosols during exhalation has been considered in human volunteers in a limited number of studies [3-5]. A common disadvantage of such *in vivo* studies is the difficulty in determining local deposition values even though imaging methods now make it possible [6]. Because of the inaccessibility of respiratory airways, modeling has long been a primary component of respiratory aerosol research. Both *in vitro* and computer models seek to simplify the complex airway geometry while still capturing the relevant physics. *In vitro* experimental studies that have evaluated expiratory resistance and aerosol deposition in human nasal replicas include Cheng *et al.* [7], Cheng *et al.* [8], and Swift *et al.* [9]. The nasal geometries used in these stud-

*Corresponding author.

ies are typically derived from medical scan data (e.g., MRI) or casts of cadavers and extend from the nostrils to the upper trachea. As with *in vivo* tests, no *in vitro* study so far has been reported in addressing the influences of larynx and pharynx on expiratory extrathoracic depositions.

A number of numerical studies have considered the transport and deposition of fine and ultrafine particles in the nasal airways [10-15]. Similar CFD studies have also evaluated the transport and absorption of vapors in the nasal passages [16-18]. Comparisons of CFD results to experimental deposition data in the nasal cavity are often difficult due to differences in the geometric models that vary in their physical realisms. Untested modeling assumptions limit the accuracy of both *in vitro* and numerical models. During exhalation, the influence of the upstream laryngeal and pharyngeal anatomy on transport and deposition in the nasal cavity may be significant. Longest and Vinchurkar [19] studied aerosol depositions in a multi-bifurcating airway geometry and showed the necessity of including the upstream effects to validate CFD predictions with experimental measurements. Xi and Longest [20] evaluated the effects of physical realism on deposition patterns for micrometer aerosols in a mouth-throat model with varying degree of geometric complexity. This study reported that geometric realism had a major effect on local inertia-based depositions and highlighted the importance of a realistic glottal aperture and angled trachea on deposition localization and particle profiles entering the lungs. Similar observations on the laryngeal effect have been reported by Li [21], Brouns *et al.* [22], Liu *et al.* [10], and Xi *et al.* [23] that the presence of larynx can significantly affect airflows and depositions downstream of the glottis. Furthermore, highly accurate respiratory models could be transformative to clinical diagnosis, treatment, and pre-surgical planning of associated respiratory disorders.

The validity of assuming simplified airway geometry to facilitate modeling is routinely adopted in previous studies but remains largely untested. The objective of this study is to systematically evaluate the effect of the laryngopharyngeal anatomical details on both airflow and aerosol depositions in the nasal airways of an adult. Starting from an image-based anatomically accurate airway model, we will progressively simplify this geometry, one anatomy at a time, to gain a more accurate understanding of the physiological roles of each anatomy in regulating airflows, breathing resistance, and particle filtering. The glottal aperture, pharynx, and uvula are especially of interest in this study.

2. Methods

2.1. Construction of the Airway Models

In order to assess the impact of the laryngopharyngeal anatomy on airflow and particle deposition in the nasal airways, four computational models with varying geometric details have been considered. We started from a highly realistic nasal-laryngeal airway model that was recently developed based on MRI images of a 53-year-old male (weight 73 kg and height 173 cm) [24]. The MRI tracings were first segmented in MIMICS (Materialise, Ann Arbor, MI) according to the contrast between osseous structures and intranasal air to convert the raw image data into a set of cross-sectional contours that define the airway of interest. Based on these contours, an internal nasal surface geometry was constructed in Gambit (Ansys, Inc.) as shown in **Figure 1**. This same MRI image set has also been used in a number of nasal particle deposition experiments [7,25-27], and therefore a direct comparison between simulated and experimental results is possible.

Bedsides the anatomically accurate nasal passages, three other key features characterize this image-based model, namely, a hanging uvula, a flat-plated pharynx, and a triangular-shaped glottal aperture. Movement of the uvula alters the airway morphology that connects the naso- and oropharynx. In this study, the uvula rests on the back of the throat and partially obstructs the inferior nasopharynx, resulting in two flow passages. This might be attributed to the supine position of the patient during image acquisition. Posterior to the uvula, the pharynx is featured by narrow and flat air channels, which converge into the wedge-shaped glottis in the larynx. In order to approximate the *in vivo* airway morphology that was captured in the images, anatomical details such as the epiglottis and the two pharyngeal sinuses on either side of the larynx were also retained, as shown in **Figure 1(a)**. In light of the glottis, several studies have reported a triangular-shaped cross-section based on bronchoscope images in living subjects [28,29]. The triangular cross section is formed by the vocal folds with one point of the triangle resting on the ventral surface and two points on the dorsal surface, as shown in the plane E-E' in the right panel of **Figure 1(a)**. In contrast to this, *in vitro* studies with larynx replica casts have reported a more elliptical shape [1]. In this study, the triangular shape of the glottis observed in the CT data has also been retained.

As a first simplification, the hanging uvula was removed in Model 1, resulting in an un-obstructed nanopharynx in Model 2 (**Figure 1(b)**). This model represents the condition when a subject takes the upright position and the soft palate bends downward due to its own weight. Further simplifications have been made to generate Model 3. This model consists of a much simpler, elliptical pharynx. The oropharynx, epiglottis, and the two laryngeal sinuses (**Figure 1(c)**), has been eliminated as most previous studies did [7-9,30]. Rodenstein *et al.* [31] reported that pharynx had a transverse to anterior-posterior

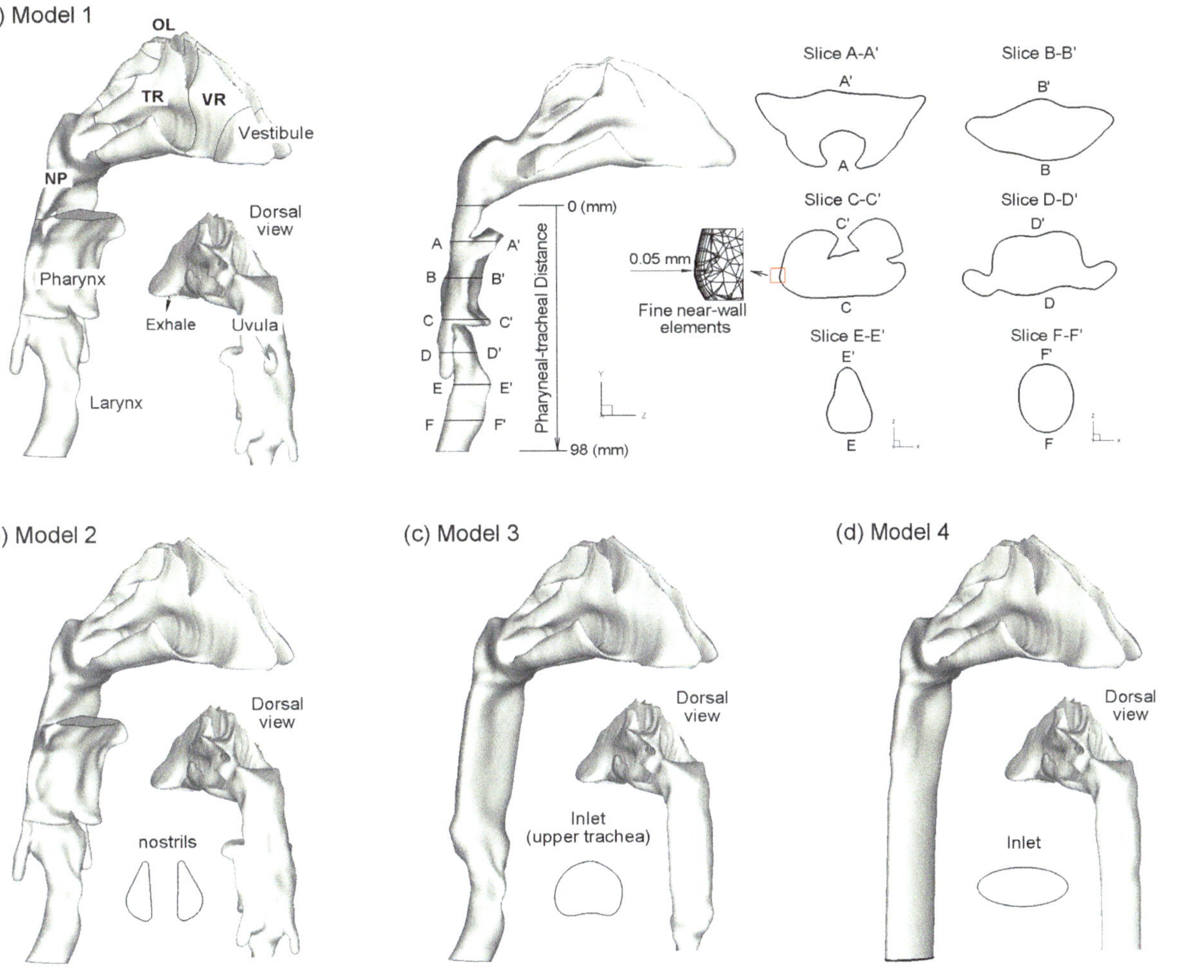

Figure 1. Surface models of the nasal-laryngeal airway with decreasing geometric complexity.

ratio of 2.5 for healthy subjects. Thus, for Model 3, we assume that the pharynx is approximately an elliptical pipe with the major axis 2.5 times as large as the minor axis. The fourth geometry model (*i.e.*, Model 4) excludes the triangular-shaped larynx and is composed of an elliptical pipe downstream of the nasal cavity (**Figure 1(d)**).

During exhalation, airflow enters the upper trachea, travels through the larynx, pharynx, and nasal turbinates, and exits the nostrils. By retaining the laryngeal-pharyngeal region, more physiologically realistic inlet flow conditions are provided to the nasal cavity. As a result, the flow field and particle deposition characteristics in the nasal cavity considered in this study will better represent *in vivo* conditions compared with some previous studies that had excluded the larynx and pharynx. Therefore, inclusion of the laryngeal-pharyngeal region allows for the results of this computational study to be directly compared with previous *in vivo* nasal deposition data. In order to characterize deposition distributions, the nose-throat airway were divided into different sub-regions that include the vestibule, valve region (VR), turbinate region (TR), olfactory region (OR), nasopharynx (NP), pharynx, and larynx (**Figure 1(a)**). In order to quantify the airflow distributions within the nasal passages, a coronal slice in the turbinate region was further divided into superior, middle, inferior meatus (abbreviated as SM, MM, and IM, respectively) and the medial passage (MP), which will be discussed in later sections.

2.2. Boundary Conditions

Steady exhalation was assumed for all simulations with uniform velocity profiles and particle distributions at the tracheal inlet (**Figure 1(a)**). Initial particle velocities were assumed to match the local fluid velocity. Atmospheric pressure conditions were assumed at the two nostrils (outlets). The airway surface was assumed smooth and rigid with no-slip ($u_{\text{wall}} = 0$) and perfect absorption conditions. In the body, the extrathoracic airway is lined with a thin layer of mucus, which captures particles at initial contact and clears them to the throat or nasal vestibule by mucocilliary movement within a time period of 10 to 15 minutes. Mass diffusion and metabolism of deposited particles may occur within the mucus layer and may change the zero-concentration conditions at the wall. However, due to the slow speed of the mucocilliary movement compared with the intranasal airflow and relatively low deposition rates, the no-slip and perfect absorption conditions are reasonable approximations in this study.

2.3. Fluid and Particle Dynamics Equations

The flow conditions considered in this study are assumed to be isothermal and incompressible. The mean inlet Reynolds number at the trachea varies from 368 to 3302. The maximum Reynolds number based on the hydraulic diameter of the glottal aperture is approximately 8037. Therefore, laminar, transitional, and fully turbulent conditions in the nasal-laryngeal model are expected. To resolve these multiple flow regimes, the low Reynolds number (LRN) k - ω model was selected based on its ability to accurately predict pressure drop, velocity profiles, and shear stress for transitional and turbulent flows. Moreover, the LRN k - ω model was shown to provide an accurate solution for laminar flow as the turbulent viscosity approaches zero [32].

The transport and deposition of the submicrometer particles are simulated with a well-tested discrete Lagrangian tracking model, which is enhanced with user-defined functions (UDFs) accounting for the finite particle inertial effects that might be significant for submitcrometer particle depositions. The aerosols evaluated in this study had a tracheal Stokes number ($St_k = r_p d_p^2 C_c U / 18 \mu D_h$) with a range of 1.68×10^{-8} to 1.0×10^{-3} and were assumed to be dilute and to not influence the continuous-phase, *i.e.*, one-way coupled particle motion. In our previous studies, the UDF-enhanced Lagrangian model with near-wall interpolation algorithm [33,34] has been shown to provide close agreement with experimental deposition data in upper respiratory airways for both submicrometer [34] and micrometer particles [35]. More details of the Largrangian tracking model can be found in [34].

2.4. Numerical Method and Convergence Sensitivity Analysis

To solve the governing mass and momentum conservation equations in each of the cases considered, the CFD package ANSYS Fluent was employed. User-supplied Fortran and C programs were implemented for the calculation of initial particle profiles, particle deposition factors, grid convergence, and deposition enhancement factors. For this study, a specific set of user-defined functions were applied that considered the Brownian force, anisotropic turbulence effect, and near-wall velocity interpolation. All transport equations were discretized to be at least second order accurate in space. A segregated implicit solver was applied to evaluate the resulting linear system of equations. This solver uses the Gauss-Seidel method in conjunction with an algebraic multigrid approach to improve the calculation performance on tetrahedral meshes. Convergence of the flow field solution was assumed when the global mass residual was reduced from its original value by five orders of magnitude and the residual-reduction-rates for both mass and momentum were sufficiently small.

The computational meshes of the four nasal-laryngeal airway models were generated with ANSYS IECM CFD (Ansys, Inc). Due to the high complexity of the model geometries, unstructured tetrahedral meshes were generated with high-resolution prismatic cells in the near-wall region (**Figure 1(a)**). A grid sensitivity analysis was conducted by testing the effects of different mesh densities with approximately 620,800, 1,140,400, 1,975,600 and 3,212,000 control volumes while keeping the near-wall cell height constant at 0.05 mm. Since the changes in both total and sub-regional depositions were less than 1% when increasing mesh size from 1,975,600 to 3,212,000, the final grid for reporting flow field and deposition conditions consisted of approximately 1,975,600 cells with a thin five-layer pentahedral grid in the near-wall region and a first near-wall cell height of 0.05 mm.

For discrete Lagrangian tracking, the number of seeded particles required to produce count-independent depositions was considered. Particle count sensitivity testing was performed by incrementally releasing groups of 10,000 particles. The number of groups was increased until the deposition rate change was less than 1%. Due to the low deposition rates, more particles were required for fine aerosols to generate count-independent results compared with ultrafine aerosols. The final number of particles tracked for 1 - 40 nm and 100 - 1000 nm aerosols were 150,000 and 600,000, respectively.

3. Results

3.1. Nasal Airway Dimensions

Dimensions of the four models considered are shown in **Figure 2** in terms of perimeter, cross-sectional area and hydraulic diameter as a function of distance downstream of the nasopharynx as denoted in **Figure 1(a)** (middle panel). As expected, Model 1 exhibits the highest variability in reported geometric parameters. The minimum hydraulic diameter at a distance of 7 mm corresponds to the airway obstruction associated with the hanging uvula. It is interesting to note that Model 1 has substantially smaller effective flow area (*i.e.*, d_h) in the oropharynx region than Model 2. A further examination of the axial cross-sections of Model 1 in this region reveals highly irregular contours as exhibited in Slice A-A' and B-B' in

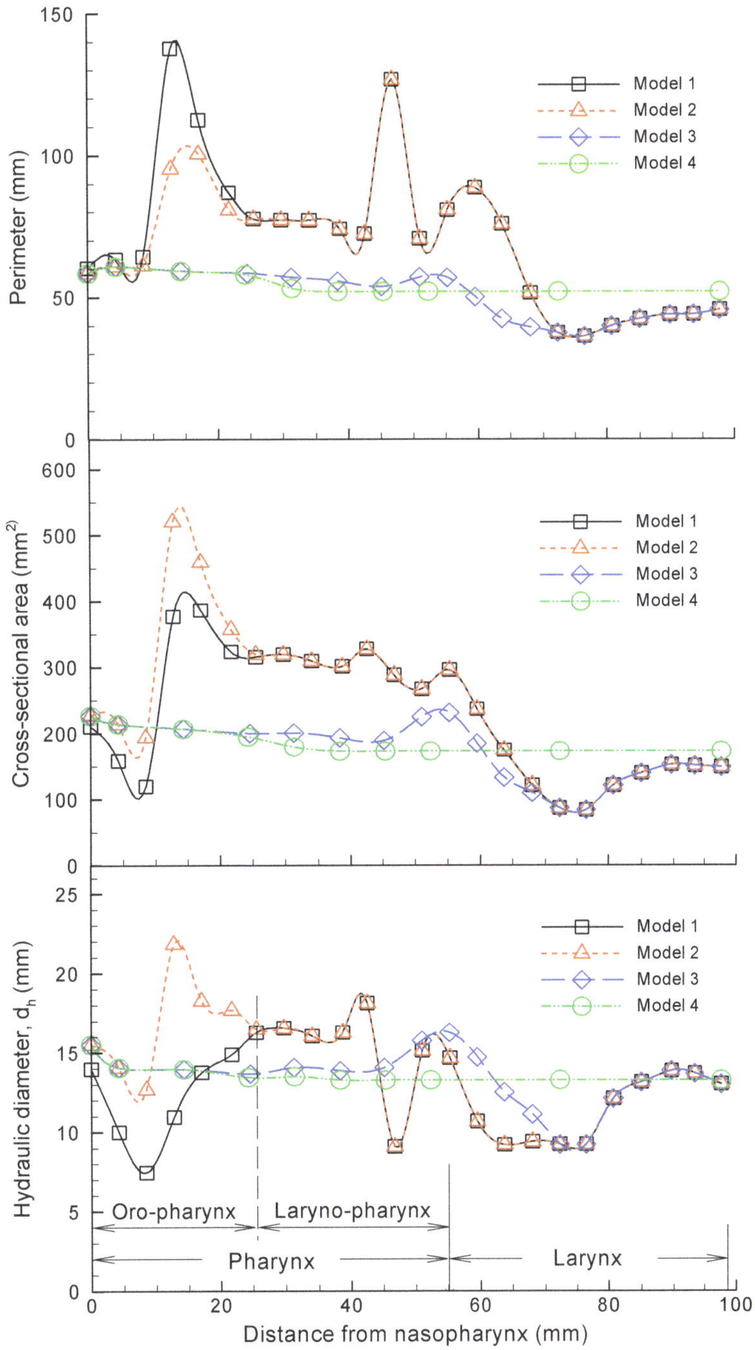

Figure 2. Airway dimensions: (a) perimeter; (b) cross sectional area; (c) hydraulic diameter.

Figure 1(a). This shape irregularity, accompanied by a much reduced effective flow area, is expected to induce significant jet effect and turbulence generation, and therefore significant energy dissipation in this region.

There is less variation in the geometric parameters of Models 3 and 4. In the laryngeal-pharyngeal region, both models (*i.e.*, 3 and 4) have smaller flow area (d_h) than Models 1 and 2. The dip in cross-sectional area at a distance of 47 mm corresponds to the epiglottis that spreads above the glottal aperture at an angle of approximately 30° to the posterior wall of the pharynx (**Figure 1(a)**, Slice C-C'). The larynx is located at a distance of 78 mm or so downstream of the nasopharynx.

3.2. Breathing Resistance

The influence of the pharyngo-laryngeal geometries on exhalation breathing resistance is illustrated in **Figure 3**. **Figure 3(a)** shows the pressure distribution in the airway

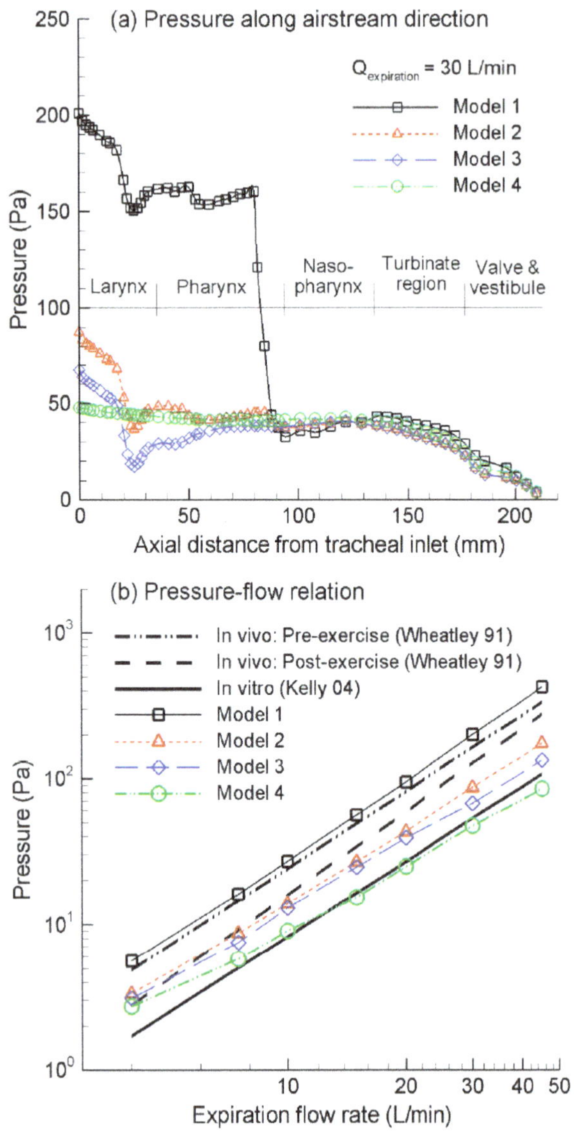

Figure 3. Expiratory pressure drop among the four nasal airway models: (a) pressure variation along the airstream direction as a function of the axial distance from the tracheal inlet; (b) expiratory pressure drop vs. flow rate with comparison to in vivo and *in vitro* measurements.

as a function of the axial distance from the tracheal inlet at an exhalation flow rate of 30 L/min for the four models considered. For Model 1, the maximum pressure drop is observed in the pharynx/nasopharynx region which is about 170 Pa in magnitude and is about 76% of the expiratory pressure drop between the upper trachea and nostrils. This abrupt pressure drop results from the airway obstruction as the hanging uvula partially blocks the pharyngeal passage. As discussed, Model 1 was based on images of a subject in a supine position during image acquisition. It is expected that changing from upright to supine position has the potential of causing pharyngeal obstruction as the uvula yields due to its own weight and that the degree of obstruction varies depending on the elasticity properties of the subject's soft palate and uvula. Considering that this pressure drop is about three times that of the laryngeal region, which is the maximum of the other three models (Models 2, 3, 4), it is possible that the patient of Model 1 has an over-flexible uvula and hence a high degree of airway obstruction. Respiratory tissues in the pharyngeal region are quite collapsible. Severe pressure drop could possibly trigger or expedite the collapse of airway walls and induce the symptoms of hyponea (*i.e.*, reduced airflow capacity) or apnea (*i.e.*, complete airway obstruction). Meanwhile, flow instability may cause vibration of the pliable pendant uvula, resulting in snoring of the supine subject. It is therefore interesting to postulate that an over-flexible uvula may be one major reason for sleep disorders such as sleep apnea, whose constriction of the pharyngeal airway and the resultant elevated pressure drop induces further pharyngeal airway collapse.

For Models 2 and 3, which corresponds to the upright position, the glottis and nasal cavity each constitutes a breathing resistance of about 50 Pa. However, there is less pressure recovery in the pharynx of Model 2 comparing to that of Model 3. Considering the fourth model, which totally excludes the anatomical details of glottis and pharynx, there is negligible pressure drop within the larynx-pharynx region, and almost all the pressure drop comes from the convoluted nasal passages.

Figure 3(b) shows the logrithmatic diagram of the pressure drop vs. exhalation flow rate in comparison to *in vitro* [26] and *in vivo* measurements [36]. Overall, the predicted pressure drop agrees well with that of a comparable physiological condition. For example, the *in vitro* nasal replica in Kelly *et al.* [26] retains the nasal cavity only and is equivalent with Model 4 in this study. As expected, the pressure drops of the above two cases also match well. Similar observations were also found in the comparison between post-exercise *in vivo* data [36] and Model 2, or the comparison between pre-exercise *in vivo* data [36] and Model 1. Therefore, the numerical predictions appear to adequately capture the nasal breathing resistances.

3.3. Turbulent Vortices and Flow Fields

Typical coherent structures (vortices) can be identified via different methods such as eigen-helicity, Q-criterion, vorticity, λ_2-criterion, swirling strength, etc. Among these methods, λ_2-criterion has been reported in better representing the topology of vortex cores for transitional flows [37], which is the case of human respiratory flows such as in this study. **Figure 4** depicts instantaneous coherent structures among the four models obtained using large eddy simulations (LES) for a steady exhalation rate

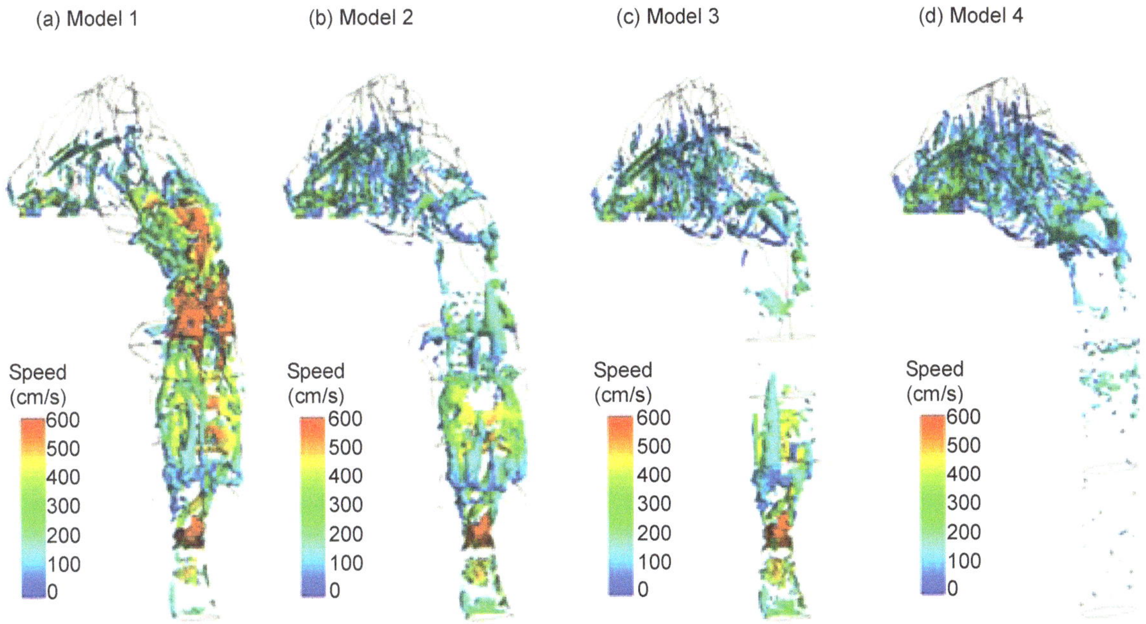

Figure 4. Instantaneous coherent structures (vortices).

of 20 L/min. These vortices are identified by the iso-surfaces of λ_2-criterion at the magnitude of 0.013 and are colored by the local airflow speed. Three observations are noteworthy in **Figure 4**. First, for Models 1-3 that retain the glottal aperture, vortex rings are observed to generate at the glottis and subsequently break into stream wise vortices with the pharynx. Second, in Model 1 with uvula constriction, a second array of strong vortex rings forms around the uvula and plays an important role in quick turbulence generation and abrupt pressure drop in this region as shown in **Figure 3(a)**. Beyond the middle nasal cavity, vortex structures decay, indicating that turbulence becomes weak and relaminarization is progressing. For the other three models without uvula constriction, similar vortex structures are observed in the nasal cavity, with standing rings that are induced by the narrow passages and streamwise vortices that are stretched along the meanstream direction. Thirdly, in Model 4 that replaces the realistic larynx and pharynx with an idealized elliptic pipe, vortices are absent in the pipe, and are limited primarily to the nasal cavity.

Representative velocity fields in the sagittal plane and at selected axial cross-sections are displayed in **Figure 5** for the exhalation flow rate of 30 L/min. Complex main and secondary flow motions are observed arising from geometrical details such as glottal constriction, epiglottal recess, pharynx, and uvula obstruction. Again, the presence of uvula obstruction in Model 1 induces remarkable variations in airflows within the nasopharynx (see Slice 1'1' in **Figure 5**), and to a lesser degree within the downstream nasal turbinate regions (a figure not shown here). Retaining the uvula in Model 1 significantly reduces the flow area, and splits the exhaled air into two jet flows that subsequently merge in the nasopharynx before entering the nasal passages. Meanwhile, these two jet flows need to negotiate through the 90° curvature in the nasopharynx. As a result, one strip of high-speed flow zone is formed near the dorsal (upper) wall of the nasopharynx in Model 1, as illustrated in **Figure 5(a)**. In contrast, for Models 2 and 3 without uvula obstruction, airflow within the pharynx interacts with the nasal septum carina and gradually bifurcates into the two nasal passages, therefore exhibiting two velocity peaks at the side walls of the Slice 1-1' (**Figures 5(b)** and **(c)**). Considering Model 4 that excludes both the glottis and pharynx, velocity patterns within the trachea are characterized by a developing laminar flow (**Figure 5(d)**). Similar as the other three models, the core flow in Model 4 is also observed to shift upward in the nasopharyngeal region due to the 90° airway curvature (Slice 1-1' in **Figure 5(d)**).

Particle profiles prior to nasal cavity (*i.e.*, slice 1-1') are compared among the four models for 200 nm particles and moderate activity conditions (Q = 30 L/min), as shown in **Figure 5**. The particles were tracked with both mean-flow and turbulent tracking algorithms after they were released at the trachea inlet, and their locations were recorded whenever a particle crossed the sample plane (Slice 1-1'). The disparities in particle profiles are evident among the four models considered. Generally speaking, the mean flow field is a good indicator of particle transport, with elevated particle concentrations at high-speed flow zones. Considering Model 1, a stripe of dense particles is observed the middle line of Slice 1-1', with even higher concentration near the upper wall and

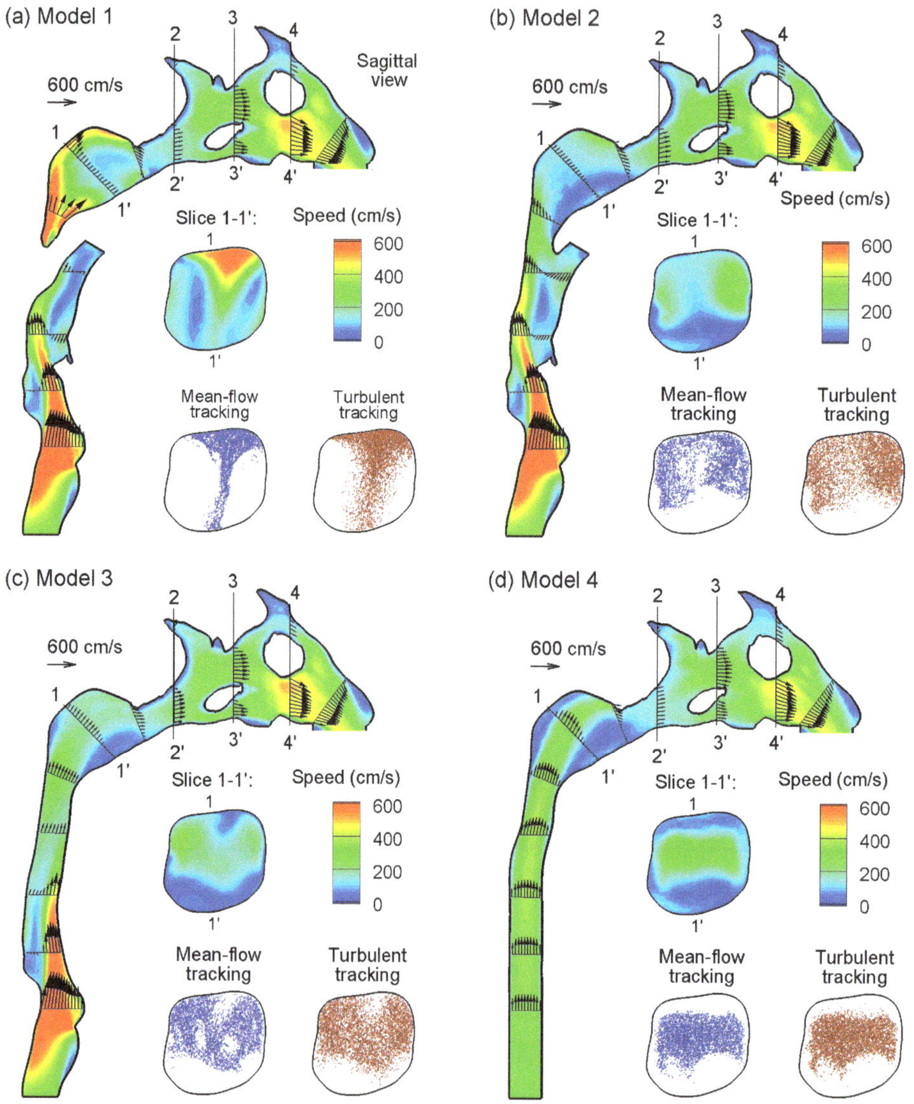

Figure 5. Velocity fields (contours and vectors) among the four models at an exhalation flow rate of 20 L/min in sagittal and cross-sectional (coronal) views.

nearly no particles at two lateral sides. As a result, stronger interactions between the particulate-laden fluid and the nasal septum carina ridge are expected for Model 1 than the other three models, leading to enhanced particle deposition in the carina and nasal septum. In the absence of uvula obstruction, quite similar particle profiles develop in Models 2 and 3, with high concentration of particles at the two lateral sides of the Slice 1-1'. Furthermore, high concentration of particles are also observed in the immediate proximity to the upper wall of Models 2 and 3, presumably due to outward secondary motions and turbulent dispersions. In contrast, the developing laminar boundary layer in the larynx region of Model 4 keeps the particle from reaching the wall surface (see **Figure 5(d)**, Slice 1-1'). In comparison to mean-flow tracking, the seemingly more random particle distributions with turbulent tracking indicates strong influences from local vortical flows in this region, as depicted in **Figure 4**. However, it should be reminded that local particle profiles are a function not just of the local flow conditions, but of the entire flow history that the particles have experienced.

Downstream of the nasopharynx, particles enter the narrow convoluted turbinate passages and start to fill up the projecting meatus driven by the secondary flow motions. To quantify the influence of the flow fields inside the nasal passages, the Slice 3-3' of the four models as defined in **Figure 5** is further divided into four sections, namely the inferior, middle, superior meatuses, and the median passage. **Figure 6** shows the flow partition percentage through each section in the right nasal passage of the four models at an exhalation flow rate of 20 L/min. It

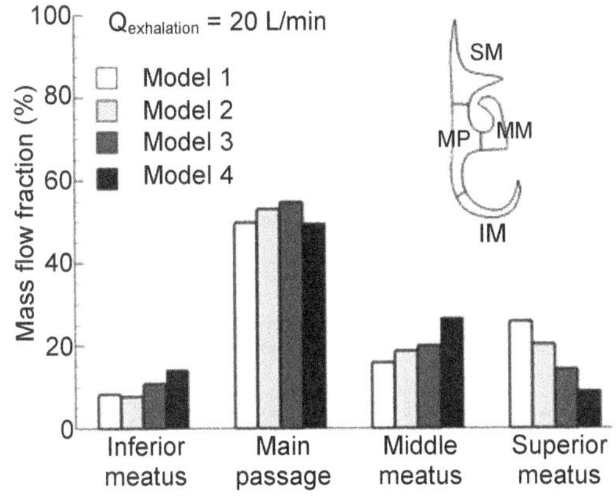

Figure 6. Flow partition into designated sections in the turbinate region. Sections of interest consist of medial passage (MP), superior meatus (SM), middle meatus (MM) and inferior meatus (IM).

is not surprising that in Model 1, whose core flow at Slice 1-1' shifts toward the upper wall, the inferior meatus receives the minimum flow partition while the superior meatus receives the maximum. On the other hand, in Model 4, the maximum flow portion is observed in the inferior meatus while the minimum in the superior meatus. Further, the middle meatus of Model 4 receives more flow than that of any other three models. It is of special interest to examine the flow partition into the superior meatus where the olfactory nerves are cached and smell can be sensed only when chemical particles or vapors reach these nerves. Taking Model 2 (*i.e.*, upright position without geometry simplification) as the reference, the flow deviation of the models 1, 3, 4 from 2 in the superior meatus is +27%, −30%, and −55%, respectively. In light of Model 4, excluding the larynx and pharynx will significantly underestimate the flow rate that could reach the olfactory nerves.

3.4. Particle Deposition

Numerically determined deposition fractions in the four models for exhalation flow rates of 10 and 20 L/min and particle sizes of 1 - 1000 nm were presented in **Figure 7**. They were also compared with *in vitro* deposition data obtained in geometrically comparable nasal replicas [7,8]. As expected, the deposition rate decreases with increasing particle size and flow rates due to decreased molecular diffusivity particle residence times, respectively. Considering that the computational geometry of this study and the nasal replica ANOT2 (Adult-Nasal-Oral-Tracheal) [7] were both constructed based on the same set of nasal images, a direct comparison of deposition data between computational and *in vitro* data is possible. As shown in **Figure** 7, the predicted deposition results appear to provide a reasonable match with *in vitro* measurements with slight under predictions by the numerical models. Specifically, for an expiratory flow rate of 10 L/min, the simulation results in the realistic model (Model 1) agree with the ANOT2 data to a high degree.

Concerning the geometric effects, deposition rates are observed to decrease progressively from Model 1 (the most complex) to Model 4 (the least complex). It is interesting to note that, for an identical particle size and flow rate in this study, the disparity in deposition (or airflow) between any two models can only result from their geometric variations. Therefore, respiratory anatomy such as uvula, pharynx, and larynx can be isolated to investigate its effect upon airflow and depositions by comparing two consecutive models. For example, the influence of the throat, which induce a turbulent laryngeal jet and subsequently enhanced convection and turbulent dispersion, can be quantified by comparing depositions between Models 3 and 4. Based on this idea, the relative impact of the uvula constriction considered in this study is found to be slightly larger than that of the larynx, which is further larger than that of the pharynx. Furthermore, considering that the pressure drop also decreases progressively in the four models with reducing geometric realism, the scatter of deposition fractions among the four models in **Figures 7(a)** and **(b)** can be largely collapsed by normalization with the pressure drop of each model (**Figures 7(c)** and **(d)**), suggesting that pressure drop (*dp*) might be a good indicator of the geometric variations among the four models.

To further evaluate the geometrical effects, the computational and experimental results are plotted as a function of two different diffusion parameters with and without including the pressure drop effect. Exhalation flow rates considered herein include 4, 7.5, 10, 15, 20, 30, and 45 L/min with particle sizes ranging from 1 to 1000 nm. The first parameter (*i.e.*, $\xi_1 = D^{0.5}Q^{-0.28}$), which was first proposed by Cheng *et al.* [38] and was well adopted in later studies [15,24], signifies the dependence of nanoparticle depositions on particle diffusive properties ("*D*" with an exponent of 0.5, molecular diffusion) and breathing conditions ("*Q*" with an exponent of 0.28, convective diffusion). In order to account for the geometric variations of the four models consider, a new correlation parameter with the form ($\xi_2 = D^{0.5}Q^{-a}dp^{b}$) was introduced, with "*dp*" signifying the geometric effects, and "*a*, *b*" being constants to be determined. Accordingly, a new correlation format, $DF = 1 - exp(-cD^{0.5}Q^{-a}dp^{b})$, was used to regress the expiratory simulation data in the four computational models. The best-fit values of *a*, *b* and *c* were 20.8, 0.62 and 0.07, respectively, suggesting insignificant influences upon the total deposition fraction from the laryngopharyngeal anatomical details. The resulting correlation for expiratory deposition for submitcrometer aerosols in the nasal-laryngeal airway is expressed,

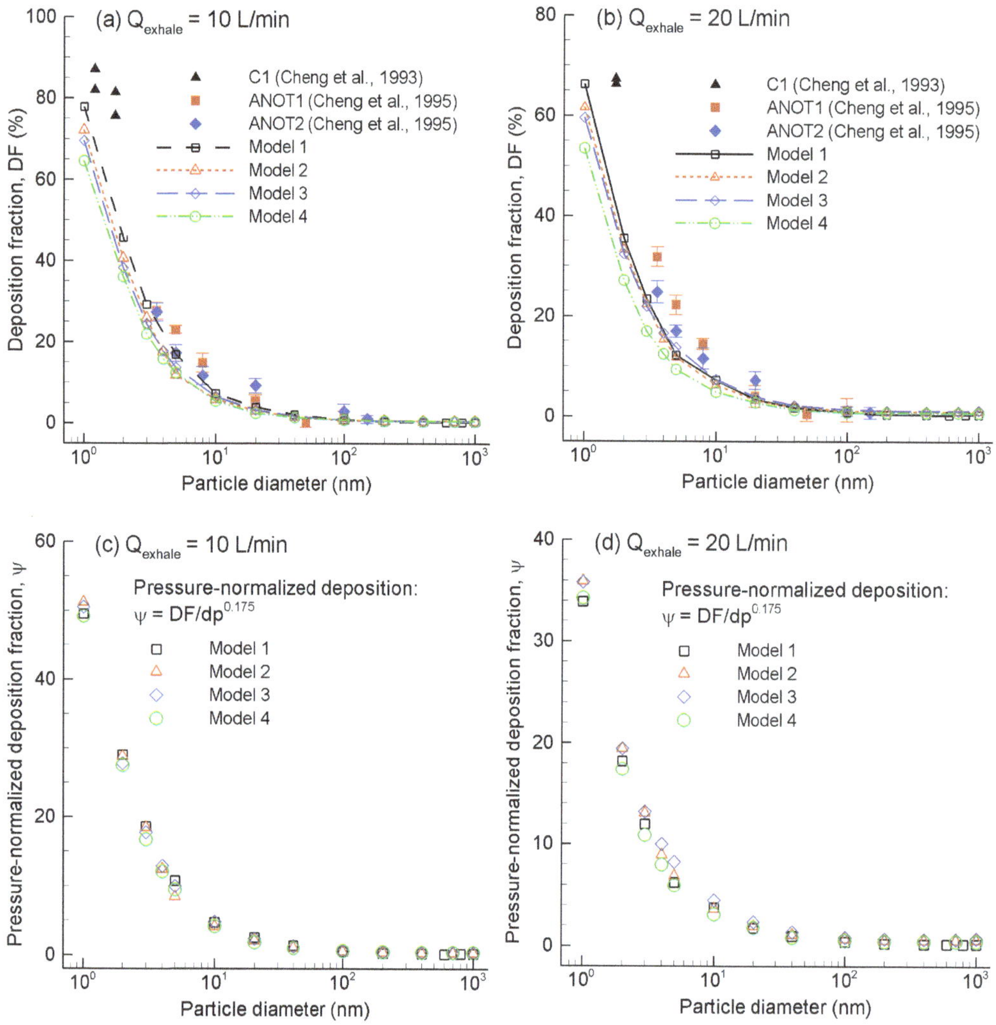

Figure 7. Predicted deposition fractions vs. particle diameter for submicrometer aerosols in the four models for an expiratory flow rate of (a) 10 L/min and (b) 20 L/min. Good agreement is obtained between predictions and in vitro measurements. The deposition rates can be collapsed by being normalized by the pressure drop (*dp*), suggesting that *dp* might be a good indictor of the geometric differences among the four models, (c) 10 L/min and (d) 20 L/min.

$$DF = 1 - \exp\left(-20.8 D^{0.5} Q^{-0.62} dp^{0.07}\right) \quad (1)$$

As shown in **Figure 8**, inclusion of the pressure drop results in a more precise approximation of the numerically predicted deposition data.

3.5. Sub-Regional and Local Depositions

The sub-regional deposition of sub-micrometer particles in different sections of the nasal cavity is shown in **Figure 9** at an exhalation flow rate of 20 L/min. The designnated sections include, in an order of the distance from the nostrils, the vestibule, nasal valve region (VR), turbinate region (TR), olfactory region (OR), and nasaopharynx (NP). Four particle sizes were considered that include particles in both ultrafine (5 nm, 40 nm) and fine (200 nm, 1000 nm) regimes. One observation from **Figure 9** is the significantly higher nasopharynx (NP) deposition of Model 1 than the other three models for particles of 40, 200, and 1000 nm. Our previous studies indicate that aerosols as small as 40 - 50 nm start to exhibit non-negligible inertial effect and will affect deposition via a concurrent mechanism of both impaction and diffusion [33,39]. In the case above, dramatic impactions are expected upon the dorsal walls of the nasopharynx in Model 1 from the particulate-laden jet flows generated by the severe uvula constriction, resulting in much elevated depositions in this region. We do not observe the same high level of NP deposition in Model 1 for 5 nm particles owing to their negligible inertial effects. Another interesting observation is that Model 3 and, to a lesser degree, Model 2 persistently receives higher dosage in the turbinate region among the four models considered. This difference may presumably be traced back to the airflow and particle profiles inside the nasopharynx

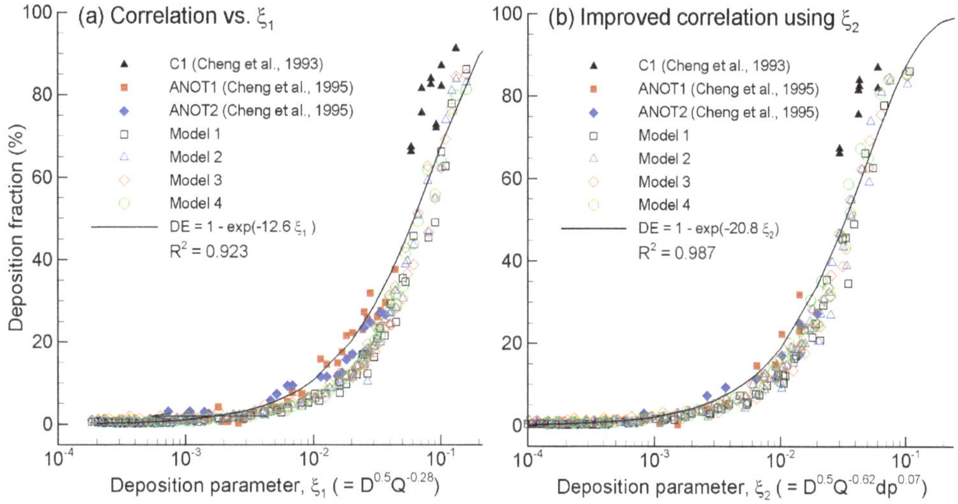

Figure 8. Improved empirical correlation by including the pressure drop in the diffusive deposition parameter: (b) $D^{0.5}Q^{-0.62}dp^{0.07}$ vs. (a) $D^{0.5}Q^{-0.28}$.

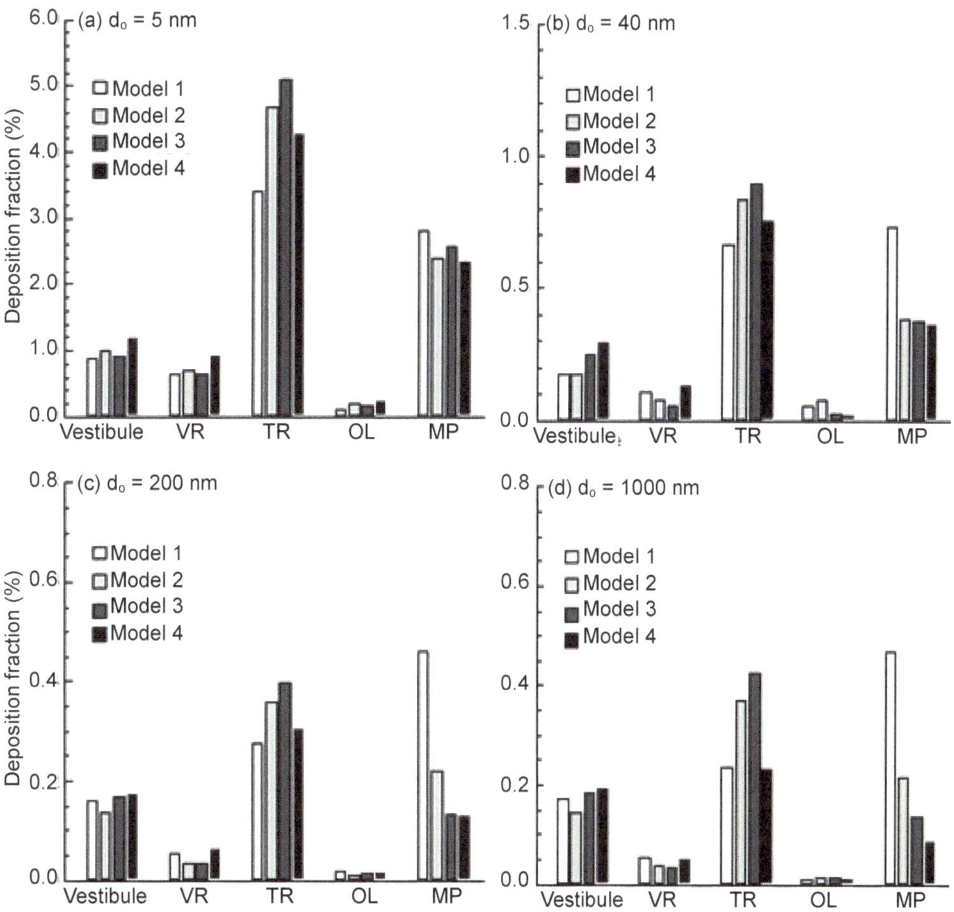

Figure 9. Sub-regional deposition fractions.

before entering the nasal turbinate region. As discussed, there are two discrete peaks of airflow speed and particle concentration occurring at both lateral side of Slice 1-1' in Models 2 and 3, which are absent in the other two models (**Figures 5(b)** and **(c)**). These two peaks, which are generated by the flow-carina interaction, convey airflows and particles into the turbinate region in a more evenly distributed manner that favors more particle contact in this region. This presumption is also corroborated by the higher flow partitions of Models 2 and 3, as shown in **Figure 6**. As regard to the nasal vestibule and valve regions, the variation of depositions is less pronounced and displays an irregular pattern as well. This result is reasonable because the upstream influences of both airflow and particles are nearly damped out when reaching these two regions after traveling through the narrow convoluted passages of the nasal turbinate.

To highlight the laryngopharyngeal anatomical effect on microdosimetry, deposition enhancement factor (DEF) values in the four models are illustrated in **Figure 10** for light activity conditions (Q_{ex} = 20 L/min) and 200 nm particles. As discussed, the DEF parameter quantifies aerosol accumulation with respect to the overall deposition rate and denotes the microdosimetry on the local tissue of interest. The maximum DEF value in Model 1 occurs at the carina ridge of the nasal septum (DEF_{max} = 269) and is nearly one order of magnitude larger than the maximum values for the other three models (DEF_{max} = 31 - 49). Minimum DEF values are observed in Model 4, representing a conservative estimate of particle localization in nasal passages. In view of the enhanced deposition locations, or hot spots, DEF patterns vary significantly among the four models considered. In general, the hot spots are observed to shift from the nasal septum outward to the lateral walls and from the nasopharynx toward the anterior turbinate. In contrast to the nasal septum (and carina ridge in particular) for Model 1, hot spots are observed mainly at the caudal turbinate regions in Models 2 and 3. Besides, the extent of particle deposition in these two models appears more widespread than that in Model 1. This deposition spreading is more pronounced in Model 4 (*i.e.*, the most simplified geometry), which displays hot spots throughout the whole turbinate region.

4. Discussion

In this study, the effects of laryngopharyngeal geometries on expiratory airflow and aerosol depositions in the nasal cavity have been assessed in four model geometries. The most realistic model considered was constructed based on medical images of two health human subjects and was intended to be physiologically accurate with minor surface smoothing and simplifications. The other three models were generated from successive simplifications of the realistic model so that functions of one particular airway anatomy could be isolated for examination, namely the uvula, pharynx, and larynx. Respiratory airflow dynamics for the four models have been considered in terms of breathing resistance and airflow distribution under different activity conditions. Particle transport and deposition was evaluated for submicron aerosols using

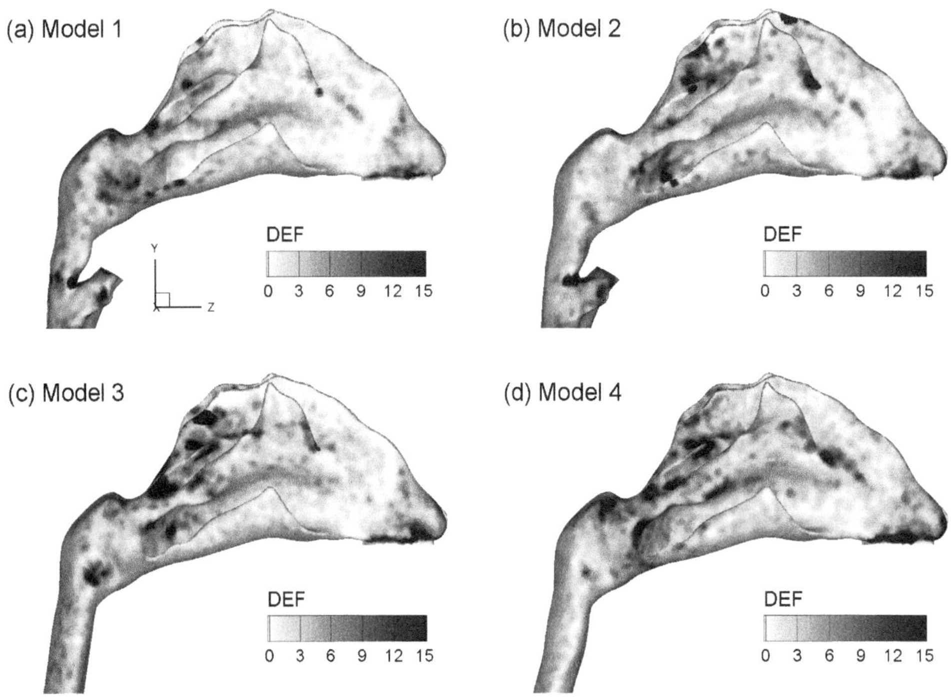

Figure 10. Deposition enhancement factors (DEF) in the four airway models for 200 nm aerosols for an exhalation flow rate of 20 L/min.

the discrete Lagrangian tracking approach on a regional, sub-regional and highly localized basis. Overall, the geometry modifications considered were found to exert substantial effects upon exhaled airflow and particle profiles within the nasal-laryngeal airways, but at varying degree of importance. Among the three anatomies studied (uvula, pharynx, and larynx), the pharynx has a much lesser impact on the airflow and particle deposition than the hanging uvula and the glottal aperture, both of which induce high velocity jet-flow and enhanced turbulent mixing. Furthermore, the variation of the uvula position remarkably alters the breathing resistance, and may constitute a major reason for sleep disorders such as snoring and apnea symptoms. For exhaled respirable aerosols, we did not observe large effects from the upstream geometrical details upon the total deposition rate in the nasal cavity. However, the local deposition patterns vary noticeably , with deposition hot spots shifting outward from nasal septum to side walls and forward from nasopharynx to the anterior turbinate as the upstream geometries being progressively simplified. In addition, the two realistic models (*i.e.*, Model 1 and 2) provided the closest prediction of the total deposition rates in comparison to experimental data, while the most simplified model (*i.e.*, Model 4) gave the worst agreement.

Breathing resistance was found to be highly sensitive to the position of uvula. Significantly elevated pressure jump was captured in the realistic model (*i.e.*, Model 1) due to the uvula obstruction in the nasopharynx. This model was developed from medical images of a 53-year-old male in supine position. Therefore, the uvula-associated flow constriction may result from the combination of an over-flexible soft palate and its own weight that make the uvula hanging over the nasopharynx. Airway tissues in the nasopharynx and pharynx region are normally flexible. Dramatic pressure change may induce further collapse of the airway walls and, in the worst scenario, causes complete obstruction of the airway (*i.e.*, sleep apnea). In addition, turbulent flows may cause fluctuations quivering of the flexible hanging uvula, producing snoring symptoms of the sleeping subject. Based on the above observations, it appears that an over-flexible uvula and weak pharynx walls are two major causes of the sleep apnea and snores [40,41]. The patient in this study (a 53-year-old male) was radiologically diagnosed because of health problems other than sleep apnea or snoring; therefore, no medical record is available at this moment to corroborate the above hypothesis. However, at senior ages as that of the patient concerned or even older, people possess higher possibilities in developing weaker soft palate and pharynx walls, and subsequently higher chances to have sleep apnea or snoring symptoms [42,43].

While multiple empirical correlations of nasal nanoparticle depositions exist, very few correlations including intersubjective variability have been reported. A critical issue in considering this variability is to determine the appropriate parameters that can adequately account for the respiratory physiological differences between different subjects. This parameter could be biological data such as age, height, weight, head circumference, respiratory rate, or nasal airway dimension itself such as volume, surface area, hydraulic diameter, or even a combination of them. From this study, pressure drop appears to be an appropriate candidate to account for the inter-subject geometric variations in light of the fact that the flow history experienced across the entire airway is inclusively characterized by its pressure drop. This does not necessarily imply that pressure drop can replace the flow rate Q in ($D^{0.5}Q^{-0.28}$) in correlating submicrometer depositions. Rather, the pressure drop is introduced as a new parameter to account for the geometry effect, whereas the flow rate Q accounts for the breathing condition (convective diffusion), and the diffusion coefficient D for the particle properties (molecular diffusion). In this study, by including the pressure drop in the diffusion factor (*i.e.*, $D^{0.5}Q^{-0.62}dp^{0.07}$), the relevant deposition mechanisms has been shown to be better captured (R^2 = 0.987 vs. 0.923 in **Figures 8(b)** and **(a)**). The resultant exponent of each parameter signifies the relative effect from that parameter, with large effect from breathing condition (Q exponent = 0.62) and particle diffusive properties (D exponent = 0.5) and small effect from geometry details (dp) for submitcrometer particles. Even though the pressure drop has been a candidate in correlating deposition data (e.g., $D^a dp^b$ for ultrafine, and $d_a^2 dp$ for micrometer particles) [38,44-47], no other study has been reported that utilizes ($D^a Q^b dp^c$) to differentiate the effects from the particle properties (D), breathing condition (Q), and geometric variation (dp), respectively.

Effects of the upstream geometrical realisms on localized deposition in the nasal airway were found to be more dramatic than on regional averages. The locations and extents of the deposition hot spots are significantly different in the four models considered. Microdosimetry is critical in accurately predicting dose-outcome relations in case of rapid absorbed components and fast acting relief medications. Results of this study demonstrate a significant difference in the occurrence of particle localization or hot spot formation in the airways of varying physical realism. As a result, models that are intended to predict cellular dose and response for rapidly absorbed components should preserve a high degree of geometric details even of the upstream anatomies. Interestingly, enhanced particle localizations were observed to be closely coincident with the regions of strong vortex structures (**Figures 4** vs. **10**).

One uncertainty that may cause the predictions to differ from *in vivo* conditions is the actual uvula position during normal expiratory breathing. The current pharynx-geal-laryngeal model was based on CT images acquired with the patient in the supine position and at the end of exhalation. It is likely that the actual upper airway geometry over the exhalation period with an upright position is different from the current model. In particular, the uvula is a highly dynamic structure whose position predominately influences the nasopharynx lumen, and subsequently the airflow and particle profiles entering the nasal cavity during exhalation. In this study, the uvula is positioned within the dorsal nasopharynx and divides the nasopharynx into two narrow channels. The resultant flow acceleration and turbulent dispersion may significantly alter the deposition in the downstream nasal cavity of particles entrained in the exhaled stream.

Other factors that limit the physiological realism of the current study include the assumptions of steady flow, simplified inlet conditions, a smooth and rigid airway surface, no humidity, and a constant glottal aperture and nasal valve for various breathing conditions. Other studies have highlighted the physical significance of transient breathing [13,48], inlet velocity profiles [49,50], nasal wall motion [51], glottal aperture variation [2], and nasal valve change during respiratory maneuvers [52,53]. Moreover, the nasal cavity model in this study is based on images of a single subject, which does not account for intersubject [54] variability.

In conclusion, the laryngopharyngeal anatomical details have been shown to dramatically alter breathing resistance and cellular-level deposition values in nasal airways during exhalation. Breathing resistance was found to be highly sensitive to the position of uvula, which might be a major cause of snoring and obstructive sleep apnea syndrome for patients with an over-flexible soft palate and weak pharynx walls. Even though the total nasal depositions was largely unaffected by the upstream geometrical details, the local deposition patterns, or hot spots, vary considerably. Results of this study indicate that the larynx and pharynx should be retained in order to reliably predict cellular-level dosages, and that patient-specific scan-based airway models are necessitated if respiratory functions in a pathological or poster-surgical situation are at stake. Further numerical analyses are needed to better evaluate the effects of tidal flows, compliable walls, variable glottal aperture and nasal valve, and intra-subject variability before the results can be transformed to clinical diagnosis, treatment and pre-surgical planning.

REFERENCES

[1] T. B. Martonen, "Mathematical-Model for the Selective Deposition of Inhaled Pharmaceuticals," *Journal of Pharmaceutical Sciences*, Vol. 82, No. 12, 1993, pp. 1191-1199.

[2] T. Brancatisano, P. W. Collett and L. A. Engel, "Respiratory Movements of the Vocal Cords," *Journal of Applied Physiology*, Vol. 54, No. 5, 1983, pp. 1269-1276.

[3] K. H. Cheng, Y. S. Cheng, H. C. Yeh, R. A. Guilmette, S. Q. Simpson, S. Q. Yang and D. L. Swift, "*In Vivo* Measurements of Nasal Airway Dimensions and Ultrafine Aerosol Depositing in Human Nasal and Oral Airways," *Journal of Aerosol Science*, Vol. 27, No. 5, 1996, pp. 785-801.

[4] Y. S. Cheng, H. C. Yeh, R. A. Guilmette, S. Q. Simpson, K. H. Cheng and D. L. Swift, "Nasal Deposition of Ultrafine Particles in Human Volunteers and Its Relationship to Airway Geometry," *Aerosol Science And Technology*, Vol. 25, No. 3, 1996, pp. 274-291.

[5] D. L. Swift and J. C. Strong, "Nasal Deposition of Ultrafine 218Po Aerosols in Humna Subjects," *Journal of Aerosol Science*, Vol. 27, No. 7, 1996, pp. 1125-1132.

[6] Z. Lee and M. S. Berridge, "PET Imaging-Based Evaluation of Aerosol Durgs and Their Delviery Devices: Nasal and Pulmonary Studies," *IEEE Transactions on Medical Imaging*, Vol. 21, No. 10, 2002, pp. 1324-1331.

[7] K. H. Cheng, Y. S. Cheng, H. C. Yeh and D. L. Swift, "Deposition of Ultrafine Aerosols in the Head Airways during Natural Breathing and during Simulated Breath-Holding Using Replicate Human Upper Airway Casts," *Aerosol Science and Technology*, Vol. 23, No. 3, 1995, pp. 465-474.

[8] Y. S. Cheng, Y. F. Su, H. C. Yeh and D. L. Swift, "Deposition of Thoron Progeny in Human Head Airways," *Aerosol Science and Technology*, Vol. 18, No. 4, 1993, pp. 359-375.

[9] D. L. Swift, N. Montassier, P. K. Hopke, K. Karpenhayes, Y. S. Cheng, Y. F. Su, H. C. Yeh and J. C. Strong, "Inspiratory Deposition of Ultrafine Particles in Human Nasal Replicate Cast," *Journal of Aerosol Science*, Vol. 23, No. 1, 1992, pp. 65-72.

[10] Y. Liu, E. A. Matida, J. Gu and M. R. Johnson, "Numerical Simulation of Aerosol Deposition in a 3-D Human Nasal Cavity Using RANS, RANS/EIM, and LES," *Aerosol Science*, Vol. 38, No. 7, 2007, pp. 683-700.

[11] T. B. Martonen, Z. Q. Zhang, G. Yue and C. J. Musante, "Fine Particle Deposition within Human Nasal Airways," *Inhalation Toxicology*, Vol. 15, No. 4, 2003, pp. 283-303.

[12] J. D. Schroeter, C. J. Musante, D. M. Hwang, R. Burton, R. Guilmette and T. B. Martonen, "Hygroscopic Growth and Deposition of Inhaled Secondary Cigarette Smoke in Human Nasal Pathways," *Aerosol Science and Technology*, Vol. 34, No. 1, 2001, pp. 137-143.

[13] H. Shi, C. Kleinstreuer and Z. Zhang, "Laminar Airflow and Nanoparticle or Vapor Deposition in a Human Nasal

Cavity Model," *Journal of Biomechanical Engineering*, Vol. 128, No. 5, 2006, pp. 697-706.

[14] G. Yu, Z. Zhang and R. Lessmann, "Fluid Flow and Particle Diffusion in the Human Upper Respiratory System," *Aerosol Science and Technology*, Vol. 28, No. 2, 1998, pp. 146-158.

[15] P. Zamankhan, G. Ahmadi, Z. Wang, P. K. Hopke, Y. S. Cheng, W. C. Su and D. Leonard, "Airflow and Deposition of Nano-Particles in a Human Nasal Cavity," *Aerosol Science and Technology*, Vol. 40, No. 6, 2006, pp. 463-476.

[16] E. A. C. Hubal, J. S. Kimbell and P. S. Fedkiw, "Incorporation of Nasal-Lining Mass-Transfer Resistance into a CFD Model for Prediction of Ozone Dosimetry in the Upper Respiratory Tract," *Inhalation Toxicology*, Vol. 8, No. 9, 1996, pp. 831-857.

[17] P. W. Scherer, K. Keyhani and M. M. Mozell, "Nasal Dosimetry Modeling for Humans," *Inhalation Toxicology*, Vol. 6, Suppl. S, 1994, pp. 85-97.

[18] K. Zhao, P. W. Scherer, S. A. Hajiloo and P. Dalton, "Effects of Anatomy on Human Nasal Air Flow and Odorant Transport Patterns: Implictions for Olfaction," *Chemical Senses*, Vol. 29, No. 5, 2004, pp. 365-379.

[19] P. W. Longest and S. Vinchurkar, "Validating CFD Predictions of Respiratory Aerosol Deposition: Effects of Upstream Transition and Turbulence," *Journal of Biomechanics*, Vol. 40, No. 2, 2007, pp. 305-316.

[20] J. Xi and P. W. Longest, "Transport and Deposition of Micro-Aerosols in Realistic and Simplified Models of the Oral Airway," *Annals of Biomedical Engineering*, Vol. 35, No. 4, 2007, pp. 560-581.

[21] Z. Li, "Particle Deposition in Oral-Tracheal Airway Models with Very Low Inhalation Profiles," *Journal of Bionic Engineering*, Vol. 9, No. 2, 2012, pp. 252-261.

[22] M. Brouns, S. Verbanck and C. Lacor, "Influence of Glottic Aperture on the Tracheal Flow," *Journal of Biomechanics*, Vol. 40, No. 1, 2007, pp. 165-172.

[23] J. Xi, P. W. Longest and T. B. Martonen, "Effects of the Laryngeal Jet on Nano- and Microparticle Transport and Deposition in an Approximate Model of the Upper Tracheobronchial Airways," *Journal of Applied Physiology*, Vol. 104, No. 6, 2008, pp. 1761-1777.

[24] J. Xi and P. W. Longest, "Characterization of Submicrometer Aerosol Deposition in Extrathoracic Airways during Nasal Exhalation," *Aerosol Science and Technology*, Vol. 43, No. 8, 2009, pp. 808-827.

[25] R. A. Guilmette, Y. S. Cheng, H. C. Yeh and D. L. Swift, "Deposition of 0.005 - 12 Micrometer Monodisperse Particls in a Computer-Milled, MRI-Based Nasal Airway Replica," *Inhalation Toxicology*, Vol. 6, Suppl. 1, 1994, pp. 395-399.

[26] J. T. Kelly, B. Asgharian, J. S. Kimbell and B. Wong, "Particle Depositon in Human Nasal Airway Replicas Manufactured by Different Methds. Part I: Inertial Regime Particles," *Aerosol Science and Technology*, Vol. 38, No. 11, 2004, pp. 1063-1071.

[27] J. T. Kelly, B. Asgharian, J. S. Kimbell and B. Wong, "Particle Depositon in Human Nasal Airway Replicas Manufactured by Different Methds. Part II: Ultrafine Particles," *Aerosol Science and Technology*, Vol. 38, No. 11, 2004, pp. 1072-1079.

[28] T. Gemci, T. E. Corcoran and N. Chigier, "A Numerical and Experimental Study of Spray Dynamics in a Simple Throat Model," *Aerosol Science and Technology*, Vol. 36, No. 1, 2002, pp. 18-38.

[29] K. Oho and R. Amemiya, "Practical Fiberopitc Bronscopy," Igaku-Shoin, Tokyo, 1980.

[30] Y. S. Cheng, Y. Yamada, H. C. Yeh and D. L. Swift, "Diffusional Deposition of Ultrafine Aerosols in a Human Nasal Cast," *Journal of Aerosol Science*, Vol. 19, No. 6, 1988, pp. 741-751.

[31] D. O. Rodenstein, G. Dooms, Y. Thomas, G. Liistro, D. C. Stanescu, C. Culee and G. Auberttulkens, "Pharyngeal Shape and Dimensions in Healthy Subjects, Snorers, and Patients with Obstructive Sleep Apnoea," *Thorax*, Vol. 45, No. 10, 1990, pp. 722-727.

[32] D. C. Wilcox, "Turbulence Modeling for CFD," 2nd Ed., DCW Industries, Inc., California, 1998.

[33] P. W. Longest and J. Xi, "Computational Investigation of Particle Inertia Effects on Submicron Aerosol Deposition in the Respiratory Tract," *Journal of Aerosol Science*, Vol. 38, No. 1, 2007, pp. 111-130.

[34] P. W. Longest and J. Xi, "Effectiveness of Direct Lagrangian Tracking Models for Simulating Nanoparticle Deposition in the Upper Airways," *Aerosol Science and Technology*, Vol. 41, No. 4, 2007, pp. 380-397.

[35] J. Xi and P. W. Longest, "Effect of Improved Near-Wall Modeling on Micro-Particle Deposition in Oral Airway Geometries," *Proceedings ASME* 2007 *Summer Bioengineering Conference*, Keystone, CO, 20-24 June 2007, pp. SBC2007-176227.

[36] J. R. Wheatley, T. C. Amis and L. A. Engel, "Oronasal Partitioning of Ventilation during Exercise in Humans," *Journal of Applied Physiology*, Vol. 71, No. 2, 1991, pp. 546-551.

[37] J. Jeong and F. Hussain, "On the Identification of a Vortex," *Journal of Fluid Mechanics*, Vol. 285, 1995, pp. 69-94.

[38] Y. S. Cheng, "Aerosol Deposition in the Extrathoracic Region," *Aerosol Science and Technology*, Vol. 37, No. 8, 2003, pp. 659-671.

[39] J. Xi and P. W. Longest, "Numerical Predictions of Submicrometer Aerosol Deposition in the Nasal Cavity Using a Novel Drift Flux Approach," *International Journal of Heat and Mass Transfer*, Vol. 51, No. 23-24, 2008, pp. 5562-5577.

[40] T. Young, L. Finn and H. Kim, "Nasal Obstruction as a Risk Factor for Sleep-Disordered Breathing," *Journal of Allergy and Clinical Immunology*, Vol. 99, No. 2, 1997, pp. S757-S762.

[41] F. Lofaso, A. Coste, M. P. d'Ortho, F. Zerah-Lancner, C. Delclaux, F. Goldenberg and A. Harf, "Nasal Obstruction as a Risk Factor for Sleep Apnoea Syndrome," *European Respiratory Journal*, Vol. 16, No. 4, 2000, pp. 639-643.

[42] E. Lindberg, A. Elmasry, T. Gislason, C. Janson, H. Bengtsson, J. Hetta, M. Nettelbladt and G. Boman, "Evolution of Sleep Apnea Syndrome in Sleepy Snorers—A Population-Based Prospective Study," *American Journal of Respiratory and Critical Care Medicine*, Vol. 159, No. 6, 1999, pp. 2024-2027.

[43] S. T. Pendlebury, J. L. Pepin, D. Veale and P. Levy, "Natural Evolution of Moderate Sleep Apnoea Syndrome: Significant Progression over a Mean of 17 Months," *Thorax*, Vol. 52, No. 10, 1997, pp. 872-878.

[44] L. Golshahi, M. L. Noga, R. B. Thompson and W. H. Finlay, "*In Vitro* Deposition Measurement of Inhaled Micrometer-Sized Particles in Extrathoracic Airways of Children and Adolescents during Nose Breathing," *Journal of Aerosol Science*, Vol. 42, No. 7, 2011, pp. 474-488.

[45] J. Heyder, J. Gebhart, G. Rudolf, C. F. Schiller and W. Stahlhofen, "Deposition of Particles in the Human Respiratory Tract in the Size Range of 0.005 - 15 Microns," *Journal of Aerosol Science*, Vol. 17, No. 5, 1986, pp. 811-825.

[46] R. F. Hounam, A. Black and M. Walsh, "The Deposition of Aerosol Particles in the Nasopharyngeal Region of the Human Respiratory Tract," *Journal of Aerosol Science*, Vol. 2, No. 1, 1971, pp. 47-61.

[47] G. J. M. Garcia, E. W. Tewksbury, B. A. Wong and J. S. Kimbell, "Interindividual Variability in Nasal Filtration as a Function of Nasal Cavity Geometry," *Journal of Aerosol Medicine and Pulmonary Drug Delivery*, Vol. 22, No. 2, 2009, pp. 139-155.

[48] Z. Li and C. Kleinstreuer, "Airflow Analysis in the Alveolar Region Using the Lattice-Boltzmann Method," *Medical & Biological Engineering & Computing*, Vol. 49, No. 4, 2011, pp. 441-451.

[49] K. Keyhani, P. W. Scherer and M. M. Mozell, "Numerical Simulation of Airflow in the Human Nasal Cavity," *Journal of Biomechanical Engineering—Transactions of the ASME*, Vol. 117, No. 4, 1995, pp. 429-441.

[50] R. P. Subramaniam, R. B. Richardson, K. T. Morgan, J. S. Kimbell and R. A. Guilmette, "Computational Fluid Dynamics Simulations of Inspiratory Airflow in the Human Nose and Nasopharynx," *Inhalation Toxicology*, Vol. 10, No. 2, 1998, pp. 91-120.

[51] R. Fodil, L. Brugel-Ribere, C. Croce, G. Sbirlea-Apiou, C. Larger, J. F. Papon, C. Delclaux, A. Coste, D. Isabey and B. Louis, "Inspiratory Flow in the Nose: A Model Coupling Flow and Vasoerectile Tissue Distensibility," *Journal of Applied Physiology*, Vol. 98, No. 1, 2005, pp. 288-295.

[52] G. P. Bridger, "Physiology of Nasal Valve," *Archives of Otolaryngology*, Vol. 92, No. 6, 1970, pp. 543-553.

[53] G. P. Bridger and D. F. Proctor, "Maximum Nasal Inspiratory Flow and Nasal Resistance," *Annals of Otology Rhinology and Laryngology*, Vol. 79, No. 3, 1970, pp. 481-488.

[54] D. N. Pickering and C. S. Beardsmore, "Nasal Flow Limitation in Children," *Pediatric Pulmonology*, Vol. 27, No. 1, 1999, pp. 32-36.

15

Direct Numerical Simulation of a Jet Issuing from Rectangular Nozzle by the Vortex in Cell Method

Tomomi Uchiyama[1], Masahiro Kobayashi[2], Shouichiro Iio[3], Toshihiko Ikeda[3]
[1]EcoTopia Science Institute, Nagoya University, Nagoya, Japan
[2]Graduate School of Information Science, Nagoya University, Nagoya, Japan
[3]Faculty of Engineering, Shinshu University, Nagano, Japan

ABSTRACT

Direct numerical simulation of a jet issuing from a nozzle having a rectangular cross-section is conducted. The vortex in cell (VIC) method, of which computational accuracy was heightened by the authors in a prior study, is used for the DNS. The aspect ratio of the nozzle cross-section is 15, and the Reynolds number based on the shorter side length of the nozzle exit is 6700. The turbulence statics, such as the mean velocity and the turbulence intensity, are favorably compared with the experimentally measured results. The behavior of the large-scale eddies as well as the development of the turbulent flow is also confirmed to agree with the measurement. These indicate that the authors' VIC method is successfully employed for the DNS of rectangular jet.

Keywords: Rectangular Jet; Vortex in Cell Method; DNS; Vortical Structure

1. Introduction

Vortex in cell (VIC) method [1] is one of the simulation methods for incompressible flows. It discretizes the vorticity field into vortex elements and computes the time evolution of the field by tracing the convection of each vortex element using the Lagrangian approach. The Lagrangian calculation markedly reduces numerical diffusion and also improves numerical stability. Thus, the VIC method is eminently suitable for direct numerical simulation (DNS) and large eddy simulation (LES) of turbulent flows, and various results have been reported. Cottet and Poncet [2] applied the VIC method for the wake simulation of a circular cylinder, and captured the streamwise vortices occurring behind the cylinder. Cocle *et al.* [3] analyzed the behavior of two vortex systems near a solid wall, and made clear the interaction between two counter-rotating vortices and the eddies induced in the vicinity of the wall. Chatelain *et al.* [4] simulated trailing edge vortices, and visualized the unsteady phenomena caused by disturbances. These studies are concerned with time-developing free shear flows. But the VIC method has not been applied to turbulent flows bounded by solid walls, which are closely related with the turbulent friction and heat transfer.

The authors [5] performed the DNS of a turbulent channel flow, which is a representative example of wall turbulent flows. When applying the existing VIC method, the oscillation of the flow increased with the progress of the computation, and eventually the computation collapsed. This was caused by the fact that the consistency among the discretized equations is not ensured. This was also because the solenoidal condition for the vorticity is not fully satisfied. To overcome such problems of the existing VIC method, the authors [5] proposed the improved VIC method. The authors [5] also applied the improved VIC method to the DNS of the turbulent channel flow. The DNS highlighted that the time evolution of the flow is fully computed and that the statistically steady turbulent flow is favorably obtained. It also demonstrated that the organized flow structures, such as streaks and streamwise vortices appearing in the near wall region, are successfully captured and that the turbulence statistics, such as the mean velocity and the Reynolds stress, agree well with the existing DNS results.

Jet is frequently observed in many engineering applications. As the jet is the representative of free shear flows governed by the convection of the eddies having various scales, the VIC method solving directly the vorticity field promises to be usefully applied to the simulation. But the VIC method has been scarcely used for the jet simulation.

The objective of this study is to search for the applicability of the authors' VIC method [5] to the DNS of jet flows. The DNS of a jet issuing from a nozzle having a rectangular cross-section is performed by the VIC method. The jet velocity field was experimentally investigated by Iio *et al.* [6]. The aspect ratio of the nozzle cross-section is 15, and the Reynolds number based on the shorter side length of the nozzle exit is 6700. The DNS results, such as the mean velocity and the turbulence intensity, are favorably compared with the measured ones. The behavior of the large-scale eddies as well as the development of the turbulent flow is also confirmed to agree with the measurement. These indicate that the authors' VIC method is successfully employed for the DNS of rectangular jet.

2. Vortex in Cell Method

2.1. Vorticity Equation and Orthogonal Decomposition of Velocity

For an incompressible flow, the vorticity equation is given by

$$\frac{\partial \boldsymbol{\omega}}{\partial t}+\nabla\cdot(\boldsymbol{\omega}\boldsymbol{u})=\nabla\cdot(\boldsymbol{u}\boldsymbol{\omega})+\nu\nabla^2\boldsymbol{\omega} \tag{1}$$

where $\boldsymbol{u}$ is the velocity and $\boldsymbol{\omega}(=\nabla\times\boldsymbol{u})$ is the vorticity.

According to the Helmholtz theorem, the velocity $\boldsymbol{u}$ is the sum of the curl of a vector potential $\boldsymbol{\psi}$ and the gradient of a scalar potential ϕ:

$$\boldsymbol{u}=\nabla\times\boldsymbol{\psi}+\nabla\phi \tag{2}$$

when $\boldsymbol{\psi}$ is postulated to be solenoidal or $\nabla\cdot\boldsymbol{\psi}=0$, the curl of Equation (2) yields the vector Poisson equation for $\boldsymbol{\psi}$:

$$\nabla^2\boldsymbol{\psi}=-\boldsymbol{\omega} \tag{3}$$

Substituting Equation (2) into the continuity equation and rewriting the resultant equation, the Laplace equation for ϕ is obtained:

$$\nabla^2\phi=0 \tag{4}$$

Once $\boldsymbol{\psi}$ and ϕ have been computed from Equations (3) and (4) respectively, the velocity $\boldsymbol{u}$ is calculated from Equation (2). The vorticity $\boldsymbol{\omega}$ in Equation (3) is estimated from Equation (1). The vortex in cell (VIC) method discretizes the vorticity field into vortex elements, and calculates the distribution of $\boldsymbol{\omega}$ by tracing the convection of each vortex element.

It is postulated that the position vector and vorticity for the vortex element p are $\boldsymbol{x}_p=(x_p,y_p,z_p)$ and $\boldsymbol{\omega}_p$, respectively. The Lagrangian form of the vorticity equation, Equation (1), is written as [7]:

$$\frac{\mathrm{d}\boldsymbol{x}_p}{\mathrm{d}t}=\boldsymbol{u}(\boldsymbol{x}_p) \tag{5}$$

$$\frac{\mathrm{d}\boldsymbol{\omega}_p}{\mathrm{d}t}=\nabla\cdot\left(\boldsymbol{u}(\boldsymbol{x}_p)\boldsymbol{\omega}(\boldsymbol{x}_p)\right)+\nu\nabla^2\boldsymbol{\omega}(\boldsymbol{x}_p) \tag{6}$$

When the position and vorticity of a vortex element are known at time t, the values at $t+\Delta t$ are computed from the Lagrangian calculations of Equations (5) and (6). In the VIC method, the flow field is divided into computational grid cells to define $\boldsymbol{\psi}$, ϕ and $\boldsymbol{\omega}$ on the grids. If $\boldsymbol{\omega}$ is defined at a position $\boldsymbol{x}_g=(x_g,y_g,z_g)$, the vorticity $\boldsymbol{\omega}$ is assigned to $\boldsymbol{x}_g$, or a vortex element with vorticity $\boldsymbol{\omega}$ is redistributed onto $\boldsymbol{x}_g$

$$\boldsymbol{\omega}=\sum_p^{N_v}\boldsymbol{\omega}_p W\left(\frac{x_g-x_p}{\Delta x}\right)W\left(\frac{y_g-y_p}{\Delta y}\right)W\left(\frac{z_g-z_p}{\Delta z}\right) \tag{7}$$

where Δx, Δy and Δz are the grid widths, and N_v is the number of vortex elements. For the redistribution function W, various forms are presented [7]. To suppress the numerical dissipation, a high-order scheme, which preserves the three first moments of the distribution, twice continuously differentiable and symmetric, is employed for W:

$$W(\varepsilon)=\begin{cases}1-2.5\varepsilon^2+1.5|\varepsilon|^3 & |\varepsilon|<1\\ 0.5(2-|\varepsilon|)^2(1-|\varepsilon|) & 1\le|\varepsilon|\le 2\\ 0 & |\varepsilon|>2\end{cases} \tag{8}$$

Equation (8) was used for the simulations of time-developing free shear flows [2-4]. The DNS of a turbulent channel flow was also successfully performed with Equation (8) by the authors [5].

2.2. Discretization by Staggered Grid

For incompressible flow simulations, the MAC and SMAC methods solve the Poisson equation, which is derived from the equation for pressure gradient and the continuity equation. These methods employ a staggered grid to ensure consistency among the discretized equations, and to prevent the numerical oscillation of the solution. The staggered grid would appear to be indispensable for discretizing the Poisson equation for $\boldsymbol{\psi}$ and the Laplace equation for ϕ, which are derived in the VIC method. However, staggered grids are not readily accommodated in the existing VIC method.

The authors proposed a VIC method using a staggered grid in their prior study [5]. **Figure 1** shows the arrangement of the variables in the grid. The scalar potential ϕ and velocity $\boldsymbol{u}$ are defined at the center and sides of a grid cell, respectively. The vorticity $\boldsymbol{\omega}$ and the vector potential $\boldsymbol{\psi}$ are defined on the edges.

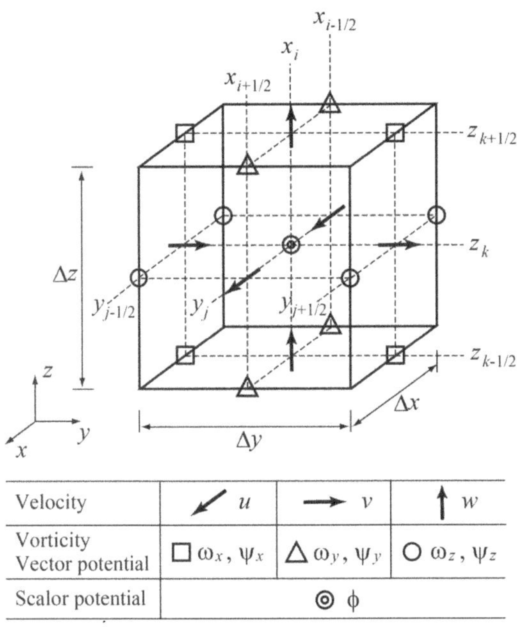

Figure 1. Staggered grid.

2.3. Correction of Vorticity Field

In the VIC method, the vorticity field is discretized into vortex elements, and the field is expressed by superimposing the vorticity distributions around each vortex element. The superposition is performed by Equation (7). The resulting vorticity field $\boldsymbol{\omega}_r$ does not necessarily satisfy the solenoidal condition [7]. Denoting the vorticity satisfying this condition by $\boldsymbol{\omega}_s = (\nabla \times \boldsymbol{u})$, $\boldsymbol{\omega}_r$ is represented as [3]

$$\boldsymbol{\omega}_r = \nabla F + \boldsymbol{\omega}_s \tag{9}$$

where F is a scalar function. Equation (9) corresponds to the Helmholtz decomposition of $\boldsymbol{\omega}_r$.

Taking the divergence of Equation (9), the Poisson equation for F is obtained:

$$\nabla^2 F = \nabla \cdot \boldsymbol{\omega}_r \tag{10}$$

Calculating F from Equation (10) and substituting into Equation (9) gives the recalculated vorticity $\boldsymbol{\omega}_s$ [3]. This correction for vorticity needs to solve the Poisson equation, which increases the computational time. To reduce this additional cost, the authors [5] have proposed a simplified correction method.

The uncorrected vorticity, $\boldsymbol{\omega}_r$, is linked to $\boldsymbol{\psi}_r$ through Equation (3). Taking the divergence of Equation (3) and substituting Equation (10) into the resultant equation, the following relations are obtained:

$$\begin{aligned} \nabla^2 (\nabla \cdot \boldsymbol{\psi}_r) &= -\nabla \cdot \boldsymbol{\omega}_r \\ &= -\nabla^2 F \end{aligned} \tag{11}$$

Unlike the assumption for the solenoidal condition of $\boldsymbol{\psi}$, the following equation for a non-solenoidal vorticity is derived from Equation (11).

$$\nabla \cdot \boldsymbol{\psi}_r = -F \tag{12}$$

Using $\boldsymbol{\omega}_r$ to calculate $\boldsymbol{\psi}_r$ from Equation (3), and determining ϕ_r from Equation (4), the curl of $\boldsymbol{u}_r$ transforms as follows:

$$\begin{aligned} \nabla \times \boldsymbol{u}_r &= \nabla \times (\nabla \phi_r + \nabla \times \boldsymbol{\psi}_r) \\ &= \nabla (\nabla \cdot \boldsymbol{\psi}_r) - \nabla^2 \boldsymbol{\psi}_r \\ &= -\nabla F + \boldsymbol{\omega}_r \\ &= \boldsymbol{\omega}_s \end{aligned} \tag{13}$$

Equation (13) demonstrates that the curl of the velocity $\boldsymbol{u}_r$ calculated from $\boldsymbol{\omega}_r$ yields a vorticity $\boldsymbol{\omega}_s$ that satisfies the solenoidal condition. If the vorticity is recalculated by Equation (13), or the vorticity is corrected immediately after calculating the velocity by Equation (2), the discretization error in the vorticity is completely removed and the flow dynamics are accurately simulated without solving the Poisson equation, Equation (10). It should be noted that the staggered grid is required for rendering the transformation in Equation (13) applicable to the corresponding discretized equations.

2.4. Simulation Procedure

Given the flow at time t, the flow at $t + \Delta t$ is simulated by the following procedure:

1) Calculate the change in the strength of each vortex element defined on the grid, or calculate the vorticity $\boldsymbol{\omega}_p$ from Equation (6).

2) Calculate the convection of each vortex element, or calculate the position $\boldsymbol{x}_p$ from Equation (5).

3) Redistribute the vortex element onto the grids, or calculate the vorticity $\boldsymbol{\omega}$ on the grids by Equation (7).

4) Calculate the vector potential $\boldsymbol{\psi}$ from Equation (3).

5) Calculate the scalar potential ϕ from Equation (4).

6) Calculate the velocity $\boldsymbol{u}$ from Equation (2).

7) Correct the vorticity, or calculate the corrected vorticity from the curl of $\boldsymbol{u}$.

3. Simulation Conditions

The DNS of an incompressible jet issuing from a nozzle having a rectangular cross-section is performed. The jet velocity field was experimentally investigated by Iio *et al.* [6]. The computational domain is shown in **Figure 2**. The shorter side length of the nozzle exit is w, and the longer side length is $15w$. The x-axis is along the jet centerline, while the y- and z-axes are parallel to the shorter and longer sides of the nozzle exit respectively. The computational domain consists of a cubic region of $40w \times 40w \times 40w$. It is resolved into uniform grid cells of $400 \times 400 \times 400$. Consequently, the shorter and longer sides of the nozzle exit are divided by 10 and 150 cells

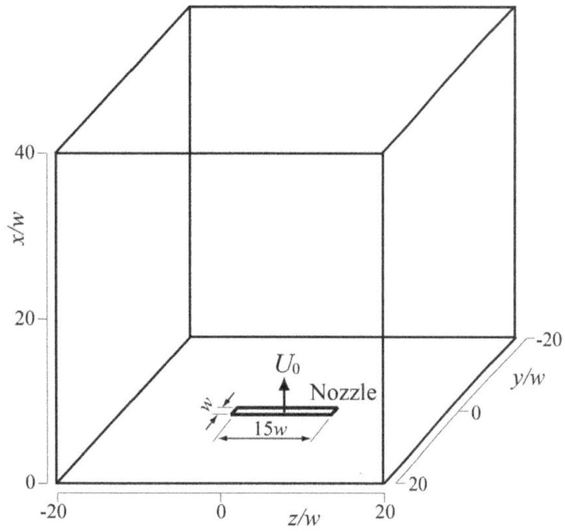

Figure 2. Rectangular nozzle and computational domain.

respectively. The jet velocity at the nozzle exit is U_0, and the Reynolds number based on w and U_0 is 6700. **Table 1** summarizes the simulation conditions.

The boundary conditions are given as follows: Within the nozzle exit, a uniform and constant velocity U_0 is imposed.

$$\begin{aligned} &(u_x, u_y, u_z) = (U_0, 0, 0) \\ &\partial \psi_x / \partial x = 0,\ \psi_y = 0,\ \psi_z = 0 \\ &\partial \phi / \partial x = U_0 \\ &\boldsymbol{\omega} = 0 \end{aligned} \tag{14}$$

The non-slip condition is adopted on the wall at $x = 0$, on which the nozzle exit is mounted.

$$\begin{aligned} &\boldsymbol{u} = 0 \\ &\partial \psi_x / \partial x = 0,\ \psi_y = 0,\ \psi_z = 0 \\ &\partial \phi / \partial x = 0 \\ &\omega_x = 0,\ \omega_y = -\partial u_z / \partial x,\ \omega_z = \partial u_y / \partial x \end{aligned} \tag{15}$$

The velocity gradient is set at zero on the other boundaries. For example, the following conditions are imposed at $y = \pm 20w$.

$$\begin{aligned} &\partial^2 \phi / \partial y^2 = 0 \\ &\partial \psi_y / \partial x = 0,\ \partial \psi_z / \partial x = 0 \\ &\partial \psi_x / \partial x = -\partial \psi_y / \partial y - \partial \psi_z / \partial z \end{aligned} \tag{16}$$

The time increment Δt is $w/(100U_0)$. A fully developed flow was obtained at a time of $200w/U_0$. The time-averaged velocity and the turbulence intensity were acquired in the period from $250w/U_0$ to $550w/U_0$.

4. Results and Discussion

4.1. Instantaneous Flow Fields

The instantaneous distribution for the absolute value of the velocity $|\boldsymbol{u}|$ for the fully developed jet is shown in **Figure 3**. The distribution on the central cross-sections of the jet is presented. **Figure 3(a)** shows the result on the x-y cross-section at $z = 0$ parallel to the shorter side of the nozzle exit. The potential core disappears at $x/w \fallingdotseq 2.6$, as mentioned later. Thus, the jet spreads in the lateral direction downstream of the disappearing point, indicating the momentum diffusion in the direction. The distribution on the x-z cross-section parallel to the longer side is shown in **Figure 3(b)**. It should be noted that the jet width becomes slightly narrower as the streamwise distance increases.

Table 1. Simulation conditions.

Direction	x	y	z
Nozzle size	-	w	$15w$
Domain length	$40w$	$40w$	$40w$
Grid cell number	400	400	400
Grid width	$w/10$	$w/10$	$w/10$
Nozzle outlet velocity	U_0	0	0
Reynolds number wU_0/ν		6700	
Time increment Δt		$w/(100U_0)$	

The vorticity components ω_z and ω_y at the same instant as **Figure 3** distribute as plotted in **Figure 4**, where the distributions on the jet central cross-sections are presented. The shear layers originate from the shorter and longer sides of the nozzle exit, and the turbulent flow rapidly develops just after the collapse of the shear layers.

Figure 5(a) shows the velocity on the jet central cross-section parallel to the shorter side of the nozzle exit, where the distribution near the nozzle is plotted. To make the distribution more understandable, the velocity $0.5U_0$ is subtracted. On the shear layers originating from the longer sides of the nozzle exit, large-scale eddies occur at $x/w \geqq 2.2$. The eddies are non-axisymmetric around the jet centerline. They change into eddies having various scales as the streamwise distance increases. **Figure 5(b)** presents the flow field visualized experimentally by illuminating an alcohol mist with a laser light sheet [6]. The image clearly visualizes the occurrence of some large-scale eddies on the shear layers as well as the rapid change of the flow into turbulence. It is confirmed that the DNS successfully resolves the flow development including the occurrence and collapse of large-scale eddies.

4.2. Time-Averaged Velocity

When the time-averaged velocity is calculated, the component in the jet axial direction U distributes on the jet

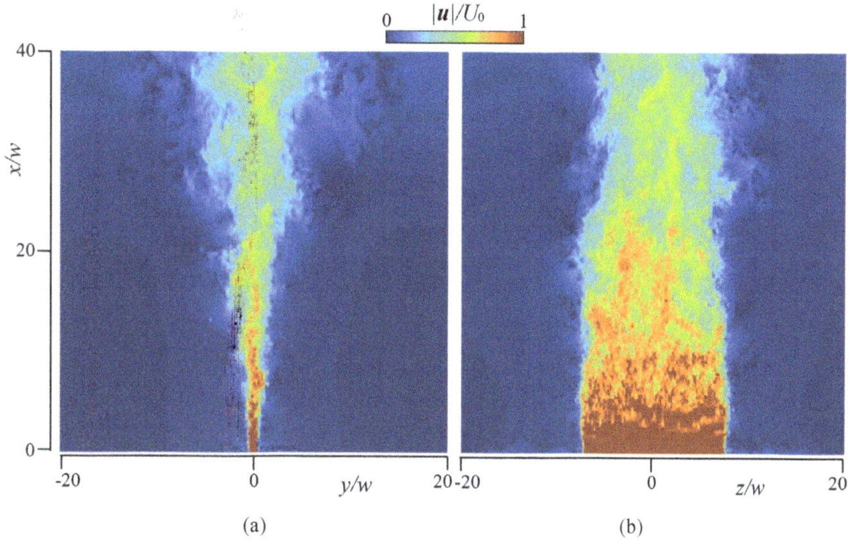

Figure 3. Instantaneous velocity distribution of fully-developed jet. (a) ***x-y*** **cross-section at** $z = 0$**; (b)** ***x-z*** **cross-section at** $y = 0$**.**

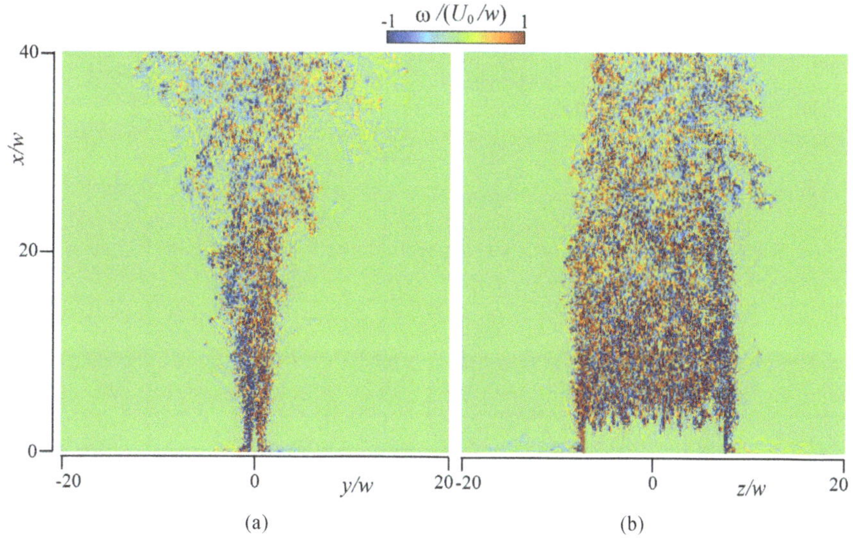

Figure 4. Instantaneous vorticity distribution of fully-developed jet. (a) ω_z **on** ***x-y*** **cross-section at** $z = 0$**; (b)** ω_y **on** ***x-z*** **cross-section at** $y = 0$**.**

central cross-sections as shown in **Figure 6**. On the *x-y* cross-section parallel to the shorter side of the nozzle exit, the jet diffuses more in the lateral direction at $x/w \geqq 3$ as the streamwise distance increases. On the *x-z* cross-section parallel to the longer side, however, the jet width becomes narrower at $x/w \geqq 3$ in the downstream direction.

Figure 7 compares the time-averaged velocity component U obtained by the DNS with the measured result of Iio *et al.* [6]. A constant temperature hot-wire anemometer was employed for the measurement. It is found that the time-averaged flow features, such as the behavior of the potential core and the change in the jet width, are favorably grasped by the DNS.

The time-averaged velocity component U distributes on the *y-z* cross-sections as shown in **Figure 8**. The sections are parallel to the nozzle exit. **Figure 8(a)** presents the DNS result. With increasing streamwise (x) distance, the jet spreads more in the y-direction. But it contracts gradually in the z-direction. These distributions agree well with the measurements depicted in **Figure 8(b)**.

The time-averaged velocity on the jet centerline U_m changes as a function of the distance from the nozzle as shown in **Figure 9**. The potential core exist until $x/w = 2.6$, where the velocity decreases markedly. Comparing with the measured result [6], the disappearing point of the potential core locates slightly upstream, and the velocity decay is smaller. It is considered that the decay of the centerline velocity is markedly affected by the appearance and collapse of the large-scale eddies [8]. The

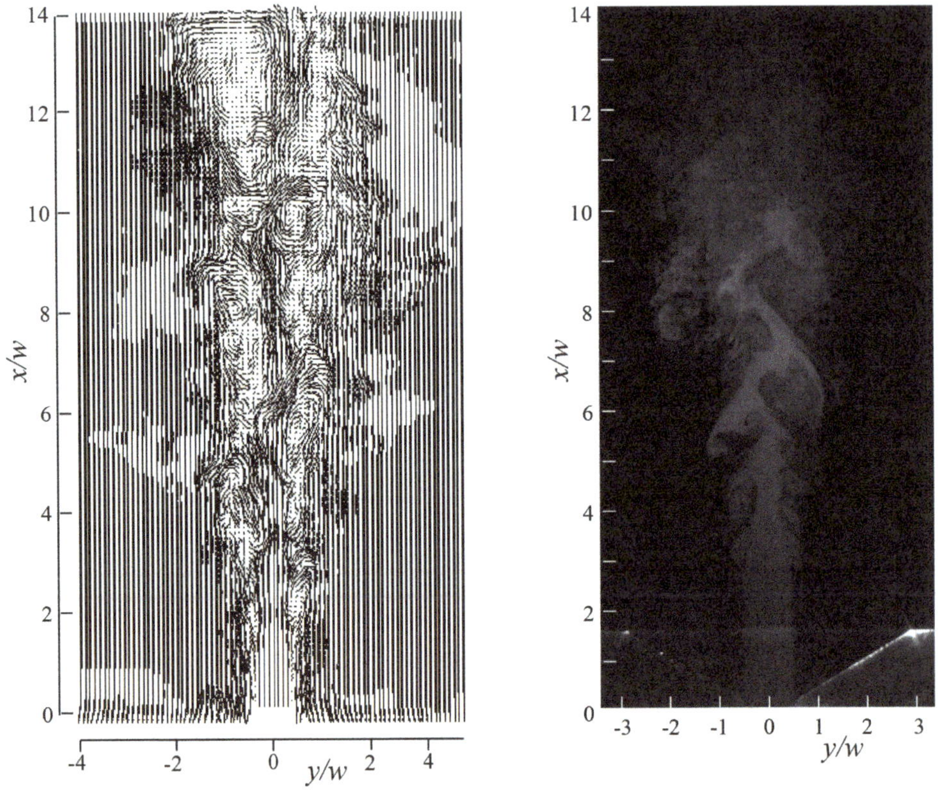

Figure 5. Flow field visualized on *x-y* cross-section at *z* = 0. (a) DNS; (b) Experiment.

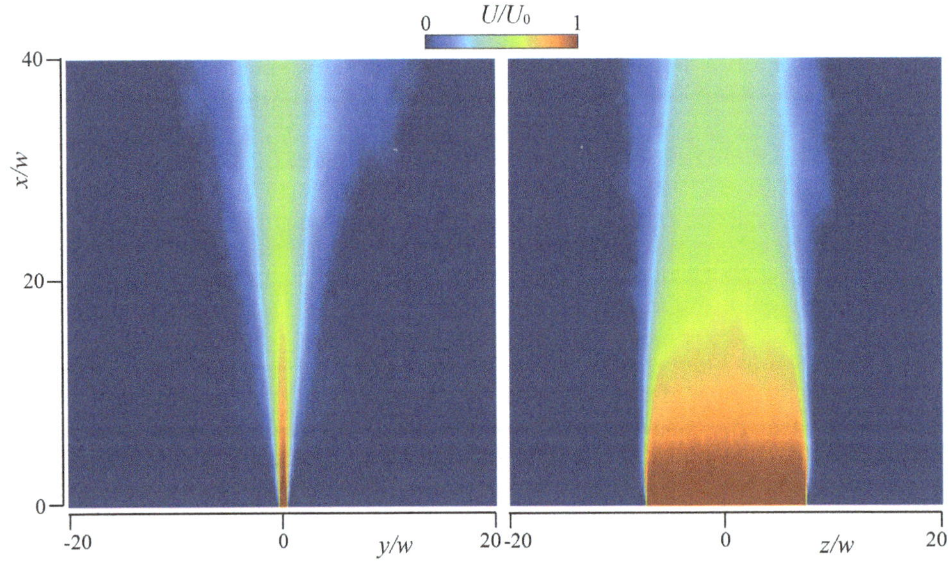

Figure 6. Distribution of time-averaged axial velocity. (a) *x-y* cross-section at *z* = 0; (b) *x-z* cross-section at *y* = 0.

behavior of the large-scale eddies depends on the velocity and turbulence intensity distributions within the nozzle exit. Some fluctuations exist in the velocity measured at the nozzle exit, and affect the velocity decay on the jet centerline through the occurrence and collapse of the large-scale eddies. The abovementioned discrepancy between the DNS and the measurement may be attributable to the fact that the boundary condition at the nozzle exit of the DNS does not accurately reproduce the velocity field of the corresponding experiments.

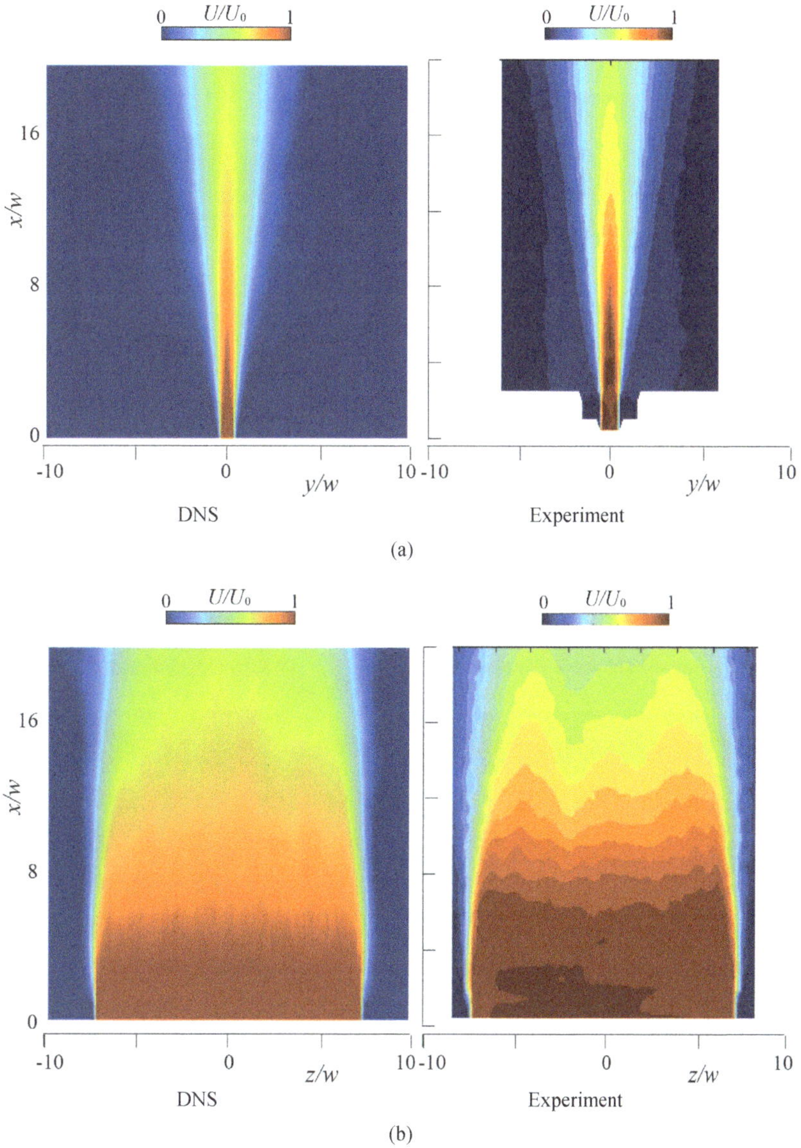

Figure 7. Comparison of time-averaged axial velocity near nozzle. (a) x-y cross-section at $z = 0$; (b) x-z cross-section at $y = 0$.

The half widths along the shorter and longer sides of the nozzle exit, b_y and b_z respectively, change as shown in **Figure 10**. b_y becomes larger as the streamwise distance increases. But b_z reduces slightly. These DNS results agree almost with the measured ones [6].

4.3. Turbulence Intensity

Figure 11 presents the jet axial component of the turbulence intensity u'_{rms}. The distributions on the jet central cross-sections are plotted. On the x-y cross-section parallel to the shorter side of the nozzle exit, u'_{rms} is larger at $x/w \leqq 10$ on the shear layers originating from the longer sides of the nozzle exit. On the x-z cross-section parallel to the longer side, u'_{rms} takes the maximum value on the shear layers from the shorter sides.

In **Figure 12**, the turbulence intensity u'_{rms} obtained by the DNS is compared with the measurement [6]. Though u'_{rms} of the DNS is larger near the nozzle as well as on the shear layers originating from the shorter side of the nozzle exit, the trend of the distribution agrees well with the measurement.

The turbulence intensity on the jet centerline varies as shown in **Figure 13**. The turbulence intensity of the DNS hardly coincides with the measurement [6] near the nozzle ($x/w \leqq 4.2$). This may be caused by the fact that the DNS ignores velocity fluctuations at the nozzle exit and the boundary layer on the nozzle wall, which exist in the corresponding experiment. But it should be noted that the turbulence intensity agrees well with the measurement at $x/w \geqq 4.2$ downstream of the disappearing point of the potential core.

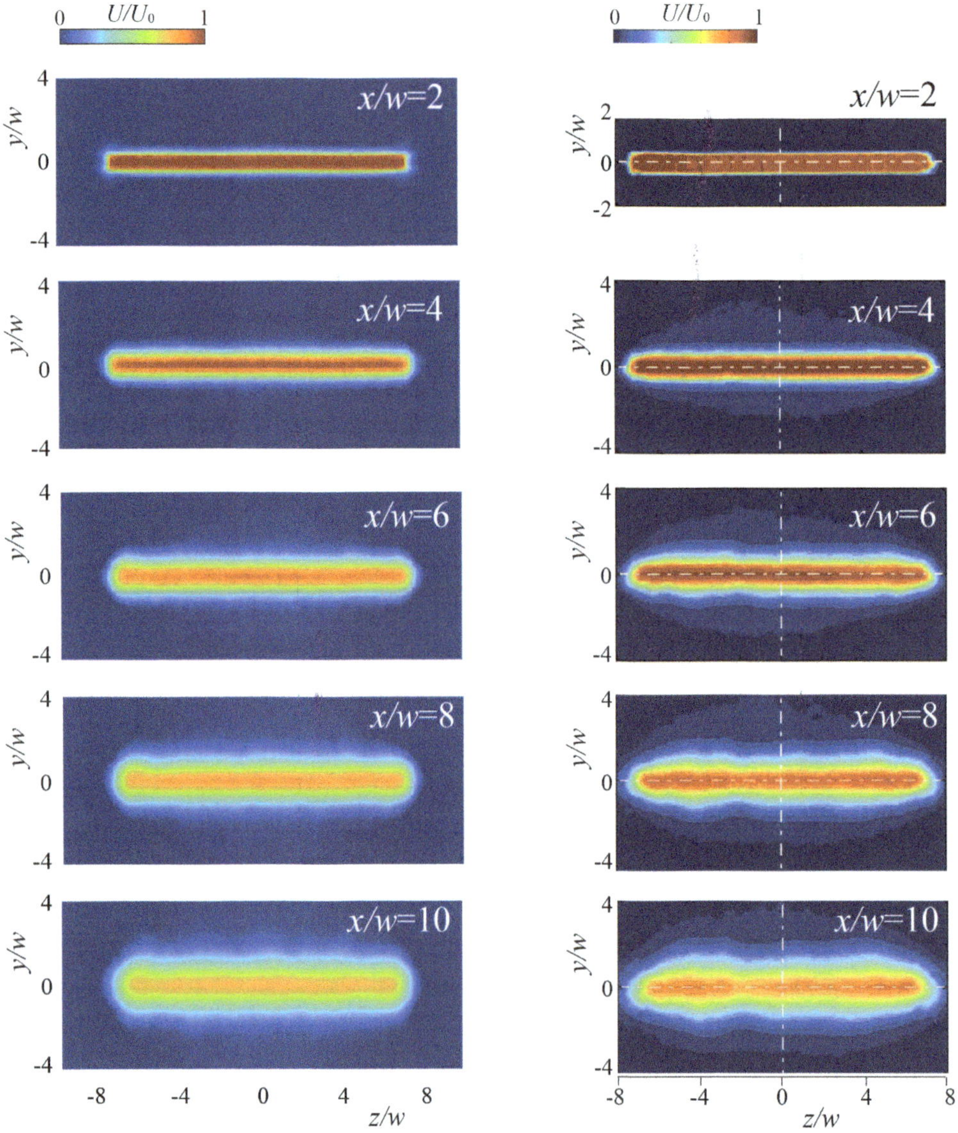

(a) (b)

Figure 8. Comparison of time-averaged axial velocity on *y-z* cross-sections. (a) DNS; (b) Experiment.

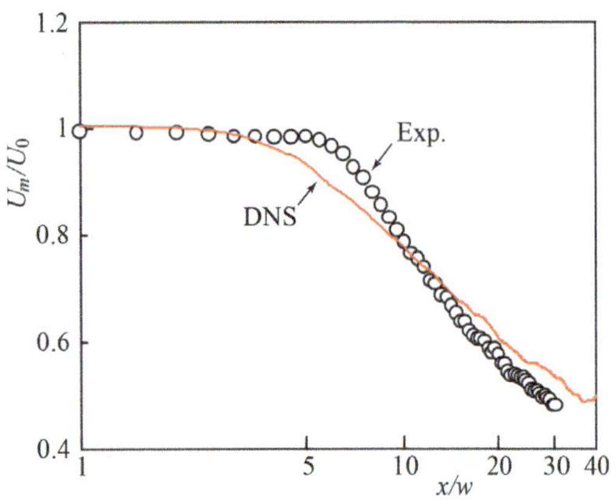

Figure 9. Axial evolution of jet centerline velocity.

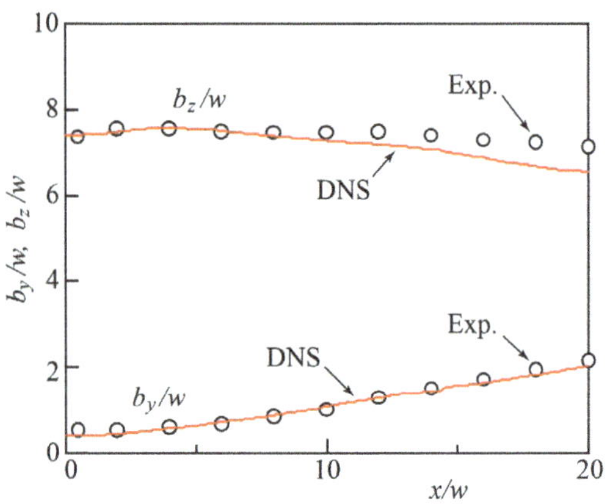

Figure 10. Axial evolution of jet half-width.

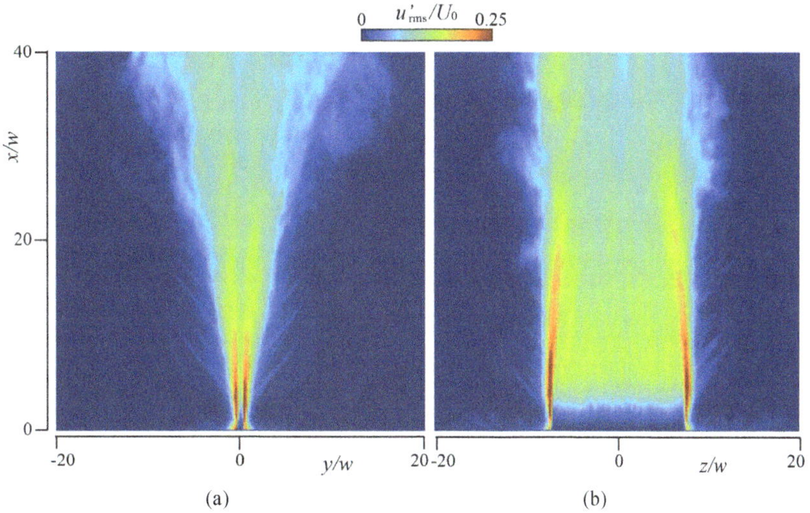

Figure 11. Distribution of axial turbulence intensity. (a) *x-y* cross-section at *z* = 0; (b) *x-z* cross-section at *y* = 0.

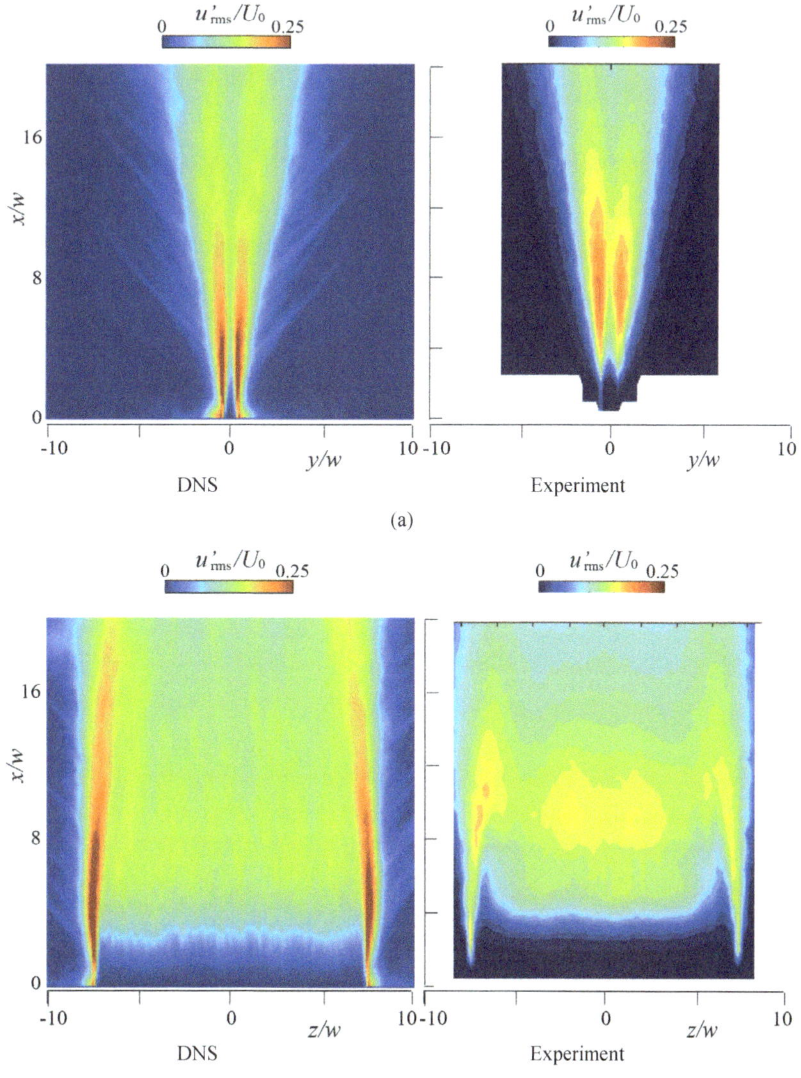

Figure 12. Comparison of axial turbulence intensity near nozzle. (a) *x-y* cross-section at *z* = 0; (b) *x-z* cross-section at *y* = 0.

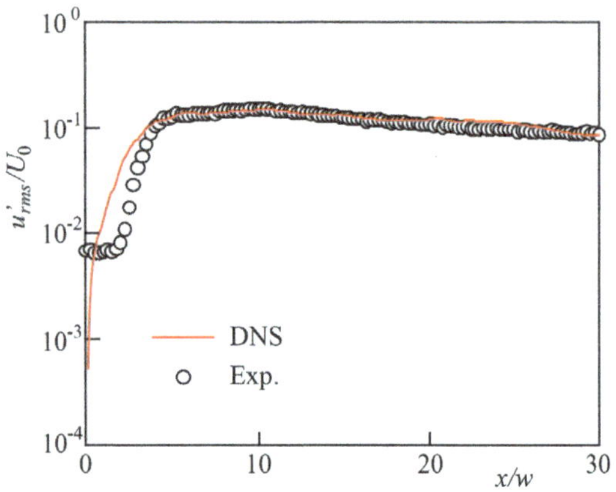

Figure 13. Axial evolution of axial turbulence intensity on jet centreline.

5. Conclusions

The DNS of a jet issuing from a nozzle having a rectangular cross-section was performed. The aspect ratio of the nozzle cross-section is 15, and the Reynolds number based on the shorter side length of the nozzle exit is 6700. The VIC method, of which computational accuracy has been heightened by the authors in a prior study, was used for the DNS. Staggered grids were employed to ensure consistency among the discretized equations, and to prevent numerical oscillations of the solution. A correction method of the vorticity was used so that the resultant vorticity field satisfies the solenoidal condition.

The DNS results, such as the mean velocity and the turbulence intensity, were favorably compared with the measured ones. The behavior of the large-scale eddies as well as the development of the turbulent flow was also confirmed to agree with the measurement. These indicated that the authors' VIC method is successfully employed for the DNS of rectangular jet.

REFERENCES

[1] I. P. Christiansen, "Numerical Simulation of Hydrodynamics by the Method of Point Vortices," *Journal of Computational Physics*, Vol. 13, No. 3, 1973, pp. 363-379.

[2] G.-H. Cottet and P. Poncet, "Advances in Direct Numerical Simulations of 3D Wall-Bounded Flows by Vortex-in-Cell Methods," *Journal of Computational Physics*, Vol. 193, No. 1, 2003, pp. 136-158.

[3] R. Cocle, G. Winckelmans and G. Daeninck, "Combining the Vortex-in-Cell and Parallel Fast Multipole Methods for Efficient Domain Decomposition Simulations," *Journal of Computational Physics*, Vol. 227, No. 21, 2008, pp. 9091-9120.

[4] P. Chatelain, A. Curioni, M. Bergdorf, D. Rossinelli, W. Andreoni and P. Koumoutsakos, "Billion Vortex Particle Direct Numerical Simulations of Aircraft Wakes," *Computer Methods in Applied Mechanics and Engineering*, Vol. 197, No. 13, 2008, pp. 1296-1304.

[5] T. Uchiyama, Y. Yoshii and H. Hamada, "Direct Numerical Simulation of a Turbulent Channel Flow by an Improved Vortex in Cell Method," *International Journal of Numerical Methods for Heat and Fluid Flow*, 2013, in Press.

[6] S. Iio, K. Hirashita, Y. Katayama, Y. Haneda, T. Ikeda and T. Uchiyama, "Jet Flapping Control with Acoustic Excitation," *Journal of Flow Control, Measurement & Visualization*, Vol. 1, No. 2, 2013, pp. 49-56.

[7] G.-H. Cottet and P. D. Koumoutsakos, "Vortex Methods: Theory and Practice," Cambridge University Press, Cambridge, 2000.

[8] S. Yuu, T. Nishioka and T. Umekage, "Direct Numerical Simulation of Three-Dimensional Navier Stokes Equations for a Slit Nozzle Free Jet and Experimental Verification," *JSME International Journal, Series B*, Vol. 36, No. 1, 1993, pp. 113-120.

MHD Transient Flow with Hall Current Past an Accelerated Horizontal Porous Plate in a Rotating System

Nazibuddin Ahmed[1], Jiwan Krishna Goswami[2], Dhruba Prasad Barua[1]
[1]Department of Mathematics, Gauhati University, Guwahati, India
[2]Department of Mathematics, ADP College, Nagaon, India

ABSTRACT

An exact solution to the problem of an MHD transient flow with Hall current past a uniformly accelerated horizontal porous plate in a rotating system has been presented. The dimensionless governing equations of the flow problem are solved by Laplace transform technique in closed form. A uniform magnetic field is assumed to be applied transversely to the direction of the flow. The expressions for velocity fields and skin-frictions are obtained in non-dimensional form. The primary and secondary velocity distributions and skin-frictions at the plate due to primary and secondary velocity field are demonstrated graphically and the effects of the different parameters namely, rotational parameter, Hartmann number, Hall parameter and acceleration parameter are discussed and the results are physically interpreted.

Keywords: MHD; Rotating System; Transient; Hall Current; Mass Transfer; Skin-Friction

1. Introduction

Many engineering problems are susceptible to MHD analysis. The study of MHD flow problems has achieved remarkable interest due to its application in MHD generators, MHD pumps and MHD flow meters etc. The study of effects of magnetic field on free convection flow is important in liquid metals, electrolytes and ionized gases. Geophysics encounters MHD phenomena in interactions of conducting fluids and magnetic fields. The rotating flow of an electrically conducting fluid in presence of magnetic field has got its importance in Geophysical problems. The study of rotating flow problems are also important in the solar physics dealing with the sunspot development, the solar cycle and the structure of rotating magnetic stars. It is well known that a number of astronomical bodies possess fluid interiors and magnetic fields. Changes that take place in the rate of rotation, suggest the possible importance of hydro magnetic spin-up. The general theory of rotating fluids has received growing interest during last decade because of its application in Cosmic and Geophysical science. In this regard, we may cite the works done by Raptis [1], Singh [2,3], Alam *et al.* [4] and Debnath [5].

MHD in the present form is due to the pioneer contri bution of several notable authors like Alfven [6], Cowling [7], Ferraro and Pulmpton [8] etc. It was emphasized by Cowling (1975) that when the strength of the applied magnetic field is sufficiently large, Ohm's law needs to be modified to include Hall current. The Hall effect is due merely to the sideways magnetic force on the drifting free charges. The electric field has to have a component transverse to the direction of the current density to balance this force. In many works of Plasma physics, it is not paid much attention to the effect caused due to Hall current. However, the Hall effect can not be completely ignored if the strength of the magnetic field is high and the number density of electrons is small as it is responsible for the change of the flow pattern of an ionized gas. Hall effect results in a development of an additional potential difference between opposite surfaces of a conductor for which a current is induced perpendicular to both the electric and magnetic field. This current is termed as Hall current. It was discovered in 1879 by Edwin Herbert Hall while working on his doctoral degree at the Johns Hopkins University in Baltimore, Maryland, USA. Pop [9], Kinyanjui *et al.* [10], Archrya *et al.* [11]

and Ahmed and Kalita [12] etc. have presented some model studies on the effect of Hall current on MHD convection flow because of its possible application in the problems of MHD generators and Hall accelerators. An unsteady MHD free convective flow past a vertical porous plate immersed in a porous medium with Hall current, thermal diffusion and heat transfer have been studied by Ahmed *et al.* [13]. Recently, Ahmed and Sarmah [14] have carried out an investigation of MHD transient flow past an impulsively started infinite horizontal porous plate in a rotating system with Hall current.

Due to the importance of studying MHD flow problems in rotating fluid with Hall current, we have proposed in the present paper to investigate an unsteady MHD transient flow with Hall currents past a uniformly accelerated porous plate in a rotating system.

2. Basic Equations

The equations governing the motion of an incompressible viscous electrically conducting fluid in a rotating system in presence of a magnetic field are as under:

Equation of continuity:

$$\nabla \cdot \boldsymbol{q} = 0 \tag{1}$$

Momentum equation:

$$\rho\left[\frac{\partial \boldsymbol{q}}{\partial t'} + 2\boldsymbol{\Omega}\times\boldsymbol{q} + \boldsymbol{\Omega}\times(\boldsymbol{\Omega}\times\boldsymbol{r}) + (\boldsymbol{q}\cdot\nabla)\boldsymbol{q}\right] = -\nabla p + \boldsymbol{J}\times\boldsymbol{B} + \mu\nabla^2\boldsymbol{q} \tag{2}$$

Kirchhoff's first law:

$$\nabla \cdot \boldsymbol{J} = 0 \tag{3}$$

General Ohm's law:

$$\boldsymbol{J} + \frac{\omega_e \tau_e}{B_0}(\boldsymbol{J}\times\boldsymbol{B}) = \sigma\left[\boldsymbol{E} + \boldsymbol{q}\times\boldsymbol{B} + \frac{1}{e\eta_e}\nabla p_e\right] \tag{4}$$

Gauss's law of magnetism:

$$\nabla \cdot \boldsymbol{B} = 0 \tag{5}$$

where $\boldsymbol{q}$ is the velocity vector, $\boldsymbol{\Omega}$ the angular velocity of the fluid, $\boldsymbol{r}$ the position vector of the fluid particle P considered, ρ the fluid density, p the pressure, $\boldsymbol{J}$ the current density, $\boldsymbol{B}$ the magnetic induction vector, μ the co-efficient of viscosity, σ the electrical conductivity, t' the time, B_0 the strength of the applied magnetic field, ω_e the electron frequency, τ_e the electron collision time, e the electron charge, η_e the number density of electron, p_e the electron pressure, $\boldsymbol{E}$ the electric field, $2\boldsymbol{\Omega}\times\boldsymbol{q}$ is the Coriolis acceleration, $\boldsymbol{\Omega}\times(\boldsymbol{\Omega}\times\boldsymbol{r})$ is the centripetal acceleration and the other symbols have their usual meanings and the other symbols have their usual meanings.

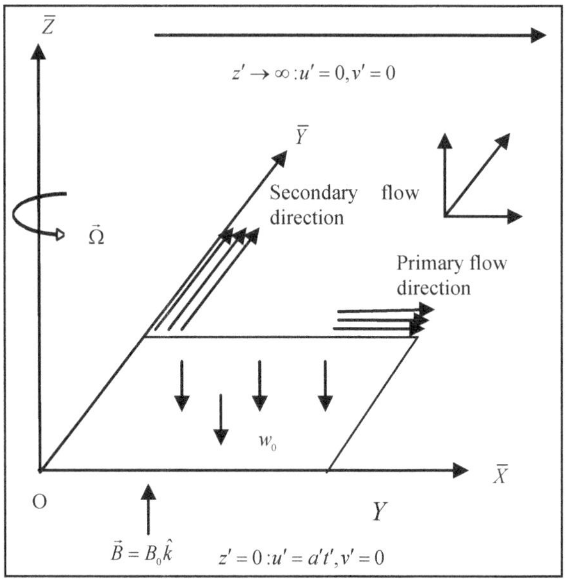

Flow configuration

We now consider an unsteady flow of an incompressible viscous electrically conducting fluid past a suddenly started infinite horizontal porous plate relative to a rotating system with constant suction in presence of a uniform transverse magnetic field taking into account the effect of Hall current. Our investigation is restricted to the following assumptions:

- All the fluid properties are constants and the buoyancy force has no effect on the flow.
- The plate is electrically non-conducting.
- The entire system is rotating with angular velocity $\boldsymbol{\Omega}$ about the normal to the plate and $|\boldsymbol{\Omega}|$ is so small that $|\boldsymbol{\Omega}\times(\boldsymbol{\Omega}\times\boldsymbol{r})|$ can be neglected.
- The magnetic Reynolds number is so small that the induced magnetic field can be neglected.
- p_e is constant.
- $\boldsymbol{E} = \boldsymbol{0}$.

Initially the plate and the fluid were rotating in unison with a constant angular velocity $\boldsymbol{\Omega}$ about the normal to the plate. At time $t' > 0$, the plate is moved in its own plane relative to the rotating system with acceleration a'. We introduce a coordinate system (x', y', z') with X-axis horizontally in the direction of the plate velocity, Y-axis horizontally perpendicular to the direction of the plate velocity and Z-axis along the normal to the plate which is the axis of rotation. Let $\boldsymbol{q} = \hat{i}u' + \hat{j}v' + \hat{k}w'$ be the fluid velocity, $\boldsymbol{J} = J'_x\hat{i} + J'_y\hat{j} + J'_z\hat{k}$ be the current density at the point $P(x', y', z', t')$ and $\boldsymbol{B} = B_0\hat{k}$ be the applied magnetic field, $\hat{i}, \hat{j}, \hat{k}$ being the unit vectors along X-axis, Y-axis and Z-axis respectively. As the plate is infinite in X-direction and Y-direction, therefore all the quantities except possibly the pressure are independent of x' and y'.

The Equation (1) gives

$$\frac{\partial w'}{\partial z'} = 0 \tag{6}$$

which is trivially satisfied by

$$\overline{w} = -w_0' \tag{7}$$

Therefore the velocity vector $\boldsymbol{q}$ is given by

$$\boldsymbol{q} = u'\hat{i} + v'\hat{j} - w_0'\hat{k} \tag{8}$$

The Equation (5) is satisfied by

$$\boldsymbol{B} = B_0\hat{k} \tag{9}$$

The Equation (3) reduces to

$$\frac{\partial J_z'}{\partial z'} = 0$$

which shows that

$$J_z' = 0 \tag{10}$$

(as the plate is electrically non-conducting).

Hence the current density is given by

$$\boldsymbol{J} = J_x'\hat{i} + J_y'\hat{k} \tag{11}$$

Under the above assumptions, the Equation (4) takes the form:

$$\boldsymbol{J} + \frac{m}{B_0}(\boldsymbol{J}\times\boldsymbol{B}) = \sigma(\boldsymbol{q}\times\boldsymbol{B}) \tag{12}$$

where $m = \omega_e\tau_e$ is the Hall parameter.

The Equations (8)-(12) yield,

$$J_x' = \frac{\sigma B_0(v' + mu')}{1+m^2} \tag{13}$$

$$J_y' = \frac{\sigma B_0}{1+m^2}(mv' - u') \tag{14}$$

With the foregoing assumptions and under the usual boundary layer approximation the Equation (2) reduces to

$$\frac{\partial u'}{\partial t'} - w_0'\frac{\partial u'}{\partial z'} - 2\Omega' v' = \upsilon\frac{\partial^2 u'}{\partial z'^2} + \frac{\sigma B_0^2(mv' - u')}{\rho(1+m^2)} \tag{15}$$

$$\frac{\partial v'}{\partial t'} - w_0'\frac{\partial v'}{\partial z'} + 2u'\Omega' = \upsilon\frac{\partial^2 v'}{\partial z'^2} - \frac{\sigma B_0^2(mu' + v')}{\rho(1+m^2)} \tag{16}$$

with $\frac{\partial p'}{\partial z'} = 0$.

where w_0' is the constant suction velocity and υ is the kinematic viscosity

The relevant initial and boundary conditions are

$$u' = 0, v' = 0 \text{ for } t' \le 0, \forall z \tag{17}$$

$$\left.\begin{aligned} u' = a't', v' = 0 \text{ at } z' = 0; \\ u' = 0, v' = 0 \text{ at } z' \to \infty \end{aligned}\right\} \forall t' > 0 \tag{18}$$

We introduce the following non-dimensional variables and parameters:

$$z = \frac{w_0' z'}{\upsilon}, t = \frac{w_0'^2 t'}{\upsilon}, u = \frac{u'}{U_0'}, v = \frac{v'}{U_0'},$$

$$M = \frac{\sigma B_0^2 \upsilon}{\rho w_0'^2}, \Omega = \frac{2\Omega' \upsilon}{w_0^2}, a = \frac{a' v}{w_0^3}$$

The non-dimensional form of the Equations (15) and (16) are

$$\frac{\partial u}{\partial t} - \frac{\partial u}{\partial z} - \Omega v = \frac{\partial^2 u}{\partial z^2} + \frac{M}{1+m^2}(mv - u) \tag{19}$$

$$\frac{\partial v}{\partial t} - \frac{\partial v}{\partial z} + \Omega u = \frac{\partial^2 v}{\partial z^2} - \frac{M}{1+m^2}(mu + v) \tag{20}$$

Subject to the initial and boundary conditions:

$$u = 0, v = 0 \text{ for } t \le 0, \forall z \tag{21}$$

$$\left.\begin{aligned} u = at, v = 0 \text{ at } z = 0 \\ u = 0, v = 0 \text{ at } z \to \infty \end{aligned}\right\} \forall t > 0 \tag{22}$$

3. Method of Solution

Let us introduce the complex variable q defined by $q = u + \mathrm{i}v$ where $\mathrm{i}^2 = -1$.

The non-dimensional forms of the equation governing the flow can be rewritten as follows:

$$\frac{\partial q}{\partial t} = \frac{\partial^2 q}{\partial z^2} + \frac{\partial q}{\partial z} - Aq \tag{23}$$

where $A = \mathrm{i}\Omega + \frac{M(1+\mathrm{i}m)}{1+m^2}$

Subject to the boundary conditions:

$$q = 0 \text{ for } t \le 0 \tag{24}$$

$$\left.\begin{aligned} q = at,\ v = 0 \text{ at } z = 0 \\ q = 0,\ v = 0 \text{ at } z \to \infty \end{aligned}\right\} \forall t \tag{25}$$

On applying Laplace Transform to the Equation (23), the following ordinary differential equation is derived

$$\frac{\mathrm{d}^2\overline{q}}{\mathrm{d}z^2} + \frac{\mathrm{d}\overline{q}}{\mathrm{d}z} - (A + s)\overline{q} = 0 \tag{26}$$

with relevant boundary conditions :

$$\overline{q} = \frac{a}{s^2} \text{ at } z = 0 \tag{27}$$

$$\overline{q} = 0 \text{ at } z \to \infty \tag{28}$$

The solution of the Equation (26) under the conditions

(27) and (28) is

$$\bar{q} = \frac{a}{s^2} e^{-\xi z} \quad (29)$$

Taking inverse Laplace transforms of the Equation (29) we derive the following:

$$q = a e^{-\eta\sqrt{t}} \left[\left(\frac{t}{2} - \eta \frac{\sqrt{t}}{2\sqrt{b}} \right) e^{-2\eta\sqrt{bt}} \operatorname{erfc}\left(\eta - \sqrt{bt}\right) + \left(\frac{t}{2} + \eta \frac{\sqrt{t}}{2\sqrt{b}} \right) e^{2\eta\sqrt{bt}} \operatorname{erfc}\left(\eta + \sqrt{bt}\right) \right] \quad (30)$$

where, $b = \frac{4A+1}{4}$, $\xi = \frac{1+\sqrt{1+4(A+s)}}{2}$, $\eta = \frac{z}{2\sqrt{t}}$

and $A = i\Omega + \frac{M(1+im)}{1+m^2}$

4. Skin Friction

The non-dimensional skin-friction at the plate is given by

$$\tau = -\frac{1}{2\sqrt{t}} \left(\frac{\partial q}{\partial \eta} \right)_{\eta=0}$$

$$= \frac{a}{2} \left[\frac{1}{\sqrt{b}} \operatorname{erf}\left(\sqrt{bt}\right) + 2t\sqrt{b} \operatorname{erf}\left(\sqrt{bt}\right) + \frac{2t}{\sqrt{\pi}} e^{-bt} + t \right]$$

$$= \tau_x + i\tau_y$$

where τ_x and τ_y are the skin-frictions at the plate due to the primary and the secondary velocity fields. The expressions for τ_x and τ_y are obtained but not presented here for the sake of brevity.

5. Results and Discussion

In order to get the physical insight into the problem we have carried out numerical calculations for the representative velocity field and skin-friction at the plate for different values of the physical parameters involved and these values have been demonstrated in different graphs. Our investigation is restricted to *t* equal to 1 and the other parameters namely, rotational parameter (Ω), Hartmann number (M), Hall parameter (m) and accelerating parameter (a) has been considered arbitrarily.

Figure 1 depicts the variation in skin-friction τ_x due to the primary velocity field versus Hall parameter m for different values of rotational parameters Ω. It is noticed that τ_x decreases with increasing values of Hall parameter m whereas a rise in the values rotational parameter Ω results a growth in τ_x.

The influence of Hartmann number M on skin-friction τ_x against Hall parameter m is presented in **Figure 2**. It is observed that a growth in the Hartmann number M leads to an increase of τ_x. Moreover, it is seen

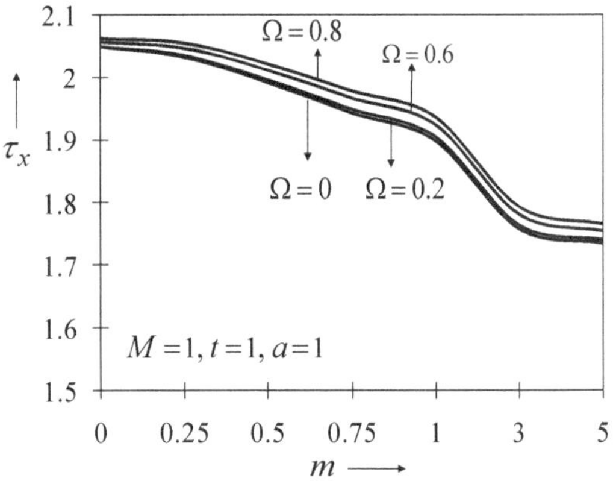

Figure 1. The behavior of τ_x versus m under the effect of Ω.

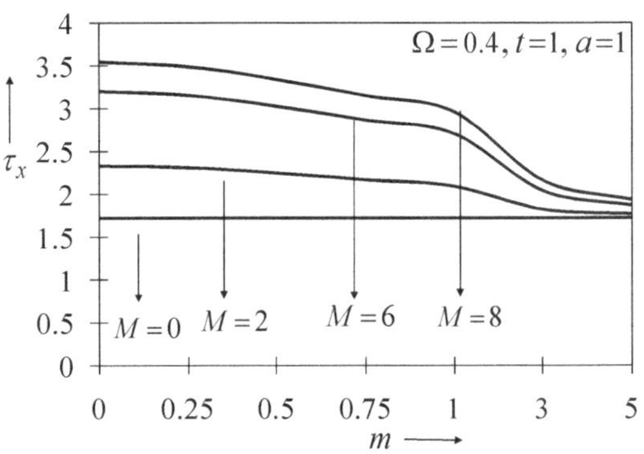

Figure 2. The behavior of τ_x versus m **under the effect of M.**

that τ_x decreases very slowly and steadily as the Hall parameter m rises. In other words, an increase in the Hall parameter or a decrease in Hartmann number results a decrease in drag force at the plate due to the primary velocity.

Figure 3 demonstrates the nature of τ_x against Hall parameter m under the effect of accelerating parameter a. **Figure 3** exhibits a substantial growth in τ_x with increasing values of accelerating parameter a. It is also seen that τ_x is unalterable for small values accelerating parameter or in absence of plate acceleration. That is, τ_x (drag force) due to the primary velocity is undisturbed whenever the plate is at rest as well for the slow movement of the plate. However, for higher values of the accelerating parameter (a) skin-friction τ_x decreases gradually with increasing values of Hall parameter m.

The behaviour of τ_y (drag force per unit area) due to the secondary velocity field versus Hall parameter m under the effects of Hartmann number M, rotational parameter Ω and acceleration parameter a respectively are depicted in **Figures 4-6**. It is noticed that a rise in M, Ω and accelerating parameter (a) results a growth in τ_y. Moreover τ_y is undisturbed by

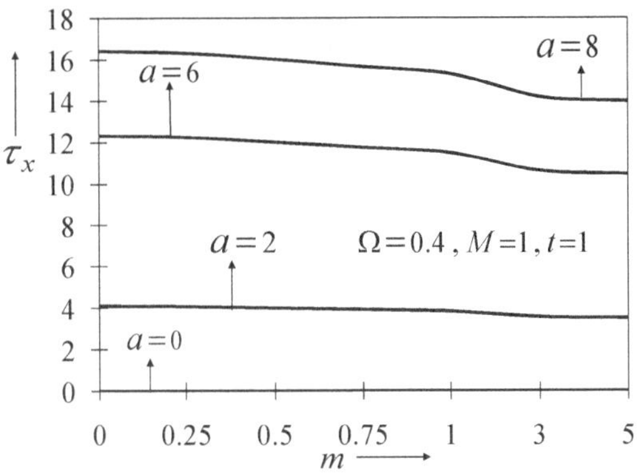

Figure 3. The behavior of τ_x versus m under the effect of a.

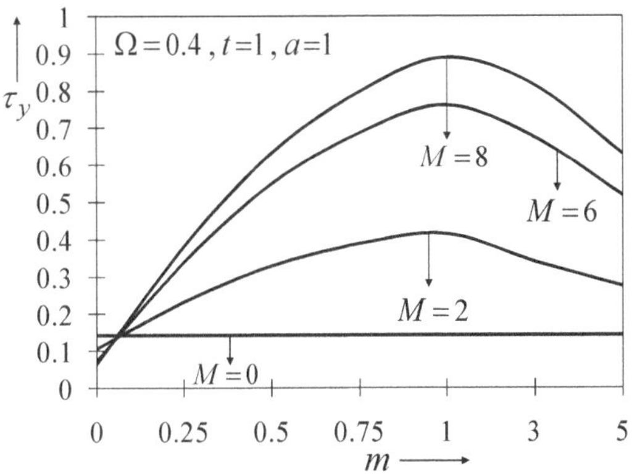

Figure 4. The behavior of τ_y under the effect of M versus m.

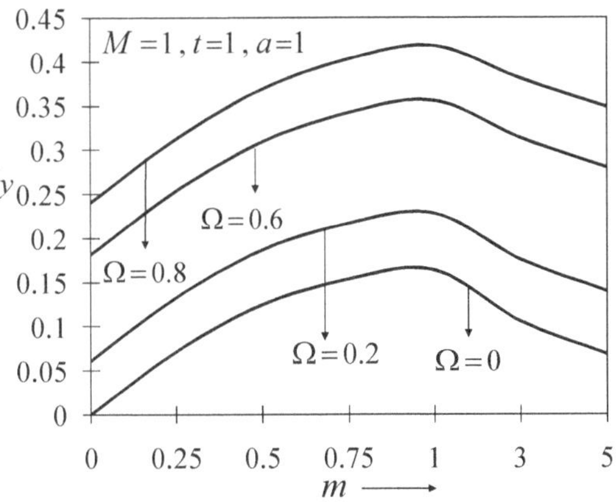

Figure 5. The behavior of τ_y versus m under the effect of Ω.

Hall effect in absence of magnetic field and acceleration of the plate. The same figures also indicate the growth of τ_y for small and moderate values of Hall parameter $(m \le 1)$ and afterwards a reversal trend on τ_y is noticed.

The effect of Hartmann number M on primary velocity field is presented in **Figure 7**. This figure clearly

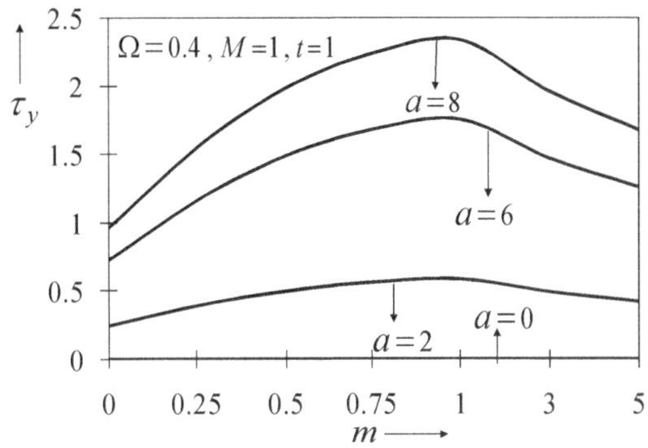

Figure 6. The behavior of τ_y versus m under the effect of a.

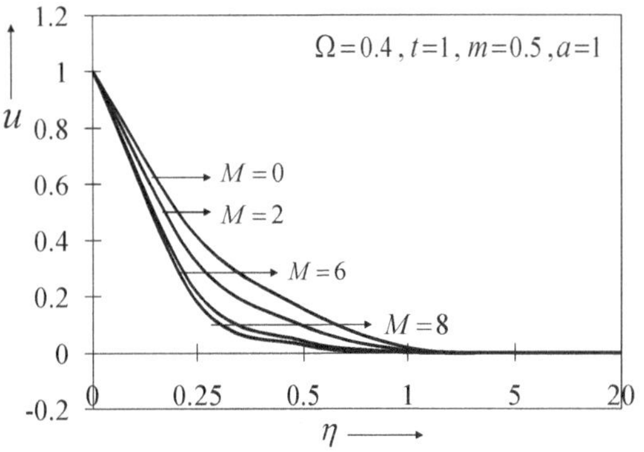

Figure 7. The variation of u versus η under the effect of M.

shows that the primary velocity sharply decreases in a thin layer of the fluid adjacent to the plate surface and then it decreases slowly and steadily to its minimum value $u=0$ as $\eta \to \infty$. The same figure also indicates that an increase in the Hartmann number M has an inhibiting effect on the primary velocity u. The primary fluid velocity u gets continuously reduced with increasing M. That is the application of the transverse magnetic field retards the primary motion. This phenomenon has an excellent agreement with the physical fact that the Lorentz force generated in present flow model due to interaction of the transverse magnetic field and the fluid velocity acts as a resistive force to the fluid flow which serves to decelerate the flow. As such the magnetic field is an effective regulatory mechanism for the flow regime.

Further, it is worthwhile to mention that the effect of Hartmann number M on primary velocity u is negligible for large η. In other words the fluid motion far away from the plate is undisturbed due to imposition of the magnetic field.

Figure 8 indicates that an increase in the plate acceleration causes the primary flow to retard comprehendsively near the plate. That is the role of accelerating parameter a on primary velocity field is almost similar to the role of Hartmann number M. However, in absence of acceleration a very small growth of primary velocity is noticed in a thin fluid region adjacent to the plate surface and afterwards it falls slowly and steadily to its free stream value as $\eta \to \infty$.

Figures 9-12 demonstrate the behaviour of the secondary velocity field (v) under the effects of rotational parameter (Ω), Hartmann number (M), Hall parameter (m) and accelerating parameter (a) respectively. From these figures, it is interesting to observe that the magnitude of v increases from its zero value at the plate surface into a fluid region adjacent to

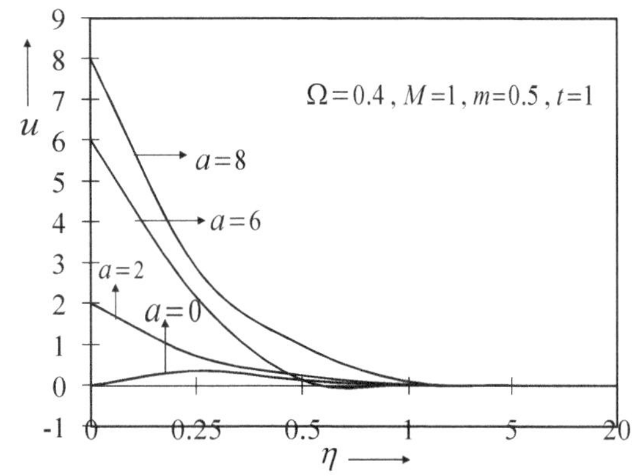

Figure 8. The variation of u versus η under the effect of a.

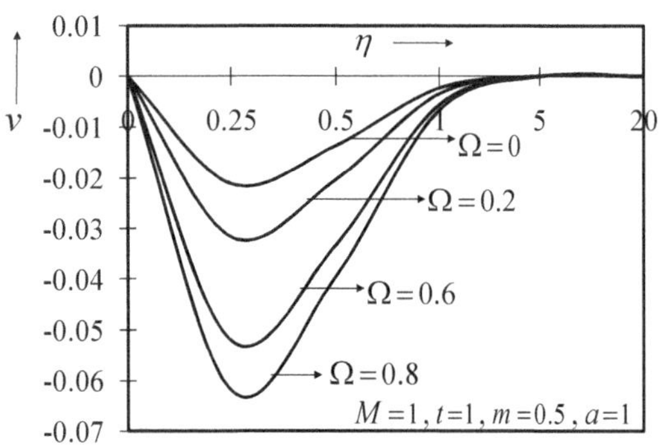

Figure 9. The behavior of v under the effect of Ω, versus η.

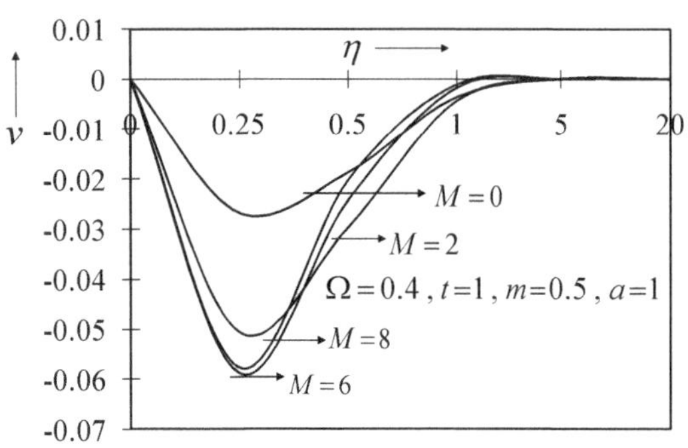

Figure 10. The behavior of v under the effect of M, versus η.

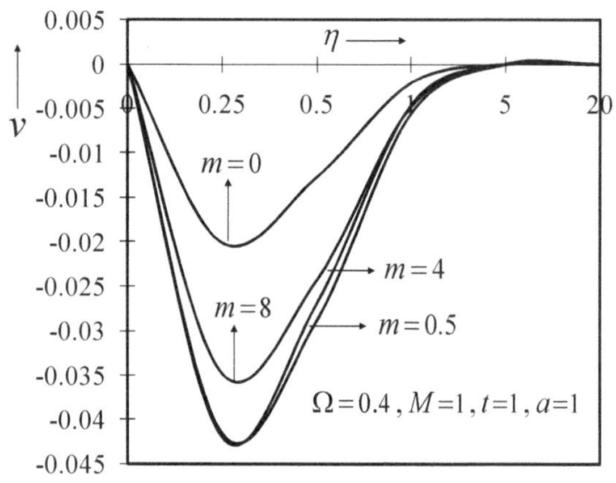

Figure 11. The behavior of v under the effect of m, versus η.

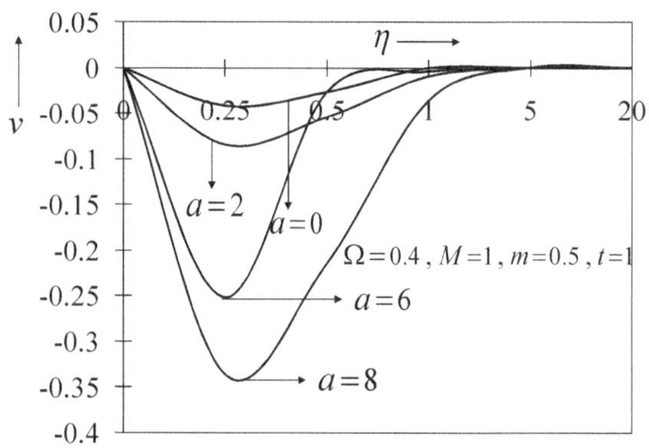

Figure 12. The behavior of v under the effect of a, versus η.

the plate and then it decreases to zero value (attainable at possibly large distances from the plate surface). This clearly agrees with the boundary conditions of the present flow problem. **Figures 9** and **12** indicate that a rise in the values of rotational parameter Ω and accelerating parameter a causes a growth in the magnitude of secondary velocity. However, this effect seems to be negligible in the fluid region far away from the plate. Further, it is worthwhile to mention that for small and moderate values of Hartmann number M and Hall parameter m, the behaviour of secondary velocity field is identical to its behaviour under the effects of Ω and accelerating parameter a whereas, for higher values of Hartmann number M and Hall parameter m the behaviour of secondary velocity v takes a reverse trend.

It is inferred from **Tables 1** and **2** that a rise in rotational parameter (Ω), and Hall parameter (m) results in a growth in the primary velocity u.

6. Conclusions

- The main flow velocity u decreases with an imposition of magnetic field.

Table 1. Behavior of u under the effect of Ω, versus η. $(M = 1, t = 1, m = 0.5, a = 1)$.

η	$u(\Omega = 0)$	$u(\Omega = 0.6)$	$u(\Omega = 0.8)$
0	1	1	1
0.25	0.3611287	0.3550634	0.3519596
0.5	0.1239504	0.1185953	0.1158712
1	0.0121564	0.0108471	0.0101901
5	0	0	0

Table 2. Behavior of u under the effect of m, versus η. $(M = 1, t = 1, \Omega = 0.4, a = 1)$.

η	$u(m = 0)$	$u(m = 4)$	$u(m = 8)$
0	1	1	1
0.25	0.3508695	0.4021751	0.4063124
0.5	0.1174215	0.1507741	0.1539457
1	0.0110595	0.0168414	0.0175416
5	0	0	0

- An increase in accelerating parameter a results in a remarkable growth in the main flow velocity u in a thin layer adjacent to the plate.
- The drag force due to primary velocity (τ_x) rises due to rotation or imposition of the transverse magnetic field or the acceleration of the plate.
- The skin friction τ_y (drag force due to secondary velocity) shows a growth for increasing each of Hartmann number M, rotational parameter Ω as well as accelerating parameter a.

7. Acknowledgements

The Authors are grateful to the CSIR-HRDG (India) for sponsoring this work through their Research-Grant-In-Aid. No. 25 (0209)/12/EMR-I.

REFERENCES

[1] A. Raptis, "Mass Transfer and Free Convection through a Porous Medium by the Presence of a Rotating Fluid," *International Communications in Heat and Mass Transfer*, Vol. 10, No. 2, 1983, pp. 141-146.

[2] A. K. Singh, "Hydromagnetic Free-Convection Flow Past an Impulsively Started Vertical Plate in a Rotating Fluid," *International Communications in Heat and Mass Transfer*, Vol. 11, No. 4, 1984, pp. 339-406.

[3] A. K. Singh, "MHD Free Convection Flow in the Stokes Problems for a Vertical Porous Plate in a Rotating System," *Astrophysics and Space Science*, Vol. 95, No. 2, 1983, pp. 283-289.

[4] M. Alam, M. A. Sattar and M. Mansur, "Similarity Solution of Steady MHD Free Convection and Mass Transfer Flow with Thermal Diffusion in a Rotating System," *Dhaka University Journal of Science*, Vol. 49, No. 2, 2001, p. 147.

[5] L. Debnath, "Exact Solutions of the Unsteady Hydrodynamic and Hydro Magnetic Boundary Layer Equations in a Rotating Fluid System," *ZAMM*, Vol. 55, No. 7-8, 1975, pp. 431-438.

[6] H. Alfven, "Existence of Electromagnetic-Hydrodynamic Waves," *Nature*, Vol. 150, No. 3805, 1942, pp. 405-406.

[7] T. G. Cowling, "Magneto Hydrodynamics," Wiley Inter Science, New York, 1957.

[8] V. C. A. Ferraro and C. Pulmpton, "An Introduction to Magneto Fluid Mechanics," Clarandon Press, Oxford, 1966.

[9] I. Pop, "The Effect of Hall Currents on Hydro Magnetic Flow near an Accelerated Plate," *Journal of Mathematical and Physical Sciences*, Vol. 5, 1971, pp. 375-379.

[10] M. Kinyanjui, J. K. Kwanza and S. M. Uppal, "Magneto Hydrodynamic Free Convection Heat and Mass Transfer of a Heat Generating Fluid Past an Impulsively, Started Infinite Vertical Porous Plate with Hall Current and Radiation Absorption," *Energy Conversion and Management*, Vol. 42, No. 8, 2001, pp. 917-931.

[11] M. Acharya, G. C. Das and L. P. Singh, "Hall Effect with Simultaneous Thermal and Mass Diffusion on Unsteady Hydro Magnetic Flow near an Accelerated Vertical Plate," *Indian Journal of Physics B*, Vol. 75, No. 1, 2001, pp. 68-70.

[12] N. Ahmed and D. Kalita, "Transient MHD Free Convection from an Infinite Vertical Porous Plate in a Rotating Fluid with Mass Transfer and Hall Current," *Journal of Energy, Heat and Mass Transfer*, Vol. 33, No. 1, 2011, pp. 271-292.

[13] N. Ahmed, H. Kalita and D. P. Barua, "Unsteady MHD Free Convective Flow Past a Vertical Porous Plate Immersed in a Porous Medium with Hall Current, Thermal Diffusion and Heat Transfer," *International Journal of Engineering, Science and Technology*, Vol. 2, No. 6, 2010, pp. 59-74.

[14] N. Ahmed and H. K. Sarmah, "MHD Transient Flow Past an Impulsively Started Horizontal Porous Plate in a Rotating System with Hall Current," *International Journal of Applied Mathematics and Mechanics*, Vol. 7, No. 2, 2011, pp. 1-15.

Units for Physical Properties

Symbol	Quantity	SI unit
a'	Acceleration of the plate	m/s^2
a	Dimensionless acceleration	
B_0	magnetic flux density, magnetic induction	Tesla
m	Hall parameter	
M	Hartmann number	
t'	time	s
t	Dimensionless time	
$(\bar{u},\bar{v},\bar{w})$	Velocity components	m/s
(u,v,w)	Dimensionless velocity components	
w_0	Suction velocity	m/s
$(\bar{x},\bar{y},\bar{z})$	Cartesian coordinates	m
Ω'	Angular velocity	s^{-1}
Ω	Dimensionless angular velocity	
ρ	Fluid density	kg/m^3
σ	Electrical conductivity	$(\text{ohm} \times \text{meter})^{-1}$
ν	Kinematic viscosity	m^2/s
(τ_x,τ_y)	Components of skin friction	

Study on Stack Effect of Stairwell by Numerical Model of Leakage Flow through Gap of Door

Jung-Yup Kim, Ji-Seok Kim
Fire Research Center, Korea Institute of Construction Technology, Goyang-Si, South Korea

ABSTRACT

Since stack effect that occurs in high-rise buildings has an effect on the indoor environment of the buildings, energy loss and smoke control in case of a fire, there is a need to conduct research on this. For an analysis of the stack effect, analysis methods on the leakage flow through gap of interior door shall be formulated. Until now, studies related to the gap leakage flow in buildings have mainly analyzed flow field and pressure in the buildings one-dimensionally using pressure difference-leakage flowrate relations of Orifice Equation and a network numerical analysis algorithm that assumes each compartment in the buildings as a single point. In this study, the Momentum Loss Model which enables pressure drop to be proportional to the flow velocity through the gap of door in computational domain of 3-dimensional numerical analysis was proposed to reflect the gap flow phenomenon effectively in 3-dimensional numerical analysis. Using the proposed model, 3-dimensional numerical analysis of the stack effect on the stairs in buildings was performed, and the effects of separation door and lobby between stair and accommodation on the stack effect were investigated.

Keywords: Stack Effect; Leakage Flow; Gap of Door; Momentum Loss Model; Separation Door; Numerical Analysis

1. Introduction

In buildings, there exist the gaps between compartments such as those of doors and windows, and a flow occurs through these gaps. Since the flow affects various heat flow phenomena that occur in buildings, analysis methods are needed for this. As for the stack effect, ventilation design, air tightness, energy analysis and smoke control in case of a fire, the gap flow has been considered as an important factor.

In particular, the stack effect has been found to affect many phenomena within high-rise buildings such as smoke flow in case of a fire, indoor environment, energy loss and elevator malfunction. The theoretical pressure difference due to the stack effect can be calculated by Equation (1).

$$\Delta P_{st} = g\Delta\rho(N-h) = g\rho_o(\Delta T/T_i)(N-h) \quad (1)$$

where g is the acceleration of gravity (m/s^2), h is the height of the measuring point (m), N is the neutral pressure level (m), T_i is the room temperature (K) and ρ_o is the outdoor air density (kg/m^3).

With respect to the stack effect in high-rise buildings, continuous research is being done internationally. The pressure fields in high-rise office buildings were presented and their characteristics were investigated using the index of Thermal Draft Coefficient [1]. Through the field measurements and analyses in two high-rise residential buildings, the characteristics of pressure distribution caused by stack effect in high-rise building were presented and solution for stack effect problems using "airlock doors" was proposed [2]. The architectural elements that affect the stack effect in a high-rise building were selected and the effect of the elements was quantitatively analyzed by simulation of network model [3].

In the analysis that includes gap leakage flow of the building that has a large influence on the stack effect, pressure difference-leakage flowrate relations around the gap of door is usually represented by Equation (2).

$$Q = f(\Delta p) \quad (2)$$

where Q is the leakage flow rate and Δp is the pressure difference before and after the gap respectively.

In relation to the pressure difference and leakage flowrate relations, a lot of researches were conducted in the 70s to 90s, focusing on derivation of theoretical equation and experimental data. Through dimensional analysis and experimental data on the gap leakage flow of compartments, equations and coefficients between related factors such as the pressure difference, flowrate and Reynolds number were derived [4]. And based on these studies and theoretical analysis, the quadratic equation between the pressure difference and flow rate was presented, and an experiment for the estimation of coefficients was performed [5].

Meanwhile, the Gap Method that can predict the leakage flow in the gap of door with various structures was presented, and equation to establish the relationship between dimensionless flow rate and dimensionless pressure difference according to three types of pressure ranges around the door was derived for a comparison with experimental data [6].

In addition, there was little difference in calculation results between the simple Orifice Equation as shown in Equation (3) and Gap Method in case of the gap flow of doors in buildings [7].

$$Q = CA\sqrt{\frac{2\Delta p}{\rho}} = C_t\sqrt{\Delta p} \quad (3)$$

where Q is the leakage flow rate, Δp is the pressure difference before and after the gap, C is the flow coefficient, A is the gap area, and ρ is the density of air respecttively.

As described above, studies related to the gap leakage flow in buildings have mainly analyzed flow and pressure field in the buildings one-dimensionally using the pressure difference-leakage flowrate relation as in Equation (3) and network numerical analysis algorithm that assumes each compartment in the buildings as a single point.

Meanwhile, 3-dimensional flow analysis technique has been greatly developed based on the computational fluid dynamics and applied to various areas of buildings. However, in case the leakage flow of the gap is included, cross-sectional area of the gap is much less compared with other areas of the building, which has led to a lot of difficulties in grid generation and 3-dimensional flow analysis. Thus, methods for 3-dimensional analysis that includes the gap are deemed necessary.

As before, the leakage flow through the gap of door can be set with the relation of the pressure difference around the door and gap flow rate, and this relationship in computational fluid dynamics can be reflected as the Momentum Loss Model in the momentum Equation [8].

That is, as shown in **Figure 1**, if the Momentum Loss Model that enables pressure drop to be proportional to the velocity of flow as in Equation (4) is applied to the gap area of door in the domain of the analytical model, the effect of the gap flow can be reflected.

$$-\frac{\partial p}{\partial x_i} = K_{loss}\frac{\rho}{2}|u|u_i \quad (4)$$

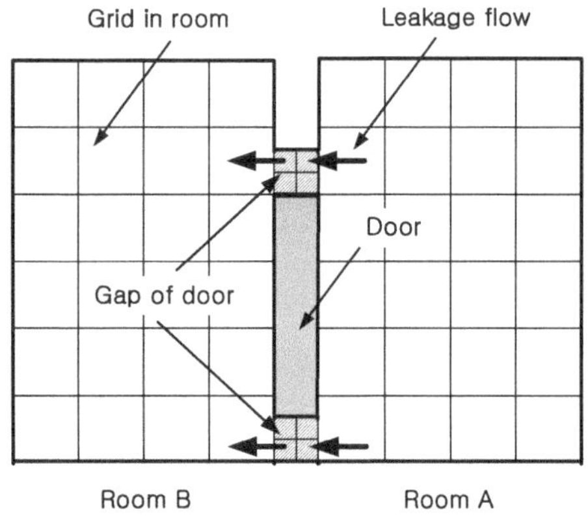

Figure 1. Schematic diagram about gap of door between rooms.

where K_{loss} is a quadratic resistance coefficient.

K_{loss} is a coefficient representing the characteristics of door gap, and a calculation can be made from the Orifice Equation of Equation (3) that contains experimental data and Momentum Loss Model of Equation (4).

In this study, a 3-dimensional flow analysis method that includes the leakage flow in the gap of the door was proposed using the Orifice Equation that is largely being applied to the gap flow analysis of buildings and Momentum Loss Model. In addition, a 3-dimensional numerical analysis of the stack effect on the stairs in buildings that include the gaps of doors was performed using this analysis method.

2. Numerical Model of Leakage Flow through Gap of Door

In this study, a 3-dimensional numerical analysis on the gap leakage flow of door that connect two rooms was performed first, and its characteristics were analyzed. For the applied pressure difference-leakage flowrate relation and geometric parameters, materials of representative publications [7] in the field of smoke design were used. As shown in **Table 1**, the door measures 2.1 m in height and 0.9 m in width, and the gap thickness at the top and side is 0.508 mm, and that at the bottom 6.35 mm. The relationship between pressure difference and leakage flowrate is determined by the Orifice Equation of Equation (2), and flow coefficient C is 0.65. **Table 2** shows the pressure difference between rooms and corresponding flow rate of the gap leakage flow obtained from the major sections in relation to characteristics of the door.

Figure 2 shows the analytical model that includes two rooms and door. As shown in the figure, there exists the closed door between Room A with an air inlet and Room B with an air outlet. The air enters Room B through the gap of the door after entering the inlet of Room A and then goes to the outside through the outlet of Room B.

In numerical analysis, a calculation is performed by

Table 1. Geometry of door.

Design factor		Value
Door	Width	0.9 m
	Height	2.1 m
Gap thickness	Top and sides	0.508 mm
	Bottom	6.35 mm
Door thickness (ΔL)		0.2 m
Flow coefficient (C)		0.65

Table 2. Flow rate induced by pressure difference through gap of door.

Pressure difference (Pa)	Flow rate (m^3/s)
10	0.022188807
20	0.031379712
30	0.038432142
40	0.044377615
50	0.049615682
60	0.054351256

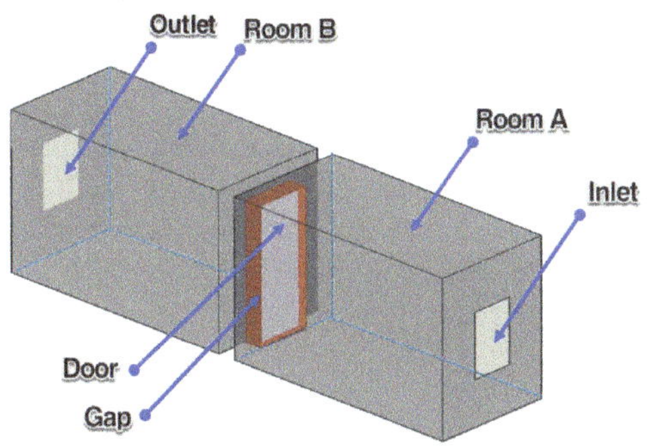

Figure 2. Analysis model of leakage flow at gap of door between rooms.

setting the leakage flow rate value of **Table 2** to the input value of boundary condition on the inlet of Room A, and the pressure difference formed between Room A and Room B was investigated. In this study, the CFX14 [8], which has widely been used in thermal and flow analysis, was used due to its verification of the accuracy and validity as a numerical analysis program.

As mentioned earlier, the door gap flow was processed with the Momentum Loss Model in the numerical analysis, and K_{loss}, the quadratic resistance coefficient of Equation (4) can be derived as shown in Equation (5) by referring to the Orifice Equation of Equation (3).

$$K_{loss} = \frac{2}{\rho} \frac{A_m^2}{\Delta L C_t^2} \quad (5)$$

where A_m is the area of the door gap in the numerical analysis model, ΔL is thickness of the door, and C_t is coefficient in the Orifice Equation respectively.

Figure 3 shows the relationship of the leakage flow rate and pressure difference between rooms obtained by the Orifice Equation and numerical analysis. As shown in the figure, result values of the Orifice Equation and numerical analysis on the flow rate values were found to be almost identical, and the 3-dimensional numerical analysis model in which the Momentum Loss Model is applied was determined to show valid results with respect to the leakage flow of the gap.

3. Effect of Separation Door in Stairwell

As a part of methods to reduce stack effect on the stairs, there is a case where a door and a wall that separate the upper stairwell and lower stairwell are installed in the intermediate floor of the stairs.

In this study, a 3-dimensional numerical analysis of the stack effect on the stairs of buildings was performed using the Momentum Loss Model of the gap leakage flow which was reviewed, and the effect of the separation door was analyzed.

Figure 4 shows the analytical model of the stairs of buildings, the target of analysis. As shown in the figure, the number of stairs is 11 floors, and the height of each floor is 3 m. A separation door was installed on the 6th floor to separate the upper stairwell and lower stairwell. Meanwhile, door and accommodation connected to the stairs were installed on the 1st and 11th floors, and

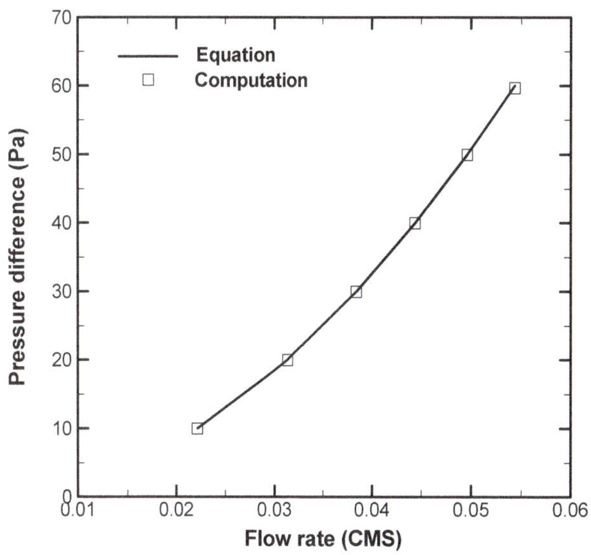

Figure 3. Pressure difference between rooms for leakage flow rate through gap of door.

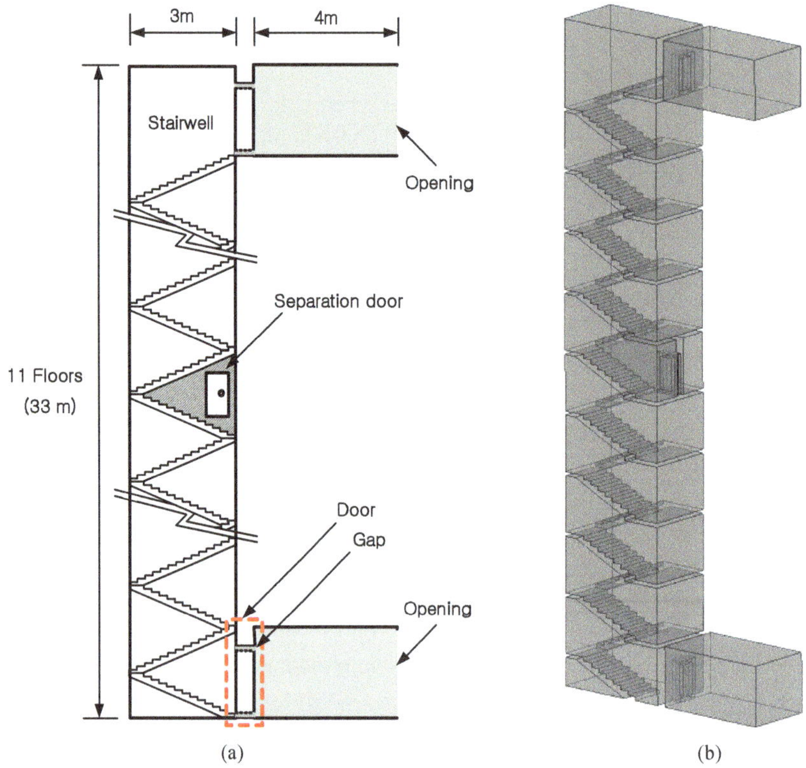

Figure 4. Analysis model of stairwell with separation door, (a) model of stairwell; (b) 3D model.

openings were made on one side of the room that comes in contact with the outside. In the door gap leakage floor, the Momentum Loss Model was applied. For analysis, the temperature of the stairs and room was set to 20˚C, and the outside temperature was set to −10˚C.

Figure 5 shows the pressure distributions on the center plane of the stairs as results of analysis with and without the installation of the separation door. **Figure 5(a)** shows the result of analysis performed under a condition where there is no separation door. On the 1st floor stairwell, a negative pressure of about −18 Pa is formed, and a positive pressure of about 20 Pa is formed on the 11th floor stairwell. Meanwhile, **Figure 5(b)** shows the result of analysis formed under a condition where there is a separation door. On the 1st floor stairwell, a negative pressure of about −10 Pa is formed, and a positive pressure of about 13 Pa is formed on the 11th floor stairwell. In addition, a positive pressure of about 3.5 Pa is formed on the 5th floor just below the separation door, and a negative pressure of about −1.0 Pa is formed on the 7th floor just above the separation door.

Figure 6 shows the result of numerical analysis performed by changing the gap area of the separation door. As shown in the figure, the less the gap area, the lower the pressure formed on the 11th stairwell and the higher the pressure formed on the 1st floor stairwell. Thus, it is adjudged that the gap area of the separation door has an effect on the stack effect on the stairs. In case of the leakage flow of the door gap due to the stack effect, as the gap area of the separation door decreases, flow rate is also reduced.

Based on the results of analysis, it can be inferred that the installation of separation door on the stairs of the building contribute to the reduction in the stack effect on the stairs to a certain extent, and the gap area of separation door has an effect on the amount of stack effect reduction.

4. Effect of Lobby between Stair and Accommodation

Meanwhile, there is a case where a lobby is installed between stair and accommodation as a part of the stack effect reduction methods.

In this study, a 3-dimensional numerical analysis on the stack effect was performed using the Momentum Loss Model after installing a lobby between stair and accommodation on the 1st and 11th floors with respect to the analytical model of the 11th stairwell presented earlier.

Figure 7 shows the pressure distribution on the center plane of the stairs as a result of analysis on the stack effect in a state where a lobby is installed. As shown in the figure, the pressure on the stairs is formed as the same result as in **Figure 5(a)** which shows the result obtained

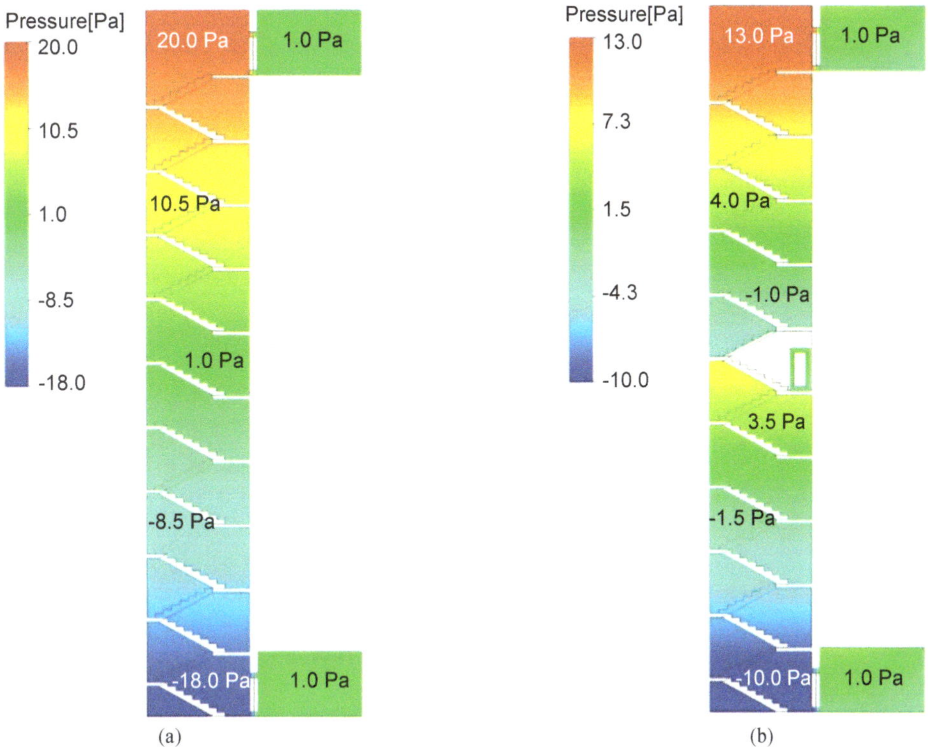

Figure 5. Pressure contours on center plane of stairwell in case of effect of separation door, (a) without separation door; (b) with separation door.

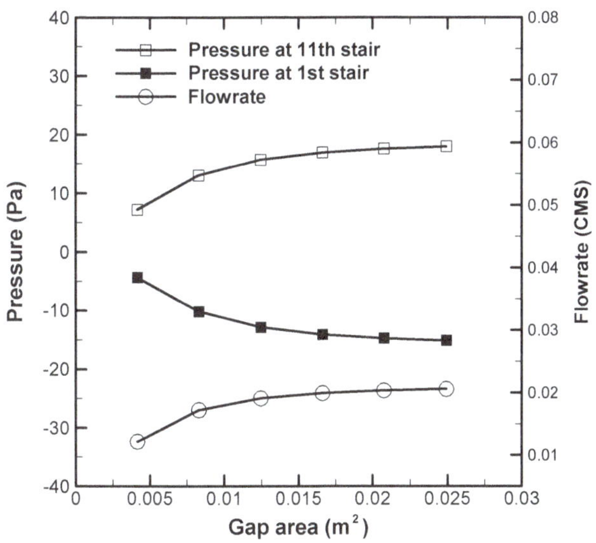

Figure 6. Pressure and flowrate for gap area in case of effect of separation door.

from case where a lobby is not installed. That is, the installation of a lobby does not contribute to the reduction in the stack effect on the stairs. Meanwhile, the pressure difference between rooms in the 1st and 11th floors was reduced considerably by installing a lobby. In case a lobby is not installed, the pressure difference between stair and accommodation of about 19 Pa is formed on the 1st floor. Whereas in case a lobby is installed, the pressure difference between stair and lobby of about 11 Pa is formed, and that between lobby and accommodation of about 8 Pa is formed respectively. As such, the installation of lobby between stair and accommodation leads to the reduction in the pressure difference between rooms due to the stack effect on each floor.

Figure 8 shows the results of numerical analysis performed by changing the door gap area between rooms. As shown in the figure, changes in the gap area of the door led to few changes in the pressure on the stair and lobby, and the leakage flow rate of the door gap was found to be increased linearly on the gap area.

5. Conclusions

In this study, a numerical analysis model was proposed with respect to the leakage flow through the gap of doors in buildings, and a 3-dimensional numerical analysis of the stack effect on the stairs of buildings was performed using the model. From the results, the following conclusions were derived.

1) In a 3-dimensional numerical analysis, the phenomenon of the gap flow can be reflected effectively through application of the Momentum Loss Model that enables the pressure drop to be proportional to the flow velocity through the gap of door.

2) As a result of the analysis on the two rooms in

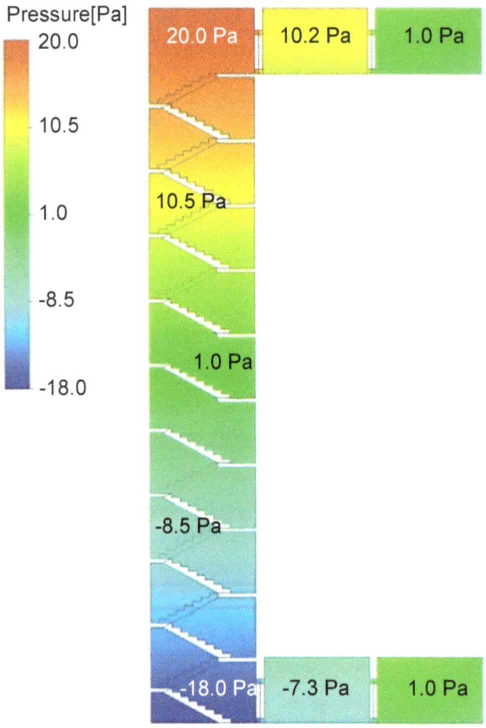

Figure 7. Pressure contours on center plane of stairwell in case of effect of lobby.

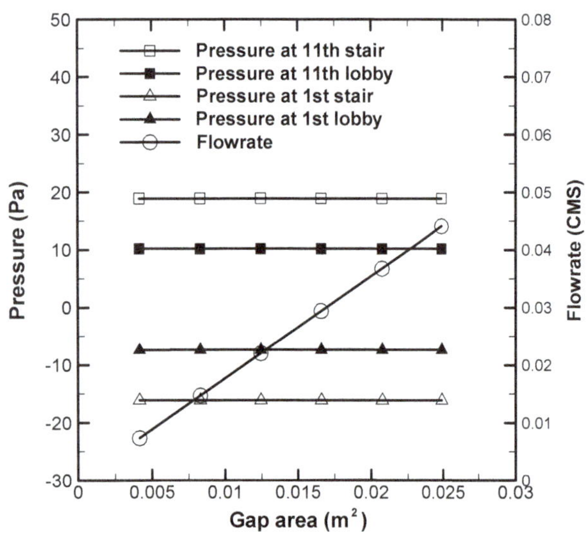

Figure 8. Pressure and flowrate for gap area in case of effect of lobby.

which a numerical model of the gap leakage flow is applied, it was found that the 3-dimensional numerical model in which the Momentum Loss Model is applied shows valid results since the results of the Orifice Equation and numerical analysis are almost identical with respect to the pressure difference corresponding to the given leakage flow rate.

3) The numerical analysis results of the stack effect on the stairs showed that a negative pressure of about −18 Pa is formed on the 1st floor stairwell, and a positive pressure of about 20 Pa on the 11th floor stairwell under the condition that the room temperature is 20˚C, and outside temperature is −10˚C. Whereas, a negative pressure of about −10 Pa is formed on the 1st floor stairwell, and a positive pressure of about 13 Pa on the 11th floor stairwell under the condition that a separation door is installed on the stairs of buildings.

4) The installation of a separation door on the stairs of buildings contributes to the reduction in the stack effect on the stairs to a certain extent, and the gap area of the separation door has an effect on the amount of stack effect reduction.

5) The installation of a lobby between stair and accommodation doesn't lead to the reduction in the stack effect on the stairs. However, the pressure difference between rooms that occurs due to the stack effect on each floor is reduced.

6. Acknowledgements

This study has been conducted by the relevant study to "(2013) Development of smoke control/evacuation schemes and over 3-hour fire rated structural members in case of fire in large-space buildings" Project, a major project of Korea Institute of Construction Technology.

REFERENCES

[1] G. T. Tamura and A. G. Wilson, "Pressure Differences Caused by Chimney Effect in Three High Buildings," *ASHRAE Transactions*, Vol. 73, Part 2, 1967, pp. II.1.1-II.1.10.

[2] J. H. Jo, J. H. Lim, S. Y. Song, M. S. Yeo and K. W. Kim, "Characteristics of Pressure Distribution and Solution to the Problems Caused by Stack Effect in High-Rise Residential Buildings," *Building and Environment*, Vol. 42, No. 1, 2007, pp. 263-277.

[3] H. Yang, M. S. Yeo, J. H. Jo and K. W. Kim, "Simulation of the Stack Effect in High-Rise Buildings," *Journal of Air Conditioning and Refrigeration*, Vol. 14, No. 6, 2002, pp. 456-467.

[4] D. W. Etheridge, "Crack Flow Equations and Scale Effect," *Building and Environment*, Vol. 12, No. 3, 1977, pp. 181-189.

[5] P. H. Baker, S. Sharples and I. C. Ward, "Air Flow through Cracks," *Building and Environment*, Vol. 22, No. 4, 1987, pp. 293-304.

[6] D. Gross and W. L. Haberman, "Analysis and Prediction of Air Leakage through Door Assemblies," *Fire Safety Science, Proceedings of the 2nd International Symposium*, Tokyo, 13 June 1988, pp. 169-178.

[7] J. H. Klote and J. A. Milke, "Principles of Smoke Management," American Society of Heating, Refrigerating and Air-Conditioning Engineers (ASHRAE), Atlanta, Society of Fire Protection Engineers (SFPE), Bethesda, 2002.

[8] ANSYS, Inc., "ANSYS CFX-Solver Theory Guide," ANSYS, Inc., Canonsburg, 2013.

About the Prospects for Passage to Instability

I. V. Lebed
Zhukovsky Central Institute of Aerohydrodynamics, Moscow, Russia

ABSTRACT

The results of the direct numerical integration of the Navier-Stokes equations are evaluated against experimental data for problem on a flow around bluff bodies in an unstable regime. Experiment records several stable medium states for flow past a body. Evolution of each of these states, after losing the stability, inevitably goes by periodic vortex shedding modes. Calculations based on the Navier-Stokes equations satisfactorily reproduced all observed stable medium states. They were, however, incapable of reproducing any of a vortex shedding modes recorded experimentally. The solutions to the classic hydrodynamics equations successfully reach the boundary of instability field. However, classic solutions are unable to cross this boundary. Most likely, the reason for this is the Navier-Stokes equations themselves. The classic hydrodynamics equations directly follow from the Boltzmann equation and naturally contain the error involved in the derivation of classic kinetic equation. Just the Boltzmann hypothesis, which closed kinetic equation, allowed us to construct classic hydrodynamics on only three lower principal hydrodynamic values. The use of the Boltzmann hypothesis excludes higher principal hydrodynamic values from the participation in the formation of classic hydrodynamics equations. The multimoment hydrodynamics equations are constructed using seven lower principal hydrodynamic values. The numerical integration of the multimoment hydrodynamics equations in the problem on flow around a sphere shows that the solutions to these equations cross the boundary and enter the instability field. The boundary crossing is accompanied by appearance of very uncommon acts in scenario of system evolution.

Keywords: Instability; Classic Hydrodynamics; Multimoment Hydrodynamics

1. Introduction

The possibility to interpret unstable phenomena becomes very topical. In particular, the reason is conditioned by visualized increase in the intensity of disasters in nature. Experiment expects that the phenomenon of vortex shedding is the example of unstable process. Vortex shedding behind bluff bodies has been studied systematically at least since the days of Strouhal. Vortex shedding is well-defined instability development regime, which is fairly extended along the Reynolds scale. Sometimes, the phenomenon of vortex shedding is called the von Karman instability. Sometimes, the phenomenon is called the Kelvin-Helmholtz instability. However, the study of evolution of solution, after losing the stability, by means of direct numerical integration of the Navier-Stokes equations became feasible comparatively recently.

The phenomenon of vortex shedding behind bluff bodies is taken for evaluation of the results of direct numerical integration of the Navier-Stokes equations against experimental data [1,2]. In Section 2, the analysis of numerous divergences between the results of numerical integration of the Navier-Stokes equations and the experiment is given. The analysis is accompanied by corresponding conclusions. In Section 3, the concepts that lie in foundation of the multimoment hydrodynamics [3] are discussed. The characteristic features of unstable solution to the multimoment hydrodynamics equations in the problem on flow around a sphere [4,5] are considered.

2. Evaluating the Results of Numerical Integration against Experiment

In each of the problems on flow around a bluff bodies experiment records several independent directions of instability development. For 3D flow past a sphere experiment records three stable medium states. The $\boldsymbol{U}_{s0}^{exp}(\boldsymbol{x})$ stable stationary flow consists of an axisymmetric toroidal recirculating zone in the near wake, which originates a single rectilinear thread in the far wake, see Figure 1(g) in [6]. The $\boldsymbol{U}_{s1}^{exp}(\boldsymbol{x})$ stable non-axisymmetric flow consists of two weakly asymmetric halves in the near wake, which originate two rectilinear threads in the far wake, see Figure 12(a) in [7]. The

$U_{s2}^{exp}(t,x)$ central-type stable state is characterized by periodic restructuring in the near non-axisymmetric wake, which causes wavy motion in the far wake without vortex shedding, see Figure 12(b) in [7].

Each of these three states, $U_{s0}^{exp}(x)$, $U_{s1}^{exp}(x)$, and $U_{s2}^{exp}(t,x)$, after losing the stability, starts to evolve in its own direction qualitatively different from the others [2]. These directions are schematically shown by three horizontal branches in **Figure 1**. Instability begins to develop upon the attainment of critical Reynolds number values, Re_0^*, Re_1^*, and Re_2^*, respectively. Instability development inevitably involves periodic vortex shedding modes. Each of the three turbulence development directions has vortex shedding features of its own only characteristic of the given direction. No matter what direction is selected by experiments, periodic vortex shedding is, however, an unavoidable, well-defined instability-development mode.

After the attainment of a certain critical Reynolds number value $\mathrm{Re}_2^* > \mathrm{Re}_2$, the $V_{s2}^{exp}(t,x)$ one-periodic vortex loop street appears in the wake behind a sphere, the upper branch in **Figure 1**. The periphery of the recirculating zone is periodically detached from the core and moves downstream in the form of hairpin vortex loop. The vortex loops move uniformly along one of the paths of the double undulated thread, which forms the far wake [7]. According to experiment, two vortex shedding modes existed at $\mathrm{Re} > \mathrm{Re}_2^{**}$. In the $W_{s2}^{exp}(t,x)$ two-periodic mode, hairpin vortex loops are shed, whereas the $Q_{s2}^{exp}(t,x)$ two-periodic mode is characterized by vortex ring shedding. As with $W_{s2}^{exp}(t,x)$, the $Q_{s2}^{exp}(t,x)$ vortex rings are alternately shed from two symmetry related recirculating zone points and rush downstream along different double undulated thread paths [7,8], the upper branch in **Figure 1**.

Above some critical Re_1^* value, the $U_{s1}^{exp}(x)$ flow loses stability, the middle branch in **Figure 1**. The periphery of the recirculating zone of the $V_{s1}^{exp}(t,x)$ one-periodic flow separates from its core and moves downstream along one of the double rectilinear thread paths in the form of hairpin vortex loop [9].

After the attainment of a certain critical Reynolds number Re_0^* the $U_{s0}^{exp}(x)$ toroidal recirculating zone in the near wake behind a sphere begins to pulsate periodically, the lower branch in **Figure 1**. The frequency of pulsations is anomalously low. The $V_{s0}^{exp}(t,x)$ pulsating flow remains axisymmetric [6,10]. After the passage of Re_0^{**}, vortex rings depart from a sphere downstream and move along the spiral path, that is, the $W_{s0}^{exp}(t,x)$ one-periodic flow becomes nonaxisymmetric [10]. The attainment of Re_0^{***} is accompanied by a change in the regime of vortex shedding from a sphere. Vortex rings penetrate into each other and form the $Q_{s0}^{exp}(t,x)$ two-periodic continuous spiral sheet in the wake behind a sphere [10], the lower branch in **Figure 1**.

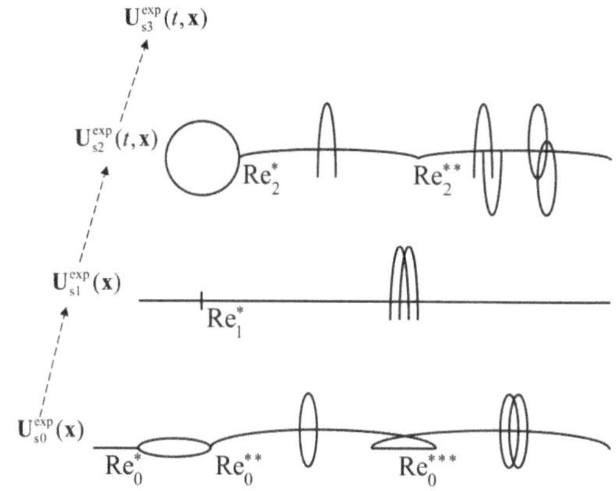

Figure 1. Three stable medium states originating three turbulence development directions for flow past a sphere. The lower branch corresponds to the evolution of stationary axisymmetric flow $U_{s0}^{exp}(x)$: $\mathrm{Re} < \mathrm{Re}_0^* - U_{s0}^{exp}(x)$; $\mathrm{Re}_0^* < \mathrm{Re} < \mathrm{Re}_0^{**}$, periodic pulsations of the axisymmetric recirculating zone in the wake behind a sphere $V_{s0}^{exp}(t,x)$, see Figure 1(h) in [6]; $\mathrm{Re}_0^{**} < \mathrm{Re} < \mathrm{Re}_0^{***}$, vortex ring shedding along a spiral path $W_{s0}^{exp}(t,x)$, see Figure 2(a) in [10]; and $\mathrm{Re} > \mathrm{Re}_0^{***}$, helicoidal vortex sheet $Q_{s0}^{exp}(t,x)$, see Figure 2(b) in [10]. The middle branch corresponds to the evolution of stable steady nonaxisymmetric flow $U_{s1}^{exp}(x)$: $\mathrm{Re}_1 < \mathrm{Re} < \mathrm{Re}_1^* - U_{s1}^{exp}(x)$; $\mathrm{Re} > \mathrm{Re}_1^*$, periodic horseshoe-shaped vortex loop shedding along a rectilinear path $V_{s1}^{exp}(t,x)$, see Figure 6 in [9]. The upper branch corresponds to the evolution of a stable central-type state $U_{s2}^{exp}(x)$: $\mathrm{Re}_2 < \mathrm{Re} < \mathrm{Re}_2^* - U_{s2}^{exp}(t,x)$; $\mathrm{Re}_2^* < \mathrm{Re} < \mathrm{Re}_2^{**}$, periodic horseshoe-shaped vortex loop shedding along one of the double undulated thread branches $V_{s2}^{exp}(t,x)$, see Figure 7(a,i) in [7]; and $\mathrm{Re} > \mathrm{Re}_2^{**}$, periodic vortex loop shedding along both double undulated thread branches $W_{s2}^{exp}(t,x)$, see Figure 4 in [8] and Figure 7(a,iii) in [7], or periodic vortex ring shedding $Q_{s2}^{exp}(t,x)$, see Figure 6 in [8].

For 2D flow past a circular cylinder experiment records two stable medium states. The $U_{c0}^{exp}(x)$ stable stationary flow is characterized by axisymmetric recirculating zone in the near wake. The $U_{c2}^{exp}(t,x)$ central-type stable state is characterized by periodic restructuring in the near nonaxisymmetric wake without vortex shedding. Experiment finds at least two independent directions of instability development. Vortex shedding along each of the two directions, $V_{c0}^{exp}(x)$ and $V_{c2}^{exp}(t,x)$ [11,12], is characterized by its own characteristic features intrinsic in it.

Evaluating the results of the direct numerical integration of the Navier-Stokes equations against experiment for unstable flows became feasible only recent twenty-five years, thanks to the vigorous development of computer facilities. As distinct from the earlier studies, the direct numerical integrations of the Navier-Stokes equations performed recently directly reproduce unstable flow conditions without artificially introducing non-stationary features into the stationary problem. Under artificial modeling a problem with time independent boundary conditions after stability loss is substituted by non-stationary stable problem [13,14]. Such substitution is need of initial conditions for non-stationary problem. The initial conditions consist of a solution for a fully developed vortex-shedding flow field. It is necessary to attract experimental data to construct such conditions. In particular, experiment must give eddy sizes and eddy positions in space, initial eddy velocities and vortex shedding frequency, beginning and completion of vortex shedding process at the Reynolds scale. To summarize, such modeling draws a picture of eddies movement. It is incapable of showing the direction of instability development.

Under the linear analysis of stability, the Navier-Stokes equations are linearized for small hydrodynamic values perturbations. Perturbations are expanded in Fourier series in time [15,16]. The linear character of studies does not allow evolution calculations to be extended to long times. That is, the linear analysis is incapable of defining the final stable saturated state. However, the linear analysis predicts characteristics of final solution, in particular, the oscillation frequencies around a stable position.

The direct numerical integration of the Navier-Stokes equations gives three stable solutions for flow around a sphere, $\boldsymbol{U}_{s0}^{cal}(\boldsymbol{x})$, $\boldsymbol{U}_{s1}^{cal}(\boldsymbol{x})$, and $\boldsymbol{U}_{s2}^{cal}(t,\boldsymbol{x})$. These solutions reproduced three stable flows observed experimentally, $\boldsymbol{U}_{s0}^{exp}(\boldsymbol{x})$, $\boldsymbol{U}_{s1}^{exp}(\boldsymbol{x})$, and $\boldsymbol{U}_{s2}^{exp}(t,\boldsymbol{x})$, correspondently. In all calculations without exception performed by the direct numerical integration of the Navier-Stokes equations, instability development occurs in strict correspondence to classic Landau-Hopf scenario [17]. The calculation leads the instability development process in the direction given by dashed slanting line in **Figure 1**. The $\boldsymbol{U}_{s0}^{cal}(\boldsymbol{x})$ solution, when loses stability, experiences bifurcation to the $\boldsymbol{U}_{s1}^{cal}(\boldsymbol{x})$ solution (regular bifurcation), and the $\boldsymbol{U}_{s1}^{cal}(\boldsymbol{x})$ solution after stability loss experiences bifurcation to the $\boldsymbol{U}_{s2}^{cal}(t,\boldsymbol{x})$ limiting cycle (Hopf bifurcation). The $\boldsymbol{U}_{s2}^{cal}(t,\boldsymbol{x})$ solution loses stability to be replaced by the $\boldsymbol{U}_{s3}^{cal}(t,\boldsymbol{x})$ chaotic solution [18-21].

The direct numerical integration of the Navier-Stokes equations gives two stable solutions for flow around a cylinder, $\boldsymbol{U}_{c0}^{cal}(\boldsymbol{x})$ and $\boldsymbol{U}_{c2}^{cal}(t,\boldsymbol{x})$. These solutions reproduced two stable flows observed experimentally, $\boldsymbol{U}_{c0}^{exp}(\boldsymbol{x})$ and $\boldsymbol{U}_{c2}^{exp}(t,\boldsymbol{x})$, correspondently. In accordance with Landau-Hopf scenario, the $\boldsymbol{U}_{c0}^{cal}(\boldsymbol{x})$ solution, when loses stability, experiences bifurcation to the $\boldsymbol{U}_{c2}^{cal}(t,\boldsymbol{x})$ limiting cycle. The $\boldsymbol{U}_{c2}^{cal}(t,\boldsymbol{x})$ solution loses stability to be replaced by the $\boldsymbol{U}_{c3}^{cal}(t,\boldsymbol{x})$ chaotic solution [22].

The bifurcation of a stationary state after stability loss to a limiting cycle is represented in **Figure 2**. The **Figure 2** was drawn for a flat plate [23]. Later, qualitatively analogous pictures were drawn for different bluff bodies [24]. The **Figure 2** gives the temporal development of the A absolute value of the velocity disturbance amplitude at an isolated point in the wake of a bluff body. After the attainment of the time moment $t=0$ lower solution loses its stability. They are the $\boldsymbol{U}_{c0}^{cal}(\boldsymbol{x})$ solution for a cylinder and the $\boldsymbol{U}_{s1}^{cal}(\boldsymbol{x})$ solution for a sphere. By the time moment $t=t_2$ increasing of disturbances results in saturation of a new stable solution, namely, the non-stationary periodic limiting cycle. They are the $\boldsymbol{U}_{c2}^{cal}(t,\boldsymbol{x})$ solution for a cylinder and the $\boldsymbol{U}_{s2}^{cal}(t,\boldsymbol{x})$ solution for a sphere. In accordance with definition [23], unstable regime is placed within the time interval between $t=0$ and $t=t_2$. So, unstable regime is limited at time. On the contrary, the visualized process of vortex shedding is not limited at time. That is, experiment records a vortex street as long as investigator wants. Moreover, unstable regime, **Figure 2**, is not a periodic one, because the A disturbance amplitude grows. As a result, time limited non-periodic unstable regime can not be put in correspondence to strictly periodic phenomenon of vortex shedding which has no time frameworks. So, classic hydrodynamics is incapable to come up the expectations of experiment on unstable nature of vortex shedding phenomenon.

The classic calculation can put only stable solution in correspondence to vortex shedding phenomenon. As noted above, the classic hydrodynamics equations in problem on flow past a sphere have four solutions, $\boldsymbol{U}_{s0}^{cal}(\boldsymbol{x})$, $\boldsymbol{U}_{s1}^{cal}(\boldsymbol{x})$, $\boldsymbol{U}_{s2}^{cal}(t,\boldsymbol{x})$ and $\boldsymbol{U}_{s3}^{cal}(t,\boldsymbol{x})$. And these four solutions must describe ten observed modes, $\boldsymbol{U}_{s0}^{exp}(\boldsymbol{x})$, $\boldsymbol{V}_{s0}^{exp}(t,\boldsymbol{x})$, $\boldsymbol{W}_{s0}^{exp}(t,\boldsymbol{x})$ and $\boldsymbol{Q}_{s0}^{exp}(t,\boldsymbol{x})$;

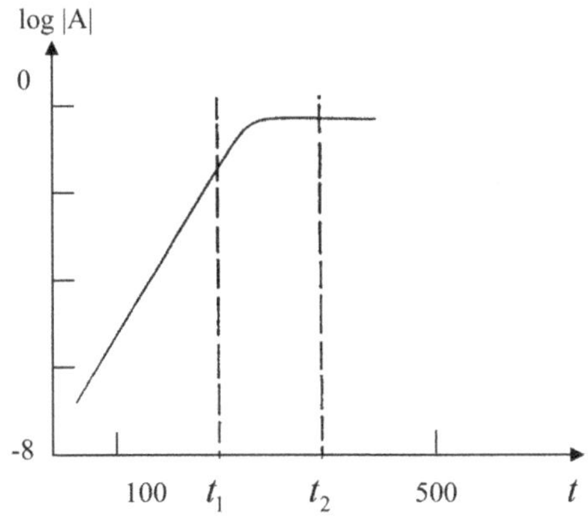

Figure 2. The temporal development of the absolute value of the velocity disturbance amplitude A: $0 < t < t_1$- exponential growth; $t_1 < t < t_2$- onset of nonlinearity; $t > t_2$- saturation state. Figure 2 was drawn using the data from [23].

$U_{s1}^{exp}(x)$ and $V_{s1}^{exp}(t,x)$; $U_{s2}^{exp}(t,x)$, $V_{s2}^{exp}(t,x)$, $W_{s2}^{exp}(t,x)$ and $Q_{s2}^{exp}(t,x)$. The $U_{s3}^{cal}(t,x)$ solution that succeeds the $U_{s2}^{cal}(t,x)$ one-period mode is multiperiodic, that is, chaotic in essence. Attempts at correlating the $U_{s3}^{cal}(t,x)$ mode with the observed strictly periodic vortex shedding modes seem to have no prospects. The $U_{s0}^{cal}(x)$ solution and the $U_{s1}^{cal}(x)$ solution are stationary. That is why, the $U_{s2}^{cal}(t,x)$ limiting cycle is the only case where it is possible to correlate the observed vortex shedding from a sphere to calculation results.

Let two experimental modes have pretensions to be described by the $U_{s2}^{cal}(t,x)$ limiting cycle. They are the $U_{s2}^{exp}(t,x)$ central-type stable state, which characterizes itself by wavy motion in the far wake without vortex shedding, and the $V_{s2}^{exp}(t,x)$ one-period vortex shedding mode, which manifests itself as a vortex loop street. Then, the $U_{s2}^{exp}(t,x)$ central-type state does not stand out as a separate regime when the experimental data are interpreted in [18-21]. The $U_{s2}^{exp}(t,x)$ central-type regime is combined with the $V_{s2}^{exp}(t,x)$ regime of periodic vortex shedding from a sphere. According to [18-21], vortex shedding does occur over the Reynolds number range $\mathrm{Re}_2 < \mathrm{Re} < \mathrm{Re}_2^*$, but its manifestations are exceedingly indistinct; the intensity of vortex shedding monotonically increases as Re grows. Earlier, this point of view was formulated in [25]. An attempt was made [18-21] to put the flow picture representing the $U_{s2}^{cal}(t,x)$ limiting process in correspondence with the observed $V_{s2}^{exp}(t,x)$ periodic vortex shedding combined with the $U_{s2}^{exp}(t,x)$ central-type mode.

Attempts at correlating the $U_{s2}^{cal}(t,x)$ solution with observed vortex shedding modes initially seem to have no prospects. The reasons are as follows. First, the suggestion of a monotonic increase in the intensity of vortex shedding [18-21] is at variance with experiment (see Figures 12(b)-12(d) in [7]). According to Figure 12(c) from [7], the vortex structure appears in the wake suddenly and is fairly well defined starting with the moment of its origination. Secondly, in accordance with calculations, the $U_{s3}^{cal}(t,x)$ chaotic solution replaces the $U_{s2}^{cal}(t,x)$ solution after the attainment of the Reynolds number value approximately equals to 500 [21]. However, in accordance with experiment, periodic vortex shedding is a well-defined mode, which is fairly extended along the Reynolds scale. In the experiments when a sphere was towed through an unperturbed medium vortex shedding was observed over the whole range of Reynolds numbers studied, up to Re = 30000 [10]. *i.e.*, calculation predicts the early transition to chaos. This prediction is, however, at variance with experiment. Thirdly, three regimes of the six vortex shedding regimes actually observed are the one-period modes. These include $W_{s0}^{exp}(t,x)$, $V_{s1}^{exp}(t,x)$, and $V_{s2}^{exp}(t,x)$ regimes. Three regimes of vortex shedding are the two-period modes. These include $Q_{s0}^{exp}(t,x)$, $W_{s2}^{exp}(t,x)$, and $Q_{s2}^{exp}(t,x)$ regimes. Attempts at correlating the one-period $U_{s2}^{cal}(t,x)$ solution with the observed two-period vortex shedding regimes seem to have no prospects. And at last, the $U_{s2}^{cal}(t,x)$ solution must be put in correspondence to several modes of vortex shedding simultaneously. For example, at some $\mathrm{Re} > \mathrm{Re}_0^{***}$, experiment records four different regimes of vortex shedding: the $Q_{s0}^{exp}(t,x)$ mode at the lower branch, the $V_{s1}^{exp}(t,x)$ mode at the middle branch, the $W_{s2}^{exp}(t,x)$ and the $Q_{s2}^{exp}(t,x)$ modes at the upper branch, **Figure 1**. However, the only $U_{s2}^{cal}(t,x)$ solution is incapable of reproducing four different modes simultaneously.

The streamline flow pictures for every quarter of the period of oscillations (Figure 25 in [20]) give an idea of the total period of recirculating zone oscillations in the $U_{s2}^{cal}(t,x)$ wake behind a sphere after the passage of the Re_2^* critical value. A detailed examination of these flow pictures was executed in [2,20]. To summarize, after the appearance in the wake behind a sphere, the size of the vortex structure becomes substantial at the surface of the sphere. Subsequently, the structure moves toward the periphery of the recirculating zone, which is accompanied by its continuous dissipation. Lastly, it fully disappears at the periphery of the recirculating zone.

This picture is qualitatively different from the observed full period of oscillation of the recirculating zone. As in calculations, the experimental vortex structure engendered begins to expand and move downstream. After reaching the periphery of the recirculating zone, this vortex structure, however, acquires a maximum size rather than dissipates as predicted by calculations. At the end of the period, the vortex localized at the periphery of the recirculating zone separates from this zone. The separation of the shed vortex from the recirculating zone is very clearly shown in Figure 40 in [20]. The periphery of the recirculating zone, which is periodically shed from the recirculating zone, rushes downstream and forms a vortex street $V_{s2}^{exp}(t,x)$ (Figure 7(a,i) in [7]).

Because of the absence of the detachment of the recirculating zone periphery in Figure 25 in the calculation [20], there is no vortex loop street in the wake behind a sphere. This is an essential difference between calculation and experiment. Analyzed Figure 25 from [20], the authors confirmed the disappearance of the vortex structure at the periphery of the recirculating zone. They note that the vortex structure "loses its spiral shape" at the periphery of the recirculating zone.

So, combining of the $U_{s2}^{exp}(t,x)$ and the $V_{s2}^{exp}(t,x)$ did not give the desired result, because it did not allow the calculated $U_{s2}^{cal}(t,x)$ mode to be put in correspondence to the observed $V_{s2}^{exp}(t,x)$ vortex shedding mode. Indeed, the calculated $U_{s2}^{cal}(t,x)$ limiting cycle satisfactorily reproduces the stable central-type position $U_{s2}^{exp}(t,x)$ at $Re < Re_2^*$ but proves a complete failure when attempts are made to reproduce the $V_{s2}^{exp}(t,x)$ vortex shedding mode at $Re > Re_2^*$ [18-21].

The $U_{s2}^{cal}(t,x)$ wake behind a sphere is drawn in [20] both by streamline flow pictures and by streakline flow pictures. On the flow pictures presented by streaklines, the lines having a form of hairpin vortices appear in near wake behind a sphere. However, a vortex street is absent too in the far wake on flow pictures presented by streaklines.

The authors of all the numerical experiments without exception discover the evident absence of a vortex street in the wake behind different bluff bodies on flow pictures. Numerical simulations were performed both for 2D and for 3D problems on flow around a circular cylinder and a flat plate, around a sphere and a disk, and so on. However, passed years gave no numerical experiments in which stability loss results in a periodic vortex shedding mode. Then, the conclusion of agreement between calculation and experiment is based in early direct numerical simulations on a comparison of the calculated vorticity distributions with visual observation results [18,21]. However, even insignificant twist in a streamline can create closed curves in the vorticity distribution pictures. And, really, the comparison of streamline flow pictures with the vorticity distributions (Figures 25 and 29 in [20]) shows that closed curves in the vorticity pictures correspond to wavy motion in the far wake behind a sphere rather than vortex structures. So, there is no reason at all for identifying the closed vorticity distribution curves in the wake with vortex structures.

That is why, Johnson and Patel [20] do not confine to drawing of streamline curves, streakline curves and vorticity distribution curves. To interpret the obvious contradiction between the observed vortex shedding and calculation results they put the regions of the supposed existence of vortex structures $\lambda_2(t,x)$ in correspondence to the observed vortex shedding. The latter direct numerical simulations select namely this way to interpret the discrepancies between calculation and experiment [26,27]. The sense is that both calculated and experimental flow pictures are known to change qualitatively depending on the system of coordinates used to observe or calculate flows. Flow pictures represented by streamlines are not invariant with respect to the Galilean transformation. Vortex structures can be absent in a certain system and well defined in another. It follows that the uniform motion of a coordinate system can mask the vortex structure which, in reality, exists.

Taking this into consideration, Jeong and Hussian developed a method for distinguishing space regions $\lambda_2(t,x)$ in which vortex structures can hypothetically exist [28]. The study performed in [28] showed that the $\lambda_2(t,x)$ regions were indeed capable of predicting the position of vortex structures in space. These predictions are, however, fairly approximate.

Namely, streamline flow pictures can contain regions with vortex structures that are undetectable by the Jeong-Hussian method. Conversely, this method can show the regions of existence for vortex structures where, according to the streamline flow pictures, vortex structures are absent. Moreover, the Jeong-Hussian method outlines the $\lambda_2(t,x)$ contours of the supposed existence of vortex structures but does not indicate the particular coordinate system in which these vortex structures are to be sought. The Jeong-Hussian method cannot be used to correct flow pictures. On the contrary, the correctness of this method is evaluated by comparing the hypothetical regions where vortex structures with calculated flow pictures exist.

Once more, the Jeong-Hussian method [28] outlines only the boundaries of the $\lambda_2(t,x)$ regions. The boundaries of the $\lambda_2(t,x)$ regions are very widespread, because vortex structures from all the reference systems without exception must find a place within these $\lambda_2(t,x)$ regions. The method is incapable of penetrating the $\lambda_2(t,x)$ region. As a consequence, the method is incapable of revealing the presence or the absence of vortex structure within the $\lambda_2(t,x)$ region. Nevertheless, namely boundaries of the $\lambda_2(t,x)$ regions are put in correspondence to observed vortex structures. The ideas of appearance of vortex structures are based on the appearance of boundaries of the $\lambda_2(t,x)$ regions. The ideas of temporal evolution of vortex structures are based on temporal evolution of boundaries of the $\lambda_2(t,x)$ regions too.

The $\lambda_2(t,x)$ region of the supposed existence of vortex structures represented in Figure 31 from [20] extends in the wake behind a sphere to distances much larger than the sphere diameter. However, in the sphere-fixed Ref_0 coordinate system the calculation does not find any vortex structures within the $\lambda_2(t,x)$ region (Figure 25 in [20]). To summarize, the $\lambda_2(t,x)$ region pattern does not remove the discrepancy between the streamline flow picture calculated in the Ref_0 system of coordinates and the flow picture observed in the same system. Really, the experiment records the hairpin vortex shedding in the wake behind a sphere (Figure 7(a,i) in [7]). The streamline flow picture draws the periodic restructuring in the near wake, which causes wavy motion in the far wake without vortex shedding (Figure 25 in [20]).

There may exist such reference system Ref_1 in which the vortex shedding appears in the calculated flow picture. Possibly, the flow picture calculated in the Ref_1 system will be identical to the photographs of the flow obtained in the same Ref_1 system. Today, calculation does not discover such reference system Ref_1.

The possibilities of variation the boundary and initial conditions of calculations to bring it in strict correspondence to the experimental boundary and initial conditions are not very narrowly, the possibilities of perfecting the numerical procedure applied in recent simulations are boundless. Nevertheless, there are almost no prospects for a cardinal change in the calculated flow pictures [2]. It is therefore very improbable that the instability development direction calculated in [15,16,18-21] can change and new solutions to the Navier-Stokes equations other than $U_{s0}^{cal}(x)$, $U_{s1}^{cal}(x)$, $U_{s2}^{cal}(t,x)$, and $U_{s3}^{cal}(t,x)$ can appear.

It may well be that the interpretation of experimental results presented in Section 2 is not ultimate. Possibly, the directions presented will be augmented by new modes that have not been recorded up to now. It is also possible that the modes that have been recorded will be differently distributed over the three directions shown in **Figure 1**, and new modes will force us to modify the concepts upon which the three turbulence development directions are based. However, irrespective of any modifications of the scheme shown in **Figure 1**, it is very improbable that one of the conclusions drawn from comparing it with experiment will change. Namely, calculations are incapable of reproducing any of the six periodical vortex shedding modes observed along the three turbulence development directions (**Figure 1**). Most likely [1,2], the reason for simulation failure is the Navier-Stokes equations themselves.

3. Increase in the Number of Principal Hydrodynamic Values

Classic hydrodynamics equations exist for about two centuries. By definition, these equations are valid for the description of arbitrary continuous media corresponding to continuity and unlimited deformability principles [29]. Statistically grounded hydrodynamics equations are, however, far from being completely established. The greatest progress in this direction was made for one of continuous medium states, namely, for the rarefied gas state, where the characteristic free path λ far exceeded the characteristic size of particles d. In a rarefied gas, that is, at $d \ll \lambda$, the path from classic mechanics equations written separately for each of the N medium particle to classic hydrodynamics equations was passed without additional assumptions. The only exception was the Boltzmann hypothesis of molecular chaos "Stosszahlansatz" (**Figure 3**).

Classic direction and pair-multimoment direction are two independent branches of the statistical deterministic approach to medium description. Classic kinetics and classic hydrodynamics are direct corollaries to the first equation of the Bogolyubov-Born-Green-Kirkwood-Yvon

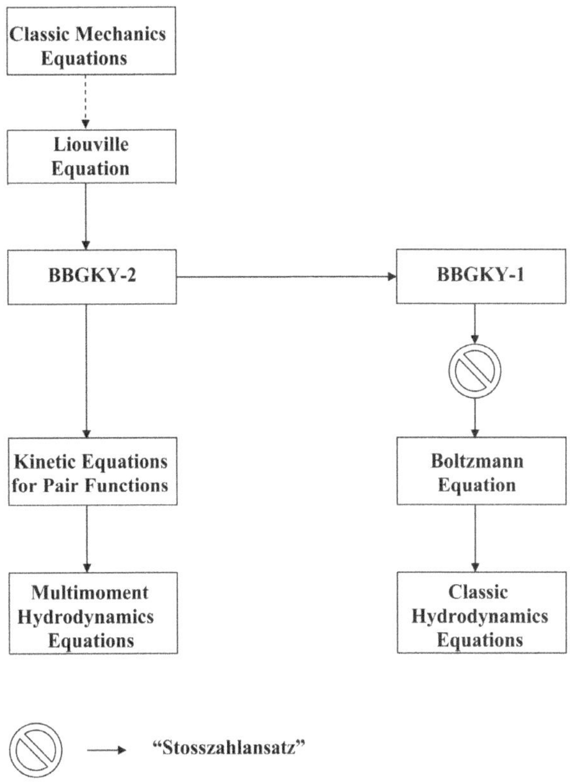

Figure 3. The scheme of the statistical deterministic approach.

chain of equations, *i.e.*, the BBGKY hierarchy [30]. The kinetic equations for pair functions and the multimoment hydrodynamics equations are derived from the second of these equations. The BBGKY hierarchy is closed by the Liouville equation (**Figure 3**). To pass from the classical mechanics equations to the Liouville equation, the concept of the Gibbs ensemble is invoked. The Gibbs concept is the connecting link between the dynamical deterministic approach and the statistical deterministic approach [30].

The sth equation of the BBGKY hierarchy has the form,

$$\begin{aligned}&\left(\frac{\partial}{\partial t}+\sum_{i=1}^{s}\xi_i\frac{\partial}{\partial x_i}+\sum_{i=1}^{s}\sum_{j\neq i,j=1}^{s}\left(\frac{\Phi_{i,j}}{m}\right)\frac{\partial}{\partial \xi_i}\right)\\&\cdot F_s\left(t,x_1,\xi_1,x_2,\xi_2,\cdots,x_s,\xi_s\right)\\&=-(N-s)\sum_{i=1}^{s}\int\left(\frac{\Phi_{i,s+1}}{m}\right)\frac{\partial}{\partial \xi_i}\\&\quad\cdot F_{s+1}\left(t,x_1,\xi_1,x_2,\xi_2,\cdots,x_{s+1},\xi_{s+1}\right)\mathrm{d}x_{s+1}\mathrm{d}\xi_{s+1}\end{aligned} \quad (1)$$

where m is the mass of the particle; $\Phi_{i,j}$ is the force of the action of the ith on j^{th} particle; and x_i, ξ_i are the spatial coordinate and velocity of the ith particle, respectively. The $F_s\left(t,x_1,\xi_1,\cdots,x_s,\xi_s\right)$ s-particle distribution function has the meaning of the probability that at time t particle 1 finds itself in unit element of the phase space near point $x_i, \xi_i, \cdots$, and particle s near point x_s, ξ_s. The $F_1(t, x_1, \xi_1)$ one-particle distribution function obeys the first equation of the BBGKY hierarchy (1). The first equation is not closed. The term on the right hand side of Equation (1) that contains the $F_2(t, x_1, \xi_1, x_2, \xi_2)$ two-particle function is responsible for the interaction of particle 1 with some particle 2.

There are several variants of the derivation of the Boltzmann equation directly from the first equation of the BBGKY hierarchy [30]. Each of them reaches the equation,

$$\begin{aligned} &\left(\frac{\partial}{\partial t} + \xi_1 \frac{\partial}{\partial x_1}\right) f_1(t, x_1, \xi_1) = N(N-1) J(t, x_1, \xi_1) \\ &J(t, x_1, \xi_1) = \int J(t, x_1, \xi_1, \xi_2) \mathrm{d}\xi_2 \\ &J(t, x_1, \xi_1, \xi_2) = \int v\Big[\bar{F}_2\left(t, x_1, \xi_1', x_1 - \rho_{v'}^{-d}, \xi_2'\right) \\ &\qquad - \bar{F}_2\left(t, x_1, \xi_1, x_1 - \rho_v^{-d}, \xi_2\right)\Big] b\mathrm{d}b\mathrm{d}\varepsilon \end{aligned} \tag{2}$$

Equation (2) is valid for a rarefied gas, that is, at $d \ll \lambda$,

$$\begin{aligned} &\bar{F}_1(t, x_1, \xi_1) = \frac{1}{l^3} \int_W F_1(t, x_1 + a, \xi_1) \mathrm{d}a \\ &f_1(t, x_1, \xi_1) = N\bar{F}_1(t, x_1, \xi_1) \\ &\bar{F}_2(t, x_1, \xi_1, x_1 - \rho, \xi_2) \\ &= \frac{1}{l^3} \int_W F_2(t, x_1 + a, \xi_1, x_1 + a - \rho, \xi_2) \mathrm{d}a \end{aligned} \tag{3}$$

The one-particle distribution function $f_1(t, x_1, \xi_1)$ in Equations (2), (3) has the meaning of the probable number of particles situated at time t in unit volume element near point x_1 and having velocities in unit interval near the ξ_1 point. The spatial integration in Equation (3) is performed within the W region having the characteristic linear size l, $d \ll l \ll \lambda$. The averaged $\bar{F}_1(t, x_1, \xi_1)$ and $\bar{F}_2(t, x_1, \xi_1, x_1 - \rho, \xi_2)$ functions weakly change on the scale of particle size d.

The $\bar{F}_2(t, x_1, \xi_1, x_1 - \rho_v^{-d}, \xi_2)$ function on the right hand side of Equation (2) corresponds to a pair of particles 1 and 2, which enter the region of their interaction C_0 at velocities ξ_1 and ξ_2, respectively; $\rho = x_1 - x_2$, and $v = \xi_1 - \xi_2$ (**Figure 4**). The ρ_v^{-d} vector has the coordinates $b, \varepsilon, -d$ in the cylindrical system of coordinates with the z axis parallel to the v vector, b is the impact parameter, and ε is the azimuthal angle. That is, the second term on the right hand side of Equation (2) is responsible for a decrease in the number of ξ_1-particles in collisions with other particles. The $\bar{F}_2(t, x_1, \xi_1', x_1 - \rho_{v'}^{-d}, \xi_2')$ function corresponds to a pair of particles 1 and 2, which enter the region of their interaction C_0 at velocities ξ_1' and ξ_2', respectively, $v' = \xi_1' - \xi_2'$. The coordinates of the $\rho_{v'}^{-d}$ vector are

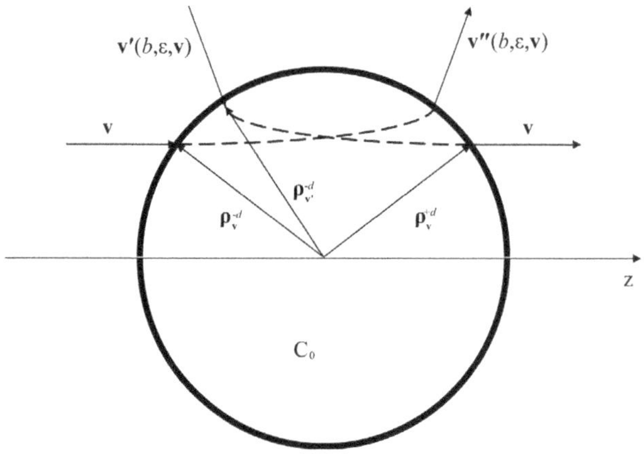

Figure 4. Interaction region C_0 of a pair of particles with characteristic radius *d*. If, at the entrance of region C_0, particles have velocity v and parameters b, ε, they acquire velocity $v''(b, \varepsilon, v)$ at its exit. If, at the entrance of region C_0, particles have relative movement velocity $v'(b, \varepsilon, v)$ and collision parameters b, ε, they acquire velocity v at its exit.

$b, \varepsilon, -d$ in the cylindrical system coordinates with the z axis parallel to v'. At the exit of the region of their interaction, particles 1 and 2 then have velocities ξ_1 and ξ_2 respectively (**Figure 4**), that is, the first term on the right hand side of (2) is responsible for an increase in the number of ξ_1-particles caused by collisions.

The factorization of two-particle distribution functions in $J(t, x_1, \xi_1, \xi_2)$ from Equation (2), that is, their representation in the form of the product of two one-particle functions,

$$\begin{aligned} &\bar{F}_2\left(t, x_1, \xi_1, x_1 - \rho_v^{-d}, \xi_2\right) = \bar{F}_1(t, x_1, \xi_1) \bar{F}_1(t, x_1, \xi_2) \\ &\bar{F}_2\left(t, x_1, \xi_1', x_1 - \rho_{v'}^{-d}, \xi_2'\right) = \bar{F}_1(t, x_1, \xi_1') \bar{F}_1(t, x_1, \xi_2') \end{aligned} \tag{4}$$

leads to a closed Boltzmann classic kinetic equation,

$$\begin{aligned} &\left(\frac{\partial}{\partial t} + \xi_1 \frac{\partial}{\partial x_1}\right) f_1(t, x_1, \xi_1) \\ &= \int v\left[f_1(t, x_1, \xi_1') f_1(t, x_1, \xi_2') - f_1(t, x_1, \xi_1) f_1(t, x_1, \xi_2)\right] \\ &\cdot b\mathrm{d}b\mathrm{d}\varepsilon\mathrm{d}\xi_2 \end{aligned} \tag{5}$$

The approximation that we use was called the molecular chaos hypothesis "Stosszahlansatz" [31]. Classic hydrodynamics equations follow directly from the Boltzmann equation (**Figure 3**) and, naturally, include errors made in the derivation of the classic kinetic equation. Let us elucidate the physical meaning of the error introduced by the Boltzmann hypothesis into hydrodynamics.

Let us pass in Equation (2) from $\bar{F}_2$ functions written in and x_1, ξ_1, x_2, ξ_2 variables to $\bar{\bar{F}}_2$ functions written in $x = (x_1 + x_2)/2$, $G = (\xi_1 + \xi_2)/2$, ρ, and v variables,

$$\tilde{J}(t,\boldsymbol{x}_1,\boldsymbol{G},\boldsymbol{v}) = \int v\left[\bar{\bar{F}}_2\left(t,\boldsymbol{x}_1,\boldsymbol{G},\boldsymbol{\rho}_{v'}^{-d},\boldsymbol{v}'\right) - \bar{\bar{F}}_2\left(t,\boldsymbol{x}_1,\boldsymbol{G},\boldsymbol{\rho}_v^{-d},\boldsymbol{v}\right)\right] b\mathrm{d}b\mathrm{d}\varepsilon \quad (6)$$

$$\tilde{J}(t,\boldsymbol{x}_1,\boldsymbol{G},\boldsymbol{v}) = J(t,\boldsymbol{x}_1,\xi_1,\xi_2)$$

$$F_2(t,\boldsymbol{x}_1,\xi_1,\boldsymbol{x}_2,\xi_2) = \tilde{F}_2(t,\boldsymbol{x},\boldsymbol{G},\boldsymbol{\rho},\boldsymbol{v})$$

The velocity of the center of mass of pair particles, $\boldsymbol{G}$, and the modulus v of the relative velocity of particles $v = |\boldsymbol{v}|$, are invariants of a binary particle collision. The enumeration of all the admissible target parameter values b and ε and relative motion velocity directions $\boldsymbol{v}$ at fixed $\boldsymbol{G}$ and v values then gives all the possible velocity directions $\boldsymbol{v}'$. It follows that a collision-caused decrease in the number of pairs of particles from a unit phase volume interval near the $\boldsymbol{x}_1,\boldsymbol{G},\boldsymbol{v}$ point characterized by all the admissible b and ε parameter values and relative motion $\boldsymbol{v}$ orientations is strictly balanced by a collision induced increase in the number of pairs of particles in this interval with these parameters,

$$\int \tilde{J}(t,\boldsymbol{x}_1,\boldsymbol{G},\boldsymbol{v})\mathrm{d}^2\Omega = 0 \quad (7)$$

Here $\mathrm{d}^2\Omega = \sin\theta\mathrm{d}\theta\mathrm{d}\varphi$, and θ and φ are the spherical coordinates of the $\boldsymbol{v}$ vector.

Suppose that $\varphi_2^{(n)}(\boldsymbol{G},v)$, $n=0,1,\cdots$, is an arbitrary weight function of velocities $\boldsymbol{G}$ and v. The properties of the $\varphi_2^{(n)}(\boldsymbol{G},v)$, $n=0,1,\cdots$, pair are invariants of a particle binary collision. Let us multiply the $\tilde{J}(t,\boldsymbol{x}_1,\boldsymbol{G},\boldsymbol{v})$ collision integral determined by Equation (6) by $\varphi_2^{(n)}(\boldsymbol{G},v)$, $n=0,1,\cdots$, and integrate the result with respect to velocities. By virtue of Equation (7), we then have,

$$\int \varphi_2^{(n)}(\boldsymbol{G},v)\tilde{J}(t,\boldsymbol{x}_1,\boldsymbol{G},\boldsymbol{v})\mathrm{d}\boldsymbol{G}\mathrm{d}\boldsymbol{v} = 0 \quad (8)$$

This means that particle collisions cannot influence the formation of hydrodynamic values constructed on the properties of the $\varphi_2^{(n)}(\boldsymbol{G},v)$, $n=0,1,\cdots$, pair. In other words, particle collisions cannot tune the distributions of all these hydrodynamic values to distributions of some other hydrodynamic values. That is, the set of hydrodynamic values constructed on the property of $\varphi_2^{(n)}(\boldsymbol{G},v)$, $n=0,1,\cdots$,

$$M^{(n)}(t,\boldsymbol{x}_1) = \frac{1}{A(t,\boldsymbol{x}_1)}\int \varphi_2^{(n)}(\boldsymbol{G},v)v\bar{\bar{F}}_2\left(t,\boldsymbol{x}_1,\boldsymbol{G},\boldsymbol{\rho}_v^{-d},\boldsymbol{v}\right)b\mathrm{d}b\mathrm{d}\varepsilon\mathrm{d}\boldsymbol{G}\mathrm{d}\boldsymbol{v} \quad (9)$$

is a set of the principal hydrodynamic values, the $A(t,\boldsymbol{x}_1)$ is local proportionality coefficient. Set (9) contains not only the lower $\bar{\bar{F}}_2$ function moments but also all the higher moments without exception. To summarize, our analysis of the properties of the $\tilde{J}(t,\boldsymbol{x}_1,\boldsymbol{G},\boldsymbol{v})$ collision integral reveals the existence of an infinite number of principal hydrodynamic values.

Under free molecular conditions, when $\lambda \gg L$, particle collisions do not participate in the formation of distributions of hydrodynamic values. These distributions are formed under the influence of the initial and boundary problem conditions, that is, each hydrodynamic value is principal. As the density of medium increases, the influence of intermolecular collisions on the formation of hydrodynamic value distributions increases. In a continuous medium, $\lambda \ll L$, this influence becomes predominant. Namely, binary particle collisions tune the distributions of all the function moments to distribution of moments (9). The velocities $\boldsymbol{G}$ and v are invariants of a particle binary collision. That is why, according to Equation (8), moments (9), as previously, do not experience the influence of binary particle collisions. As in the free molecular mode, that is, at $\lambda \ll L$, moments (9) remain principal hydrodynamic values.

The equation for the $f_1(t,\boldsymbol{x}_1,\xi_1)$ one-particle distribution function is written in a six-dimensional phase space of one particle (μ space). The dimensionality of the μ space allows only the properties of particle $\varphi_1^{(n)}(\xi_1)$, $n=0,1,\cdots$, to be accommodated in it; binary particle collision invariants, that is, $\varphi_2^{(n)}(\boldsymbol{G},v)$, $n=0,1,\cdots$, do not fit into the μ space. The symmetry properties of the $J(t,\boldsymbol{x}_1,\xi_1,\xi_2)$ collision integral strictly correspond to the symmetry properties of the Boltzmann collision integral (5), [31]. It follows that only three lower particle properties, namely $1,\xi_1,\xi_1^2$, can lead to the disappearance of the $J(t,\boldsymbol{x}_1,\xi_1)$ collision integral. Let us sequentially accommodate three lower properties of a particle in the μ space. We then have,

$$\begin{aligned}
0 &= \int J(t,\boldsymbol{x}_1,\xi_1)\mathrm{d}\xi_1 = \int \tilde{J}(t,\boldsymbol{x}_1,\boldsymbol{G},\boldsymbol{v})\mathrm{d}\boldsymbol{G}\mathrm{d}\boldsymbol{v} \\
0 &= \int \xi_1 J(t,\boldsymbol{x}_1,\xi_1)\mathrm{d}\xi_1 \\
&= \int\left(\boldsymbol{G}+\frac{\boldsymbol{v}}{2}\right)\tilde{J}(t,\boldsymbol{x}_1,\boldsymbol{G},\boldsymbol{v})\mathrm{d}\boldsymbol{G}\mathrm{d}\boldsymbol{v} \\
&= \int \boldsymbol{G}\tilde{J}(t,\boldsymbol{x}_1,\boldsymbol{G},\boldsymbol{v})\mathrm{d}\boldsymbol{G}\mathrm{d}\boldsymbol{v} \qquad (10)\\
0 &= \int \xi_1^2 J(t,\boldsymbol{x}_1,\xi_1)\mathrm{d}\xi_1 \\
&= \int\left(\boldsymbol{G}^2+\frac{\boldsymbol{v}^2}{4}+\boldsymbol{G}\boldsymbol{v}\right)\tilde{J}(t,\boldsymbol{x}_1,\boldsymbol{G},\boldsymbol{v})\mathrm{d}\boldsymbol{G}\mathrm{d}\boldsymbol{v} \\
&= \int\left(\boldsymbol{G}^2+\frac{\boldsymbol{v}^2}{4}\right)\tilde{J}(t,\boldsymbol{x}_1,\boldsymbol{G},\boldsymbol{v})\mathrm{d}\boldsymbol{G}\mathrm{d}\boldsymbol{v}
\end{aligned}$$

Equations (10) are valid because of symmetry of the partial distribution function $\tilde{F}_2(t,\boldsymbol{x},\boldsymbol{G},\boldsymbol{\rho},\boldsymbol{v})$ with respect to the permutation of the phase coordinates of two particles [3]. The symmetry properties allow all the moments of the two-particle distribution functions odd with respect to $\boldsymbol{v}$ to be removed from (10), [32].

It follows that the integrals that contain the lower particle properties $1,\xi_1,\xi_1^2$, as weight functions are strictly reduced to the integrals with the $1,\boldsymbol{G},\left(\boldsymbol{G}^2+\boldsymbol{v}^2/4\right)$ weight functions belonging to $\varphi_2^{(n)}(\boldsymbol{G},\boldsymbol{v})$ $n=0,1,\cdots$, invariants. Three lower particle properties $1,\xi_1$, and ξ_1^2 give three lower $f_1(t,\boldsymbol{x}_1,\xi_1)$ one-particle distribution function moments, the density of the number of particles $n(t,\boldsymbol{x})$, medium velocity $\boldsymbol{U}(t,\boldsymbol{x})$, and medium pressure $p(t,\boldsymbol{x})$. It follows that, when we pass to the hydrodynamic stage from the phase space of one particle, such hydrodynamics equations cannot be constructed using more than three lower principal hydrodynamic values corresponding to the $1,\xi_1,\xi_1^2$ particle properties. The transition to the hydrodynamic stage from the phase space of one particle excludes higher principal hydrodynamic values (9) from participation in the construction of hydrodynamics equations. However, there is no rigorous passage to hydrodynamics from the μ space. This passage is closed because Equation (2) is not closed.

The use of the Boltzmann hypothesis ("Stosszahlansatz") (4) opens up the possibility of approximate passage to hydrodynamics. So, the physical meaning of the error introduced by the Boltzmann hypothesis into hydrodynamics is as follows. It follows that just Boltzmann hypothesis allows us to construct hydrodynamics on only three lower principal hydrodynamic values. It follows that the use of the Boltzmann hypothesis excludes higher principal hydrodynamic values (9) from the participation in the formation of classic hydrodynamics equations. Since the classic three-moment hydrodynamics is constructed without the use of values (9), its applicability range is limited to states that are only weakly removed from the statistical equilibrium state. To substantiate this conclusion, we must find passage to hydrodynamics from the phase space capable of accommodateing the whole set of binary particle collision $\varphi_2^{(n)}(\boldsymbol{G},\boldsymbol{v})$ $n=0,1,\cdots$, invariants.

The $F_2(t,\boldsymbol{x}_1,\xi_1,\boldsymbol{x}_2,\xi_2)$ two-particle distribution function obeys the second equation of the BBGKY hierarchy (1). The second equation is valid for arbitrary positions of particles 1 and 2 in space with respect to each other. It is not closed. There are, however, positions of particles 1 and 2 for which the second equation is considerably simplified.

The left hand side of the second equation of the BBGKY hierarchy (1),

$$\begin{aligned}&\left[\frac{\partial}{\partial t}+\xi_1\frac{\partial}{\partial \boldsymbol{x}_1}+\xi_2\frac{\partial}{\partial \boldsymbol{x}_2}+\left(\frac{\boldsymbol{\Phi}_{1,2}}{m}\right)\frac{\partial}{\partial \xi_1}+\left(\frac{\boldsymbol{\Phi}_{2,1}}{m}\right)\frac{\partial}{\partial \xi_2}\right]\\&\cdot F_2(t,\boldsymbol{x}_1,\xi_1,\boldsymbol{x}_2,\xi_2)\\&=-(N-2)\int\left[\left(\frac{\boldsymbol{\Phi}_{1,3}}{m}\right)\frac{\partial}{\partial \xi_1}+\left(\frac{\boldsymbol{\Phi}_{2,3}}{m}\right)\frac{\partial}{\partial \xi_2}\right]\\&\cdot F_3(t,\boldsymbol{x}_1,\xi_1,\boldsymbol{x}_2,\xi_2,\boldsymbol{x}_3,\xi_3)\mathrm{d}\boldsymbol{x}_3\mathrm{d}\xi_3\end{aligned}\tag{11}$$

describes the free movement of particles 1 and 2 and the interaction between the particles 1 and 2. The integral terms containing the $F_3(t,\boldsymbol{x}_1,\xi_1,\boldsymbol{x}_2,\xi_2,\boldsymbol{x}_3,\xi_3)$ three-particle distribution function are responsible for the interaction of particles 1 and 2 with some third particle 3. Presuming that a particle may be present at all phase space locations with equal probabilities, introduce the dimensionless quantities,

$$\begin{aligned}&F_2=\left(1/V^2c^6\right)\hat{F}_2,\quad F_3=\left(1/V^3c^9\right)\hat{F}_3,\\&\xi_i=c\hat{\xi}_i,\quad\left(\frac{\boldsymbol{\Phi}_{i,j}}{m}\right)=\left(c^2/d\right)\left(\frac{\hat{\boldsymbol{\Phi}}_{i,j}}{\hat{m}}\right),\quad i,j=1,2,3\end{aligned}\tag{12}$$

In Equation (12) the hat appears above the dimensionless quantities, c is the characteristic velocity of the particle, the ratio of the system volume V to the characteristic volume $d^2\lambda$ occupied by one particle is used as an estimation for N. In the thermodynamic limit, $N\to\infty$, $V\to\infty$, yet, N/V is finite.

Let us bring the convective terms in the left hand side of Equation (11) into dimensionless form on the scale of particle size, $t=(d/c)\hat{t}$, $\boldsymbol{x}_i=d\hat{\boldsymbol{x}}_i$. In this case, the ratio of order of the right hand side terms to order of the left hand side terms of Equation (11) is of the order of d/λ, $d\ll\lambda$. It means that the interaction with third particle has a very weak influence on the free movement of particles 1 and 2 and the interaction between the particles 1 and 2 at times, proportional to d/c. The right hand side terms of Equation (11) receive the fundamental order only at times, proportional to λ/c.

Let us substitute the force term of the left hand side of the third equation of the BBGKY hierarchy (1) into the first term of the right hand side of Equation (11). Then,

$$\begin{aligned}&-(N-2)\int\left(\frac{\boldsymbol{\Phi}_{1,3}}{m}\right)\frac{\partial}{\partial \xi_1}F_3(t,\boldsymbol{x}_1,\xi_1,\boldsymbol{x}_2,\xi_2,\boldsymbol{x}_3,\xi_3)\mathrm{d}\boldsymbol{x}_3\mathrm{d}\xi_3\\&=(N-2)\int w\Big[\tilde{\tilde{F}}_3\left(t,\boldsymbol{x}_1,\xi_1,\boldsymbol{x}_2,\xi_2,\boldsymbol{r}_w^{+d},\boldsymbol{w}\right)\\&\qquad-\tilde{\tilde{F}}_3\left(t,\boldsymbol{x}_1,\xi_1,\boldsymbol{x}_2,\xi_2,\boldsymbol{r}_w^{-d},\boldsymbol{w}\right)\Big]b\mathrm{d}b\mathrm{d}\varepsilon\mathrm{d}\boldsymbol{w}\end{aligned}\tag{13}$$

Here,
$\tilde{\tilde{F}}_3(t,\boldsymbol{x}_1,\xi_1,\boldsymbol{x}_2,\xi_2,\boldsymbol{r},\boldsymbol{w})=F_3(t,\boldsymbol{x}_1,\xi_1,\boldsymbol{x}_2,\xi_2,\boldsymbol{x}_3,\xi_3)$, $\boldsymbol{r}=\boldsymbol{x}_1-\boldsymbol{x}_3$, $\boldsymbol{w}=\xi_1-\xi_3$. In the right hand side of Equation (13), the terms responsible for triple collisions of particles are omitted. The convective terms responsible for displacement of particles 1 and 2 during the time of interaction of particles 1 with particle 3 are omitted too. The omitted terms are negligible with respect to saved terms in the right hand side of Equation (13). The estimation of orders is carried out in accordance with Equation (12). As noted above, the order of the convective terms is estimated at times, proportional to λ/c.

Positions of particles 1 and 3, described by the first and second terms of the right hand side of Equation (13), are depicted in **Figure 4**. However, when interpreting the collision integral (13) the vectors $\boldsymbol{v}$, $\boldsymbol{\rho}_v^{+d}$, $\boldsymbol{\rho}_v^{-d}$ in **Figure 4**, should be replaced respectively by the vectors $\boldsymbol{w}$, $\boldsymbol{r}_w^{+d}$, $\boldsymbol{r}_w^{-d}$. The first term of the right hand side of Equation (13) is responsible for an increase in the number of ξ_1-particles caused by collisions with some third ξ_3-particles $\left(\boldsymbol{r}=\boldsymbol{r}_w^{+d}\right)$. If at time t particles 1 and 3 leave the region of their interaction C_0, then later, within a time interval not exceeding λ/c, they will be placed with respect to each other within the cylindrical volume oriented in the direction of vector $\boldsymbol{w}$ with bases area of πd^2 and height of λ. *I.e.*, these particles within a time interval of λ/c will be in the stage of departure from each other. The second term of the right hand side of Equation (13) is responsible for a decrease in the number of ξ_1-particles in collisions with some third ξ_3-particles $\left(\boldsymbol{r}=\boldsymbol{r}_w^{-d}\right)$. If at time t particles 1 and 3 enter the region of their interaction C_0, then earlier, within a time interval not exceeding λ/c, they were placed with respect to each other within the cylindrical volume oriented in the direction of vector $\boldsymbol{w}$ with bases area of πd^2 and height of λ. *i.e.*, these particles within a time interval of λ/c were in the stage of approaching with each other. Thus, the terms of the right hand side of Equation (13) at times, proportional to λ/c, describe groups of three particles, in which two of three particles (1 and 3) find itself either at the stage of approaching before the collision with each other, or in the stage of departure from each other after collision. The second term of the right hand side of Equation (11) has the identical physical meaning.

The idea that leads to the concept of pair distribution functions is as follows. For instance, in a rarefied gas at each time moment, each particle after its last collision moves toward the next collision. This means that every particle 1 in a rarefied gas simultaneously flies away from some particle 2 with which it collided last at point *D* (**Figure 5**), and approaches some particle 2 with which it is to collide next at point *B*.

It follows that it is particle 2 plays the role of a third particle 3 for particle 1, with which particle 1 has already experienced a collision and will be in the stage of departure within a time interval not exceeding λ/c (**Figure 5(a)**). At the same time, it is particle 1 plays the role of a third particle 3 for particle 2, with which the particle 2 has already experienced a collision and will be in the stage of departure within a time interval not exceeding λ/c (**Figure 5(a)**). From the other hand, it is a particle 2 plays the role of a third particle 3 for particle 1, with which particle 1 should collide within a time interval not exceeding λ/c (**Figure 5(b)**). At the same time, it is particle 1 plays the role of a third particle 3 for particle 2,

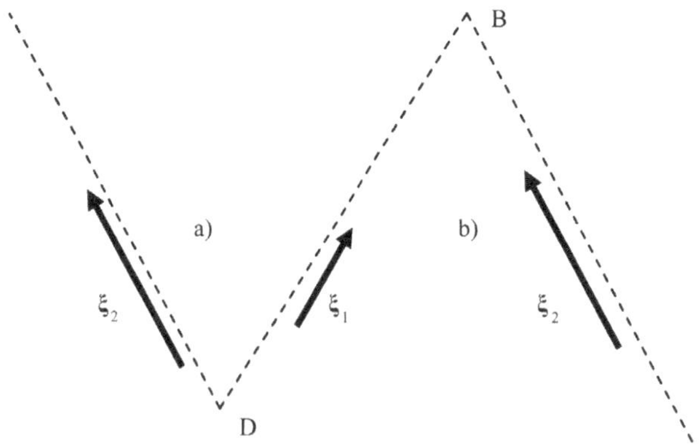

Figure 5. Graphic representation of a pair of particles.

with which the particle 2 should collide within a time interval not exceeding λ/c (**Figure 5(b)**).

The motion and interaction of particles 1 and 2 are, however, described by the left-hand side terms of the second equation of hierarchy (11) that are proportional to $F_2\left(t,\boldsymbol{x}_1,\xi_1,\boldsymbol{x}_2,\xi_2\right)$. For this reason, the terms containing $F_3\left(t,\boldsymbol{x}_1,\xi_1,\boldsymbol{x}_2,\xi_2,\boldsymbol{x}_3,\xi_3\right)$ must be excluded for particles 1 and 2. The second equation of the BBGKY hierarchy (11) written for particles 1 and 2 (**Figure 5**), is therefore considerably simplified,

$$\left(\frac{\partial}{\partial t}+\xi_1\frac{\partial}{\partial \boldsymbol{x}_1}+\xi_2\frac{\partial}{\partial \boldsymbol{x}_2}+\frac{\boldsymbol{\Phi}_{12}}{m}\frac{\partial}{\partial \xi_1}+\frac{\boldsymbol{\Phi}_{21}}{m}\frac{\partial}{\partial \xi_2}\right)\cdot F_2^p\left(t,\boldsymbol{x}_1,\xi_1,\boldsymbol{x}_2,\xi_2\right)=0 \tag{14}$$

Generally, any medium particle forms a pair with every other particle. A medium therefore contains $N(N-1)$ pairs of particles. All these pairs are described by the $F_2\left(t,\boldsymbol{x}_1,\xi_1,\boldsymbol{x}_2,\xi_2\right)$ function, which obeys the second equation of hierarchy (11). If a single particle 2, which either flies away from (**Figure 5(a)**), or approaches (**Figure 5(b)**), some particle 1, is selected as a partner of this particle, this pair is described by the $F_2^p\left(t,\boldsymbol{x}_1,\xi_1,\boldsymbol{x}_2,\xi_2\right)$ function that obeys Equation (14). Note that Equation (14) is valid for an arbitrary gas particle rather than some particular particle 1 (**Figure 5**). For this reason, Equation (14) is capable of describing the gas as a whole. A rigorous derivation of Equation (14) is given in [33]. There is an analogy between the derivation [33] and Grad method used in deriving the Boltzmann equation for a gas consisting of rigid spheres [30].

The arguments of the $F_2^p\left(t,\boldsymbol{x}_1,\xi_1,\boldsymbol{x}_2,\xi_2\right)$ function are $\boldsymbol{x}_1$ and $\boldsymbol{x}_2$, which are the spatial coordinates of particles 1 and 2. Information about the position of a separate particle in space, however, becomes lost in kinetics and hydrodynamics. Classic kinetics and hydrodynamics deal with the place in space near which a set of particles are situated rather than with particle coordinates. It follows that a necessary condition for the transition from (14) to kinetics and hydrodynamics equations is getting rid of one of the spatial coordinates of the F_2^p function. It was suggested in [33] that F_2^p be integrated in one of its spatial coordinates within the cylinder of collisions. The Δ_1 and Δ_2 cylinders of collisions are oriented in the direction of the velocity $\boldsymbol{v}$, the areas of their bases equal the collision cross section area πd^2, and the heights of the cylinders equal the characteristic free path $\lambda(t,\boldsymbol{x})$. Let us pass from the 12-dimensional phase space of particles 1 and 2 constructed on the $\boldsymbol{x}_1,\boldsymbol{\xi}_1,\boldsymbol{x}_2,\boldsymbol{\xi}_2$ vectors to the 12-dimensional space constructed on the $\boldsymbol{x}$, $\boldsymbol{G}$, $\boldsymbol{\rho}$, $\boldsymbol{v}$ vectors and produce pair distribution functions $f_p^{app}(t,\boldsymbol{x},\boldsymbol{G},\boldsymbol{v})$ and $f_p^{div}(t,\boldsymbol{x},\boldsymbol{G},\boldsymbol{v})$,

$$\begin{aligned} & f_p^{app}(t,\boldsymbol{x},\boldsymbol{G},\boldsymbol{v}) \\ & = \frac{N(N-1)}{2l^3} \int_W \int_{\Delta_1} \tilde{F}_2(t,\boldsymbol{x}+\boldsymbol{a},\boldsymbol{G},\boldsymbol{\rho},\boldsymbol{v})\mathrm{d}\boldsymbol{\rho}\mathrm{d}\boldsymbol{a} \\ & f_p^{div}(t,\boldsymbol{x},\boldsymbol{G},\boldsymbol{v}) \\ & = \frac{N(N-1)}{2l^3} \int_W \int_{\Delta_2} \tilde{F}_2(t,\boldsymbol{x}+\boldsymbol{a},\boldsymbol{G},\boldsymbol{\rho},\boldsymbol{v})\mathrm{d}\boldsymbol{\rho}\mathrm{d}\boldsymbol{a} \\ & F_2(t,\boldsymbol{x}_1,\boldsymbol{\xi}_1,\boldsymbol{x}_2,\boldsymbol{\xi}_2) = \tilde{F}_2(t,\boldsymbol{x},\boldsymbol{G},\boldsymbol{\rho},\boldsymbol{v}) \end{aligned} \tag{15}$$

Integration (15) collects all pairs of particles that either will inevitably collide during the time interval not exceeding the characteristic time between collisions $\tau = \lambda/v$ or fly away from each other after a collision that occurs during the time interval not exceeding τ. The integration with respect to $\boldsymbol{a}$ within W region removes strong spatial dependence of the functions on the scale having the characteristic size of particles d. The pair distribution function $f_p^{app}(t,\boldsymbol{x},\boldsymbol{G},\boldsymbol{v})$ $\left(f_p^{div}(t,\boldsymbol{x},\boldsymbol{G},\boldsymbol{v})\right)$ has the meaning of the probable number of pairs of particles approaching (diverging from) each other whose centers of mass are concentrated at time t in a unit volume element near point $\boldsymbol{x}$ and the velocities of the centers of mass and relative particle velocities, in unit velocity elements near $\boldsymbol{G}$ and $\boldsymbol{v}$, respectively. The relations between pair distribution functions (15) and $v\bar{\tilde{F}}_2(t,\boldsymbol{x}_1,\boldsymbol{G},\boldsymbol{\rho}_v^{-d},\boldsymbol{v})$, and $f_1(t,\boldsymbol{x}_1,\boldsymbol{\xi}_1)$ from [3] can be used to find the proportionality coefficient (9),

$$A = N(N-1)\tau \tag{16}$$

Heuristic derivation of a system of kinetic equations for pair distribution functions $f_p^{app}(t,\boldsymbol{x},\boldsymbol{G},\boldsymbol{v})$ and $f_p^{div}(t,\boldsymbol{x},\boldsymbol{G},\boldsymbol{v})$ was given in [34]. In [33], equations for pair distribution functions (15) are derived directly from the fundamental statistical mechanics concepts. An analysis of the system of kinetic equations [33,34] reveals the most important property of $f_p(t,\boldsymbol{x},\boldsymbol{G},v)$ pair distribution function constructed on the velocity of the center of mass $\boldsymbol{G}$ and the relative velocity modulus v,

$$\left(\frac{\partial}{\partial t} + \boldsymbol{G}\frac{\partial}{\partial \boldsymbol{x}}\right) f_p(t,\boldsymbol{x},\boldsymbol{G},v) = 0 \tag{17}$$

Here,

$$\begin{aligned} & f_p(t,\boldsymbol{x},\boldsymbol{G},v) = f_p^{app}(t,\boldsymbol{x},\boldsymbol{G},v) = f_p^{div}(t,\boldsymbol{x},\boldsymbol{G},v) \\ & f_p^{app}(t,\boldsymbol{x},\boldsymbol{G},v) = \int f_p^{app}(t,\boldsymbol{x},\boldsymbol{G},\boldsymbol{v})\sin\theta\mathrm{d}\theta\mathrm{d}\varphi \\ & f_p^{div}(t,\boldsymbol{x},\boldsymbol{G},v) = \int f_p^{div}(t,\boldsymbol{x},\boldsymbol{G},\boldsymbol{v})\sin\theta\mathrm{d}\theta\mathrm{d}\varphi \end{aligned} \tag{18}$$

and v,θ, and φ are the spherical coordinates of the $\boldsymbol{v}$ vector. According to Equation (17), the $f_p(t,\boldsymbol{x},\boldsymbol{G},v)$ function remains unchanged at time along the trajectory of the center of mass of the pair.

The lower moments of the pair distribution functions (18) are the local density of the number of pairs $n^G(t,\boldsymbol{x})$, the mean velocity of the centers of mass of pairs $\boldsymbol{U}^G(t,\boldsymbol{x})$, the $P_{ij}^G(t,\boldsymbol{x})$ tensor of stresses created as a result of movement of the centers of mass of pairs of particles, the $P_{ij}^v(t,\boldsymbol{x})$ tensor of stresses created because of relative movement of pair particles, the $\boldsymbol{q}^G(t,\boldsymbol{x})$ heat flux vector corresponding to the transfer of thermal energy because of movement of the centers of mass of pairs of particles, $\boldsymbol{q}^v(t,\boldsymbol{x})$ and $\boldsymbol{q}^{Gv}(t,\boldsymbol{x})$ the and heat flux vectors corresponding to thermal energy transfer because of relative movement of particles in pairs. Along with $P_{ij}^G(t,\boldsymbol{x})$ and $P_{ij}^v(t,\boldsymbol{x})$ let us determine small stress tensors $p_{ij}^G(t,\boldsymbol{x})$ and $p_{ij}^v(t,\boldsymbol{x})$, and $p^G(t,\boldsymbol{x})$ and $p^v(t,\boldsymbol{x})$ pressure, and $T^G(t,\boldsymbol{x})$ and $T^v(t,\boldsymbol{x})$ temperature, $p^G = nkT^G$, $p^v = nkT^v$. The lower moments of the pair distribution functions (18) are determined by Equations (21)-(26) in [3]. The set of hydrodynamic values, namely, $n^G(t,\boldsymbol{x})$, $\boldsymbol{U}^G(t,\boldsymbol{x})$, $p^G(t,\boldsymbol{x})$, $p^v(t,\boldsymbol{x})$, $p_{ij}^G(t,\boldsymbol{x})$, $\boldsymbol{q}^G(t,\boldsymbol{x})$ and $\boldsymbol{q}^v(t,\boldsymbol{x})$ is constructed on the property of $\varphi_2^{(n)}(\boldsymbol{G},v)$ $n = 0,1,\cdots$ It is a set of the principal hydrodynamic values. The lower moments $p_{ij}^v(t,\boldsymbol{x})$ and $\boldsymbol{q}^{Gv}(t,\boldsymbol{x})$ are not the principal hydrodynamic values.

The relations between the $f_1(t,\boldsymbol{x}_1,\boldsymbol{\xi}_1)$ one-particle distribution function and the pair distribution functions (18) from [3] can be used to find a relation between the one-particle distribution function and the pair distribution functions moments,

$$\begin{aligned} & n = n^G \quad \boldsymbol{U} = \boldsymbol{U}^G \\ & p = \frac{1}{2}p^G + \frac{1}{2}p^v \qquad T = \frac{1}{2}T^G + \frac{1}{2}T^v \\ & P_{ij} = \frac{1}{2}P_{ij}^G + \frac{1}{2}P_{ij}^v \quad p_{ij} = \frac{1}{2}p_{ij}^G + \frac{1}{2}p_{ij}^v \\ & \boldsymbol{q} = \frac{1}{2}\boldsymbol{q}^G + \frac{5}{6}\boldsymbol{q}^v + \boldsymbol{q}^{Gv} \end{aligned} \tag{19}$$

Here, $n = n(t,\boldsymbol{x})$ denotes the number densities of particles, $\boldsymbol{U} = \boldsymbol{U}(t,\boldsymbol{x})$ is the hydrodynamic velocity, $p = p(t,\boldsymbol{x})$ is the pressure, $T = T(t,\boldsymbol{x})$ is the temperature, $p = nkT$, $P_{ij} = P_{ij}(t,\boldsymbol{x})$ and $p_{ij} = p_{ij}(t,\boldsymbol{x})$ are the viscous stress tensors, and $\boldsymbol{q} = \boldsymbol{q}(t,\boldsymbol{x})$ is the heat flux.

A derivation of equations determining lower moments of the pair distribution functions is given in [3]. In [32], these hydrodynamics equations are written in the form,

$$\frac{\partial}{\partial t}[mn] + \frac{\partial}{\partial x_i}[mnU_i] = 0 \quad (20a)$$

$$\frac{\partial}{\partial t}[mnU_i] + \frac{\partial}{\partial x_j}\left[\frac{1}{2}p^G\delta_{ij} + \frac{1}{2}p_{ij}^G + mnU_iU_j\right] = 0 \quad (20b)$$

$$\frac{\partial}{\partial x_j}\left[\frac{1}{2}p^v\delta_{ij} + \frac{1}{2}p_{ij}^v\right] = 0 \quad (20c)$$

$$\frac{\partial}{\partial t}\left[\frac{3}{4}nkT^G + \frac{1}{2}mn\boldsymbol{U}^2\right] + \frac{\partial}{\partial x_i}\left[\frac{1}{2}q_i^G + \frac{5}{4}U_i nkT^G + \frac{1}{2}U_j p_{ij}^G + \frac{1}{2}mnU_i\boldsymbol{U}^2\right] = 0 \quad (20d)$$

$$\frac{\partial}{\partial t}\left[\frac{3}{4}nkT^v\right] + \frac{\partial}{\partial x_i}\left[\frac{1}{2}q_i^v + \frac{3}{4}U_i nkT^v\right] = 0 \quad (20e)$$

$$\frac{\partial}{\partial x_i}\left[q_i^{Gv} + \frac{1}{3}q_i^v + \frac{1}{2}U_i p^v + \frac{1}{2}U_j p_{ij}^v\right] = 0 \quad (20f)$$

Taking (19) into account, we find that the sum of Equations (20b) and (20c) is none other than the equation of the conservation of momentum, and the sum of Equations (20d), (20e) and (20f) is none other than the equation of conservation of energy. Equation (20a) is the continuity equation. So, the equation for conservation of momentum separates into two equations, and the equation for conservation of energy, into three separate equations. In Equation (20), nonprincipal hydrodynamic values p_{ij}^v and q_i^{Gv} are given by the equations,

$$p_{ij}^v = -\tilde{\eta}\left(\frac{\partial U_i}{\partial x_j} + \frac{\partial U_j}{\partial x_i} - \frac{2}{3}\delta_{ij}\frac{\partial U_k}{\partial x_k}\right) - \frac{2\tilde{\eta}}{3p^v}\left(\frac{\partial q_i^v}{\partial x_j} + \frac{\partial q_j^v}{\partial x_i} - \frac{2}{3}\delta_{ij}\frac{\partial q_k^v}{\partial x_k}\right) \quad (21a)$$

$$q_i^{Gv} = -\tilde{\lambda}\frac{\partial T}{\partial x_i} - \frac{\tilde{\lambda}}{5p^v}\frac{\partial\left(p_{ij}^G T^v\right)}{\partial x_j} - \frac{4\tilde{\lambda}}{15p^v}\frac{m}{k}\left(q_i^v\delta_{jk} + q_j^v\delta_{ik} + q_k^v\delta_{ij}\right)\frac{\partial U_k}{\partial x_j} \quad (21b)$$

Here, $\tilde{\eta}$ is the dynamic viscosity coefficient and $\tilde{\lambda} = 15k\tilde{\eta}/4m$ is the heat conductivity coefficient.

Relations (19) and Equations (20), (21) contain seven lower principal hydrodynamic values, namely, $n^G(t,\boldsymbol{x})$, $\boldsymbol{U}^G(t,\boldsymbol{x})$, $p^G(t,\boldsymbol{x})$, $p^v(t,\boldsymbol{x})$, $p_{ij}^G(t,\boldsymbol{x})$, $\boldsymbol{q}^G(t,\boldsymbol{x})$ and $\boldsymbol{q}^v(t,\boldsymbol{x})$. Just these values dominate in equations for measurable one-particle function moments (19). All the other higher principal hydrodynamic values (9) were omitted in Equations (19)-(21). An increase in the number of principal hydrodynamic values taken into account will cause the additional appearance of derivatives of all the higher principal moments (9) taken into account in both definitions (19) and Equations (20), (21). The lower one particle distribution function moments n, $\boldsymbol{U}$, and p are written through the principal hydrodynamic values n^G, U^G, $p^G(t,\boldsymbol{x})$ and $p^v(t,\boldsymbol{x})$. The higher measurable moments p_{ij} and $\boldsymbol{q}$ are represented by linear combinations of principal and non-principal hydrodynamic values. The p_{ij}^G stress tensor component is a principal hydrodynamic value. The first term of the p_{ij}^v component is proportional to the tensor of deformation velocities U_{ij} from Equation (21a), the remaining term of the component appears because of an increase in the number of principal hydrodynamic values. The first two components of the $\boldsymbol{q}$ heat flux vector, $\boldsymbol{q}^G$ and $\boldsymbol{q}^v$, are principal hydrodynamic values. The first term of the third component, $\boldsymbol{q}^{Gv}$, is proportional to the temperature gradient from Equation (21b), and the other terms of the third component appear because of an increase in the number of principal hydrodynamic values.

The upper indexes "*div*" and "*app*" were omitted in the p_{ij}^v and q_i^{Gv} hydrodynamic values because the right hand sides of Equations (19) and (21) do not experience any changes in calculations of moments with pair functions corresponding either to particles that approach each other, $f_p^{app}(t,\boldsymbol{x},\boldsymbol{G},\boldsymbol{v})$, or to particles that fly away from each other, $f_p^{div}(t,\boldsymbol{x},\boldsymbol{G},\boldsymbol{v})$ [3].

The analysis performed in [35] showed that hydrodynamic Equations (20), (21) could be rigorously reduced to classic hydrodynamics equations only when the state of the system weakly deviated from thermodynamic equilibrium. In particular, in the $\mathrm{Re} \ll 1$ limit, the $p_{ij}^G(t,\boldsymbol{x})$ main hydrodynamic value and the $p_{ij}^v(t,\boldsymbol{x})$ value are equal. Moreover, the second term on the right-hand side of Equation (21a), which is responsible for viscous stresses caused by heat fluxes, becomes negligibly small compared with the first term. It follows that, as is shown in [35], the Newton law [29] is only valid in the $\mathrm{Re} \ll 1$ limit.

An indefinite $\lambda(t,\boldsymbol{x})$ parameter is the characteristic free path. An indefinite $\lambda(t,\boldsymbol{x})$ parameter is introduced at the stage of the formation of pair distribution functions (15). The characteristic time between collisions $\tau(t,\boldsymbol{x}) = \lambda(t,\boldsymbol{x})/v$ is contained in resulting kinetic equations and hydrodynamics equations [3]. The limit of weak non-equilibrium allows us to find the proportionality factor between the $\tilde{\eta}$ dynamic viscosity coefficient and τ,

$$\tau = 2\tilde{\eta} / p^{v} \tag{22}$$

The $\tilde{\eta}$ coefficient can be calculated with arbitrary accuracy by solving the Boltzmann equation with the use of the Chapman–Enscog method for an arbitrary law of interparticle interactions [31].

In calculations of the right hand sides of the equations for the pair functions (18), collision integrals were calculated in [3] for a gas consisting of rigid spheres with diameter d. This case has exceptional advantages, because, in integration in velocities $\boldsymbol{G}$ and $\boldsymbol{v}$ with the $\varphi_2^{(n)}(\boldsymbol{G}, v)$ $n = 0, 1, \cdots$, weight function, the collision integrals of kinetic equations are expressed exactly in terms of the moments of pair distribution functions. The classic Boltzmann collision integral offers similar advantages to the case of so-called Maxwell molecules [36]. The structure of the derived hydrodynamics Equations (20) and (21) does not depend on the form of the law of structureless particle interactions. The interaction law only influences the values of transfer coefficients. It follows that system (20) and (21) is valid for the description of gas flows with an arbitrary law of interaction of structureless particles.

In the derivation of hydrodynamics Equations (19)-(21), only terms linear in τ are retained. At the hydrodynamic description stage, the omitted terms correspond to so-called Barnett accuracy [31,36]. In follows that hydrodynamics Equations (20) and (21) presented above correspond to Navier-Stokes accuracy of description.

The system of equations determining lower moments of the pair distribution functions (20,21) is not closed. Let us multiply the Equation (17) for the $f_p(t, \boldsymbol{x}, \boldsymbol{G}, v)$ pair function by lower invariants of binary particle collision, $\varphi_2^{(n)}(\boldsymbol{G}, v)$, and integrate the result with respect to velocities $\mathbf{G}$ and v. By virtue of definition of lower moments of pair functions [3], we then have hydrodynamic Equations (20a, b, d, e). Thus, Equation (17) is equivalent to several equations, namely, Equations (20a, b, d, e). According to (17), the $f_p(t, \boldsymbol{x}, \boldsymbol{G}, v)$ function remains unchanged at time along the trajectory of the center of mass of the pair. This property creates the additional relations between the principal hydrodynamic values. The $\boldsymbol{U}^G$, $\boldsymbol{q}^G$ and $\boldsymbol{q}^v$ vectors, and the p_{ij}^G tensor are multicomponent principal hydrodynamic values. It turned out that the use of the conservation property (17) is accompanied by the appearance of additional relations between each vector and tensor components. Generally, just these relations must close the system (20,21) of hydrodynamics equations.

However, a problem on flow around a fixed solid sphere was solved in [4,5] by another manner. In [4], the solution to Equation (17) was built. The first integrals of Equation (17) for a stationary system are the $F_{xy}(\boldsymbol{G}, \boldsymbol{x})$, $F_{yz}(\boldsymbol{G}, \boldsymbol{x})$, and $F_{zx}(\boldsymbol{G}, \boldsymbol{x})$ functions,

$$\begin{gathered} F_{xy} = G_x y - G_y x \quad F_{yz} = G_y z - G_z y \\ F_{zx} = G_x z - G_z x \\ \boldsymbol{G} = (G_x, G_y, G_z), \quad \boldsymbol{x} = (x, y, z) \end{gathered} \tag{23}$$

According to [4], a solution to Equation (17) in the problem with time-independent boundary conditions should be sought in the form of a series of the products of the first integrals,

$$f_p(t, \boldsymbol{x}, \boldsymbol{G}, v) = \left(\frac{m}{2\pi k T_0}\right)^3 \exp\left(-\frac{m\boldsymbol{G}^2}{kT_0}\right) \cdot \exp\left(-\frac{mv^2}{4kT_0}\right) \sum_{i=1}^{S} c_i \varphi_2^{(j)} F_{xy}^l F_{yz}^r F_{zx}^s \tag{24}$$

$i = (j, l, r, s)$

Here, T_0 is the temperature of the unperturbed medium and the c_i coefficients generally depend on t. Substituting the distributions of the principal hydrodynamic values calculated by Equation (24) into the hydrodynamic Equations (20c, f) yields a nonlinear system of differential equations for the C_i coefficients,

$$\frac{\partial C_i}{\partial t} = F_i(C_1, \cdots, C_S) \quad i = 1, \cdots, S \tag{25}$$

Relations between the C_i dimensionless coefficients and the c_i dimensional coefficients are given in [4].

4. Conclusions

Figure 1 represents observed flow pictures around a sphere. Experiment records three stable flows: $\boldsymbol{U}_{s0}^{exp}(\boldsymbol{x})$, $\boldsymbol{U}_{s1}^{exp}(\boldsymbol{x})$, and $\boldsymbol{U}_{s2}^{exp}(t, \boldsymbol{x})$. Each of these three flows after losing the stability starts to evolve in its own direction qualitatively different from the others. These directions are schematically shown by three horizontal branches. At these branches experiment records at least six vortex shedding modes: $\boldsymbol{W}_{s0}^{exp}(t, \boldsymbol{x})$ and $\boldsymbol{Q}_{s0}^{exp}(t, \boldsymbol{x})$; $\boldsymbol{V}_{s1}^{exp}(t, \boldsymbol{x})$; $\boldsymbol{V}_{s2}^{exp}(t, \boldsymbol{x})$, $\boldsymbol{W}_{s2}^{exp}(t, \boldsymbol{x})$ and $\boldsymbol{Q}_{s2}^{exp}(t, \boldsymbol{x})$.

The integration of the Navier-Stokes equations for the problem of flow past a sphere [15,16,18-21] gives two stable solutions, $\boldsymbol{U}_{s0}^{cal}(\boldsymbol{x})$ and $\boldsymbol{U}_{s1}^{cal}(\boldsymbol{x})$, and the $\boldsymbol{U}_{s2}^{cal}(t, \boldsymbol{x})$ stable limiting cycle. These solutions satisfactorily reproduce three stable flows observed experimentally, $\boldsymbol{U}_{s0}^{exp}(\boldsymbol{x})$, $\boldsymbol{U}_{s1}^{exp}(\boldsymbol{x})$, and $\boldsymbol{U}_{s2}^{exp}(t, \boldsymbol{x})$. Calculations were, however, incapable of reproducing any of the six vortex shedding modes. The calculation leads the instability development process in the direction given by dashed slanting line (**Figure 1**). The dashed slanting line separates stable solutions from unstable ones, that is, the dashed line gives the boundary of the instability field. So, solutions to the classic hydrodynamics equations successfully reach the boundary of the instability field. As Reynolds grows, these solutions move along the boundary of the field. However, the solutions to the classic hydrodynamics equations are unable to cross this boundary and to pass to the instability field. There are no numerical experiments on flow around different bluff bodies in which this boundary is crossed. Most likely, the reason for simulation failure is the Navier-Stokes equations themselves.

Analysis of the multimoment hydrodynamics equations [35] substantiates the suggestion that the applicability scope of classic hydrodynamics is limited, in particular, by small Reynolds number values, $\mathrm{Re} \ll 1$. Nevertheless, the calculated stable solutions to the Navier-Stokes equations, $\boldsymbol{U}_{s0}^{cal}(\boldsymbol{x})$, $\boldsymbol{U}_{s1}^{cal}(\boldsymbol{x})$ and $\boldsymbol{U}_{s2}^{cal}(t,\boldsymbol{x})$, satisfactorily reproduce stable flows around a sphere $\boldsymbol{U}_{s0}^{exp}(\boldsymbol{x})$, $\boldsymbol{U}_{s1}^{exp}(\boldsymbol{x})$, and $\boldsymbol{U}_{s2}^{exp}(t,\boldsymbol{x})$ recorded experimentally at moderately high Reynolds number values, reaching a few hundreds. This means that the error in stable solutions to the Navier-Stokes equations is not large and cannot distort calculated flow pictures noticeably. This error, however, grows very rapidly after the loss of stability. This is explained by the tendency of nonlinear equations toward causing the divergence of close solutions even in a limited phase space region. This sensitivity to initial conditions was called the Lorentz butterfly effect. The loss of instability development direction is the result of error growth, *i.e.*, the classic hydrodynamics equations become helpless in producing of regular nonstationary periodic flows appearing after the attainment of a certain critical Reynolds number value.

The numerical integration of the multimoment hydrodynamics equations in the problem on a flow past a sphere [4,5] shows that the solutions to these equations cross the boundary and enter the instability field. Let $\overline{C}_i^{(\alpha)}$ be a stationary $\alpha-$ solution of the system (25) and $\delta C_i^{(\alpha)}(t)$ be a perturbation of the $\overline{C}_i^{(\alpha)}$ solution, $C_i(t)=\overline{C}_i^{(\alpha)}+\delta C_i^{(\alpha)}(t)$. The retention of three-lower axisymmetric terms in expansion (24) leads to the classic Stokes solution valid in the $\mathrm{Re} \ll 1$ limit [29]. For advancement up the Re scale, twenty lower axisymmetric expansion (24) terms are retained in [4]. The $\overline{C}_i^{(0)}, i=1,\cdots,20$, stationary solution to the nonlinear algebraic system of twenty equations gives the stationary distribution of the hydrodynamic values (19). According to the $\overline{C}_i^{(0)}, i=1,\cdots,20$, solution, an axisymmetric recirculating zone is formed in the wake behind a sphere at Re ~ 20. This recirculating zone has the shape of an axisymmetric toroidal ring. It expands as Re grows but its shape remains unchanged. The calculated distribution $\boldsymbol{U}_{s0}^{cal}(\boldsymbol{x})$ satisfactorily reproduces the $\boldsymbol{U}_{s0}^{exp}(\boldsymbol{x})$ axisymmetric flow.

The $\overline{C}_i^{(0)}, i=1,\cdots,20$, solution remains stable up to a certain critical Re_0^* value. This means that the $\delta C_i^{(0)}(t)$ small perturbations of the solution $\overline{C}_i^{(0)}, i=1,\cdots,20$, are damped at $\mathrm{Re} < \mathrm{Re}^*$. The passage of Re_0^* is accompanied by the $\overline{C}_i^{(0)}, i=1,\cdots,20$, solution stability loss. Starting with $t=0$, the $\delta C_i^{(0)}(t), i=1,\cdots,20$ small axisymmetric perturbations begin to increase exponentially at $\mathrm{Re} > \mathrm{Re}_0^*$. The $\delta C_i^{(0)}(t)$ perturbations grow up to time $t_* > 0$. At $t=t_*$, the $\delta C_i^{(0)}(t)$ solution is cut off. The reason for this is analyzed in [4].

It turned out that another solution to the nonlinear system of differential Equations (25) existed in the neighborhood of the cut-off point. It is the $\delta C_i^{+(0)}(t), i=1,\cdots,20$, solution of the so-called system of the reversed hydrodynamics equations [35]. According to [37], the law of large numbers is violated near singular points (bifurcations, regions of the coexistence of several stable solutions, etc.), and large fluctuations may appear in the system. In conformity with [37], a large fluctuation causes the transfer of the system from the cut-off point to the $\delta C_i^{+(0)}(t), i=1,\cdots,20$, solution at time $t=t_*$. At time $t=2t_*$, the $\delta C_i^{+(0)}(t)$ solution reaches the neighborhood of the $\overline{C}_i^{(0)}, i=1,\cdots,20$, stationary solution. At time $t=2t_*$ further evolution of the $\delta C_i^{+(0)}(t), i=1,\cdots,20$, solution is finished. The $\delta C_i^{(0)}(t), i=1,\cdots,20$, solution exists at $0 \le t \le t_*$, the $\delta C_i^{+(0)}(t), i=1,\cdots,20$, solution exists at $t_* \le t \le 2t_*$. A fluctuation causes the transfer of the system from the $\overline{C}_i^{(0)}+\delta C_i^{+(0)}(t), i=1,\cdots,20$, solution to the $\overline{C}_i^{(0)}+\delta C_i^{(0)}(t), i=1,\cdots,20$, solution at time $t=2t_*$. The $\overline{C}_i^{(0)}, i=1,\cdots,20$ solution is unstable. It follows that small axisymmetric $\delta C_i^{(0)}(t), i=1,\cdots,20$, perturbations begin to grow starting with time $t=2t_*$. This process is repeated periodically.

The evolution of the $\overline{C}_i^{(0)}+\delta C_i^{(0)}(t)$ and $\overline{C}_i^{(0)}+\delta C_i^{+(0)}(t), i=1,\cdots,20$, solution satisfactorily reproduces the $\boldsymbol{V}_{s0}^{exp}(t,\boldsymbol{x})$ periodic axisymmetric pulsations from [6,10]. According to the $\overline{C}_i^{(0)}+\delta C_i^{(0)}(t)$ and $\overline{C}_i^{(0)}+\delta C_i^{+(0)}(t), i=1,\cdots,20$, solution, toroidal recirculating zone in the near wake behind a sphere begins to pulsate periodically after the attainment of the Reynolds number Re_0^*. The recirculating zone remains toroidal during pulsations, and its forefront is firmly fixed on the sphere. The pulsating flow remains axisymmetric. Because of the absence of the detachment of the recirculating zone periphery, there is no vortex loop street in the far wake behind a sphere.

The scenario of instability appearance, $\mathrm{Re} > \mathrm{Re}_0^*$, [4,5], has no analogues. Currently, we have no alternative to this scenario, because Landau's scenario based on solutions to the Navier-Stokes equations [15,16,18-21] fails to describe experiment.

The crossing of the instability field boundary is accompanied by the appearance of uncommon acts in the scenario of instability development. It turned out that the exponential growth of small perturbations of stationary unstable solution is not infinite. At some point the nonstationary solution is cut off and the growth of perturbations is completed. After the break the system finds extremely unlike direction of evolution. It starts a returning motion in the direction of the unstable stationary solution. As a result, the system after losing the stability does not find a new stable position, *i.e.*, after crossing the boundary of the instability field, the evolution of the system takes place only within the instability field. Such a scenario differs from the ideas of classic hydrodynamics, which interprets the development of instability in terms of bifurcations from one stable state to another stable state.

The importance of the entropy in interpretation of instability is revealed in [5]. It turns out that the explanation of instability appearance lies in change in entropy balance. Namely, an open system with time-independent boundary conditions has a steady state while entropy production in it exceeds entropy outflow through the surface confining the system. Then, any fluctuation generated by the system fades out. As soon as the entropy is removed through the confining surface faster than it is produced in the system, any fluctuation generated by the system begins to grow. The system becomes unstable.

Further studies [38] show that reproduction of the $\boldsymbol{W}_{s0}^{exp}(t,\boldsymbol{x})$ and $\boldsymbol{Q}_{s0}^{exp}(t,\boldsymbol{x})$ vortex shedding regimes, $\mathrm{Re} > \mathrm{Re}_0^{**}$, leads to new scenario acts that do not have analogues. Moreover, the $S_p(t)$ pair entropy of system begins to play the key role to predict the instability development direction. The calculation of the system evolution becomes impossible without participation of the entropy. The multimoment hydrodynamics equations find a large number of ways for system evolution. And the $S_p(t)$ pair entropy shows the single way, in which the system evolution moves.

So, the perspectives for removal the discrepancies between calculations and experiment for unstable regimes are revealed along the direction on increase of the principal hydrodynamic values number when deriving the hydrodynamics equations.

REFERENCES

[1] I. V. Lebed, "On the Inapplicability of Navier-Stokes Equations to Interpreting the Turbulence," *Physica A*, Vol. 315, No. 1-2, 2002, pp. 228-235.

[2] I. V. Lebed and S. Y. Umanskii, "The Appearance and Development of Turbulence in a Flow Past a Sphere: Problems and the Existing Approaches to Their Solution," *RJ Physical Chemistry B*, Vol. 1, No. 1, 2007, pp. 52-73.

[3] I. V. Lebed, "Method of Two-Particle Distribution Functions. Hydrodynamic Equations," *Chemical Physics Reports*, Vol. 15, No. 6, 1996, pp. 861-883.

[4] I. V. Lebed, "The Method of Pair Functions as Applied to the Problem of a Flow around a Quiescent Solid Sphere," *Chemical Physics Reports*, Vol. 16, No. 7, 1997, pp. 1263-1301.

[5] I. V. Lebed, "About the Behavior of the Entropy of a Gas Flow Losing Its Stability," *Chemical Physics Reports*, Vol. 17, No. 3, 1998, pp. 411-439.

[6] S. Taneda, "Experimental Investigation of the Wake behind a Sphere at Low Reynolds Numbers," *Journal of the Physical Society of Japan*, Vol. 11, 1956, pp. 1104-1108.

[7] H. Sakamoto and H. Haniu, "The Formation Mechanism and Shedding Frequency of Vortices from a Sphere in Uniform Shear Flow," *Journal of Fluid Mechanics*, Vol. 287, 1995, pp. 151-171.

[8] R. H. Magarvey and R. L. Bishop, "Transition Ranges for Three-Dimensional Wakes," *Canadian Journal of Physics*, Vol. 39, No. 10, 1961, pp. 1418-1422.

[9] E. Achenbach, "Vortex Shedding from Spheres," *Journal of Fluid Mechanics*, Vol. 62, No. 2, 1974, pp. 209-221.

[10] J. M. Chomaz, P. Bonneton and E. J. Hopfinger, "The Structure of the Near Wake of a Sphere Moving Horizontally in a Stratified Fluid," *Journal of Fluid Mechanics*, Vol. 254, 1993, pp. 1-21.

[11] J. Gerrard, "The Wakes of Cylindrical Bluff Bodies at Low Reynolds Numbers," *Philosophical Transactions of the Royal Society of London A*, Vol. 288, No. 1354, 1978, pp. 351-382.

[12] M. M. Zdravkovich, "Smoke Observation of the Formation of a Karman Vortex Street," *Journal of Fluid Mechanics*, Vol. 37, No. 3, 1969, pp. 491-496.

[13] B. E. Eaton, "Analysis of Laminar Vortex Shedding behind a Circular Cylinder by Computer-aided Flow Visualization," *Journal of Fluid Mechanics*, Vol. 180, 1987, pp. 117-145.

[14] J. Cohen, I. G. Shukhman, M. Karp and J. Philip, "An Analytical-Based Method for Studying the Nonlinear Evolution of Localized Vortices in Planar Homogenous Shear Flows," *Journal of Computational Physics*, Vol. 229, No. 20, 2010, pp. 7765-7773.

[15] I. Kim and A. J. Pearlstein, "Stability of the Flow Past a Sphere," *Journal of Fluid Mechanics*, Vol. 211, 1990, pp. 73-83.

[16] R. Natarajan and A. Acrivos, "The Instability of the Steady Flow Past Spheres and Disks," *Journal of Fluid Mechanics*, Vol. 254, 1993, pp. 323-344.

[17] H. G. Schuster, "Deterministic Chaos," Physik-Verlag, Weinheim, 1984.

[18] A. G. Tomboulides, S. A. Orszag and G. E. Karniadakis, "Direct and Large-Eddy Simulations of Axisymmetric Wakes," AIAA Paper, No. 93-9546, 1993.

[19] R. Mittal, "Planar Symmetry in the Unsteady Wake of a Sphere," *AIAA Journal*, Vol. 37, No. 3, 1999, pp. 388-400.

[20] T. A. Johnson and V. S. Patel, "Flow Past a Sphere Up to a Reynolds Number of 300," *Journal of Fluid Mechanics*, Vol. 378, 1999, pp. 19-70.

[21] A. G. Tomboulides and S. A. Orszag, "Numerical Investigation of Transitional and Weak Turbulent Flow Past a Sphere," *Journal of Fluid Mechanics*, Vol. 416, 2000, pp. 45-73.

[22] C. P. Jackson, "A Finite-Element Study of the Onset of Vortex Shedding in Flow Past Variously Shaped Bodies," *Journal of Fluid Mechanics*, Vol. 182, 1987, pp. 23-45.

[23] K. Hannemann and H. Oertel Jr., "Numerical Simulation of the Absolutely and Convectively Unstable Wake," *Journal of Fluid Mechanics*, Vol. 199, 1989, pp. 55-88.

[24] P. J. Strykowski and K. Hannemann, "Temporal Simulation of the Wake behind a Circular Cylinder in the Neighborhood of the Critical Reynolds Number," *Acta Mechanica*, Vol. 90, No. 1-4, 1991, pp. 1-20.

[25] G. K. Batchelor, "An Introduction to Fluid Dynamics," Cambridge at the University Press, Cambridge, 1970.

[26] V. Gushchin and P. Matyushin, "Vortex Formation Mechanisms in the Wake behind a Sphere for $200 < Re < 380$," *Fluid Dynamics*, Vol. 41, 2006, pp. 795-809.

[27] S. Baghery, P. Schlatter, P. Schmid and D. Henningson, "Global Stability of a Jet Cross-Flow," *Journal of Fluid Mechanics*, Vol. 624, 2009, pp. 33-44.

[28] J. Jeong and F. Hussain, "On the Identification of a Vortex," *Journal of Fluid Mechanics*, Vol. 285, 1995, pp. 69-94.

[29] L. G. Loitsyanskii, "Mechanics of Liquids and Gases," Pergamon, Oxford, 1966.

[30] R. L. Liboff, "Introduction to the Theory of Kinetic Equations," Wiley, New York, 1969.

[31] J. H. Ferziger and H. G. Kaper, "Mathematical Theory of Transport Processes in Gases," North-Holland Publishing Company, Amsterdam, 1972.

[32] I. V. Lebed and S. Y. Umanskii, "On the Possibility of Improving Classic Hydrodynamics Equations by an Increase in the Number of Hydrodynamic Values," *RJ Physical Chemistry B*, Vol. 6, No. 1, 2012, pp. 149-162.

[33] I. V. Lebed, "Derivation of the Equations for Pair Distribution Functions," *Chemical Physics Reports*, Vol. 14, No. 5, 1995, pp. 599-615.

[34] I. V. Lebed, "Equations of Pair Distribution Functions," *Chemical Physics Letters*, Vol. 165, No. 2-3, 1990, pp. 226- 228.

[35] I. V. Lebed, "Hydrodynamic Equations Stemming from Two Particle Distributions in the Limit of Weak Non-equilibrium. Analysis of Invertibility of Equations," *Chemical Physics Reports*, Vol. 15, No. 12, 1996, pp. 1725-1750.

[36] H. Grad, "About Kinetic Theory of Rarefied Gases," *Communications on Pure and Applied Mathematics*, Vol. 2, No. 4, 1949, pp. 331-407.

[37] I. Prigogine, "From Being to Becoming," Freeman, San Francisco, 1980.

[38] I. V. Lebed, "The Development of Instability in Problem on Flow around a Sphere," *RJ Physical Chemistry B*, Vol 8, No. 1, 2014, in Press.

A Comparison of the Ghost Cell Technique for Front Tracking Method

Donghong Wang[1,2], Yanqing Wang[3], Ning Zhao[1]
[1]College of Aerospace Engineering, Nanjing University of Aeronautics and Astronautics, Nanjing, China
[2]College of Science, Nanjing University of Aeronautics and Astronautics, Nanjing, China
[3]College of Science, PLA University of Science and Technology, Nanjing, China

ABSTRACT

The treatment of moving material interfaces and their vicinity is very important for compressible multifluids. In this paper, we propose one type of ghost fluid method based on Riemann solutions for front tracking method. The accuracy of the interface boundary condition is discussed for the gas-gas Riemann problem. It is shown that the solution of the ghost fluid method approximates the exact solution to second-order accuracy in the sense of comparing to the exact solution of a Riemann problem at the material interface. Numerical examples suggest that the present scheme is able to handle multifluids problems with large density differences and has the property of reduced conservation error.

Keywords: Compressible Multifluid; Front Tracking Method; Riemann Problem

1. Introduction

The dynamics of interfaces separating different fluids in compressible flows is of interest in several scientific fields as diverse as astrophysics and geophysics. It is also of significant importance in many engineering applications. A relatively dominant difficulty for simulating compressible multi-medium flow is the treatment of moving material interfaces and their vicinity. In general, there are two basic issues that need to be taken into account. One is to capture the interface location and topological changes accurately. Researchers can take all kinds of effective measures such as volume of fluid method [1] or level set technique [2] or front tracking technique [3] to deal with it. The other is to faithfully simulate the interface state including physical nonlinear wave interaction occurring at the interface.

The ghost fluid method (GFM) based techniques [4-8] provide us simple and flexible ways for handling multi-medium flows with immiscible material interfaces. A closely related ghost cell level set method [4] was proposed by Fedkiw. But the GFM is problem-related and not suitable for some cases like high speed jet impacting [5]. To take into consideration the influence of both, wave interaction and material properties on the interfacial evolution led to the development of a modified ghost fluid method (MGFM) [5,7]. Wang [8] also presented a real ghost fluid method (RGFM) by predicting the flow states for the real fluid nodes just next to the interface and the ghost fluid nodes using the Riemann problem solver.

For the GFM [4-8], the level set technique [2] is employed to capture the moving interface. However, they can be used with other techniques for tracking the interface. Recently, Hao [9] and Terashima [3] extend Tryggvason's method [10] using Fedkiw's ghost fluid method [4] to handle compressible flows. Based on Riemann solutions, front tracking method combined with MGFM and RGFM had been discussed [11].

In this paper, a relatively class of GFM based on Riemann solutions is developed. One solves a system of a two-shock approximation to the Riemann problem for prediction of the condition at the interface. Then one uses the predicted pressure and velocity as those for the ghost fluid and the real fluid nodes just next to the interface. The isobaric entropy is employed at the interface to fix the real fluid density at node just next to the interface to suppress the possible "overheating" and also to assign density for the ghost fluid. The RGFM is one of this type of GFM. We combine the GFM with a front tracking method (GFM-FT). The front tracking method [12] is used here. The accuracy of the interface boundary condition is discussed for the gas-gas Riemann problem.

The conservation errors of this class of GFM and MGFM are analyzed for FT method. Numerical tests show that the GFM-FT has the property of reduced conservation error comparing to MGFM-FT.

2. Equations

The Euler equations governing one-dimensional compressible flows are written as

$$\frac{\partial U}{\partial t}+\frac{\partial F(U)}{\partial x}=0 \tag{1}$$

where

$$U=(\rho,\rho u,E)^{\mathrm{T}},\ F(U)=\left(\rho u,\rho u^2+p,(E+p)u\right)^{\mathrm{T}},$$

Here ρ is the density, u is the velocity, p is the pressure, and E is the total energy per unit volume. The total energy is the sum of internal energy and kinetic energy,

$$E=\rho e+\frac{1}{2}\rho u^2, \tag{2}$$

where e is the internal energy per unit mass.

For closure of Equation (1), the equation of state (EOS) is required. The EOS for compressible gases and water can be expressed in the following consistent form as

$$p=(\gamma-1)\rho e-\gamma B, \tag{3}$$

where γ and B are treated as constants. For the air, $\gamma=1.4$, $B=0$. For the water medium (Tait's equation), $\gamma=7.15$, $B=3309$.

3. Tracking Fluid Interfaces

Suppose that the interface lies between node *i* and node *i* + 1. To define the Riemann problem at the interface, the two initial constant states of the Riemann problem are simply given as $U_L=U_{i-1}$ and $U_R=U_{i+2}$. Once the Riemann problem is defined, the two shock Riemann problem solver [5] can then be employed to provide intermediate interfacial states— p_I (pressure), u_I (velocity), and ρ_{IL} and ρ_{IR} (the densities on the left and right sides of the interface). Now, having determined the interface velocity u_I and its spatial location, the interface is moved between t_n and t_{n+1}.

4. Ghost Fluid

The initial conditions for the governing Equation (1) are given as

$$U=\begin{cases}U_L & x<x_0\\ U_R & x>x_0\end{cases}. \tag{4}$$

Regardless of the numerical scheme used, an algorithm based on the GFM essentially consists of solving two separate Riemann problems in the two respective single media with one-sided ghost fluid. One is in medium 1 (on the left side of the interface) with the initial conditions of

$$U=\begin{cases}U_L & x<x_0\\ U_L^* & x>x_0\end{cases} \tag{5}$$

and it solves from the grid point 1 on the left-end to the ghost point, The other is in medium 2 (on the right side of the interface) with the initial conditions of

$$U=\begin{cases}U_R^* & x<x_0\\ U_R & x>x_0\end{cases} \tag{6}$$

and it solves from the ghost point to the end point on the right. Here, "*" indicates the fluid states have been replaced by the ghost fluid states.

With the method of characteristics, we have the relations

$$u_I=u_L^*+\int_{p_L^*}^{p_I}\frac{1}{\rho_L^* c_L^*}\mathrm{d}p \tag{7}$$

$$u_I=u_R^*-\int_{p_R^*}^{p_I}\frac{1}{\rho_R^* c_R^*}\mathrm{d}p \tag{8}$$

where ρ_L^*,c_L^* and ρ_R^*,c_R^* are the densities and sound speeds of the ghost fluid.

The ghost fluid states can be chosen by solving Equation (7) and (8). The simplest case is that the ghost fluid pressure be defined as the interface pressure, *i.e.*

$$p_L^*=p_I,\ p_R^*=p_I. \tag{9}$$

Hence the integral in Equation (7) and (8) become zero and the ghost fluid velocity is

$$u_L^*=u_I,\ u_R^*=u_I. \tag{10}$$

Furthermore, one can find that any ghost fluid density can satisfy Equation (5) or (6).

For Equation (5) or (6), Hu [6] had considered two algorithms. The focus is on defining the ghost fluid states while the pressure and velocity in the real fluid side are taken for granted, except for the correction made to the density at the real fluid nodes next to the interface to overcome the overheating. Indeed, the real fluid states next to the interface instead of the ghost fluid states should be predicted by solving a Riemann problem [8]. This will result in the complete redefinition of the real fluid next to the interface. In doing so, one will find that the nonphysical reflection for the shock impedance matching problem can be greatly and further suppressed.

For medium 1, we take p_I and u_I as the pressure and velocity at nodes i, $i+1$, $i+2,\cdots$, For medium 2, we also take p_I and u_I as the pressure and velocity at nodes $i+1$, $i,i-1,\cdots$. The ghost fluid density is based on isentropic fixing. Here, we shall consider two cases:

Case 1: For medium 1, the predicted isobaric entropy $S_{IL} = \frac{p_I + B_L}{\rho_{IL}^{\gamma_L}}$ at the interface is employed to fix the real fluid density at point *i* and the ghost fluid density at point $i+1$, $i+2, \cdots$. A similar procedure is used for computation in medium 2. This is the RGFM [8] which can be seen from **Figure 1**.

Case 2: For medium 1, one employs isobaric entropy $S_{i-1} = \frac{p_{i-1} + B_L}{\rho_{i-1}^{\gamma_L}}$ to fix the real and ghost fluid density. The rest of the procedures are the same as those for Case 1. We take this method as SGFM which can be seen from **Figure 2**.

5. The Accuracy of SGFM and RGFM

We only consider the gas–gas Riemann problem. The accuracy of MGFM had been discussed [13]. It may be noted that the accuracy discussed should be interpreted as how accurate the boundary conditions are implicitly imposed at the material interface and how accurate the interface states are approximated by the GFM technique; it should not be explained or analyzed as the accuracy of final numerical solution and/or the numerical scheme used because the detailed numerical scheme is never involved in the discussion. An approximate Riemann problem solver (ARPS) based on a doubled shock structure is used for the Riemann problem. For the SGFM applied to the Riemann problem (4), the following error estimations are held [13]

$$\begin{cases} u_I^S - u_I^e = O\left[\max\left(\left|\frac{p_I^e}{p_L} - 1\right|, \left|\frac{p_I^e}{p_R} - 1\right|\right)\right]^3 \\ p_I^S - p_I^e = O\left[\max\left(\left|\frac{p_I^e}{p_L} - 1\right|, \left|\frac{p_I^e}{p_R} - 1\right|\right)\right]^3 \end{cases} \tag{11}$$

Here, p_I^e and u_I^e are the exact interfacial pressure and velocity, u_I^S and p_I^S are the interfacial pressure and velocity from ARPS.

For SGFM, we have $p_R^* = p_I^S$, $u_R^* = u_I^S$, $p_L^* = p_I^S$ and $u_L^* = u_I^S$, the pressure and velocity at the real fluid node next to the interface are also redefined. Following conclusions are held for the SGFM when applied to the Riemann problem (4).

Theorem 1. *The following error estimates are held for the respective GFM Riemann problems* (5) *and* (6) *using the SGFM*

(A) $u_I^{SA} - u_I^e = O\left(\frac{p_I^{SA}}{p_I^S} - 1\right)^2 + O\left(\frac{p_I^e}{p_L} - 1\right)^3$

(B) $p_I^{SA} - p_I^e = O\left(\frac{p_I^{SA}}{p_I^S} - 1\right)^2 + O\left(\frac{p_I^e}{p_L} - 1\right)^3$

(C) $u_I^{SB} - u_I^e = O\left(\frac{p_I^{SB}}{p_I^S} - 1\right)^2 + O\left(\frac{p_I^e}{p_R} - 1\right)^3$

(D) $p_I^{SB} - p_I^e = O\left(\frac{p_I^{SB}}{p_I^S} - 1\right)^2 + O\left(\frac{p_I^e}{p_R} - 1\right)^3$

$u_I^{SA}\left(u_I^{SB}\right)$ *and* $p_I^{SA}\left(p_I^{SB}\right)$ *are the exact interfacial velocity and pressure of the GFM Riemann problem* (5) *and* (6) *using the SGFM.*

Proof: Applying Theorem 2.1 [13] to the GFM Riemann problem (5), one gets

$$\begin{cases} u_I^{SA} = \frac{\rho_L c_L u_L + \rho_R^* c_R^* u_R^* + p_L - p_R^*}{\rho_L c_L + \rho_R^* c_R^*} + O\left(\frac{p_I^{SA}}{p_L} - 1\right)^2 \\ p_I^{SA} = \frac{\rho_L c_L \rho_R^* c_R^* \left(u_L - u_R^*\right) + \rho_R^* c_R^* p_L + \rho_L c_L p_R^*}{\rho_L c_L + \rho_R^* c_R^*} + O\left(\frac{p_I^{SA}}{p_L} - 1\right)^2 \end{cases} \tag{12}$$

According to SGFM, one has $p_R^* = p_I^S$, $u_R^* = u_I^S$ and the pressure and velocity at the real fluid node next to the interface are $u_L = u_I^S$, $p_L = p_I^S$. Equation (12) can be expressed as

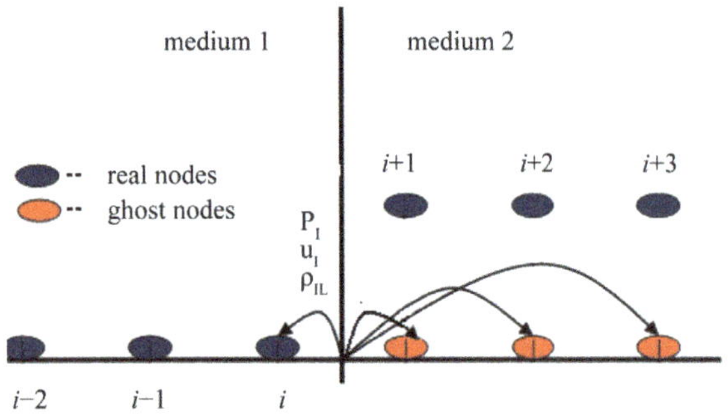

Figure 1. Isobaric fixing for RGFM.

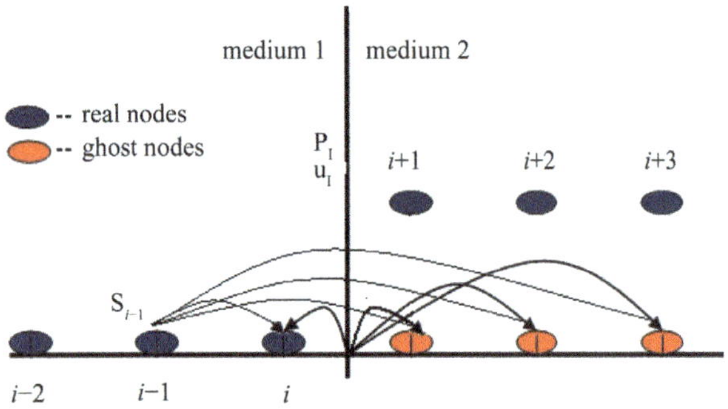

Figure 2. Isobaric fixing for SGFM.

$$\begin{cases} u_I^{SA} = u_I^S + O\left(\dfrac{p_I^{SA}}{p_I^S} - 1\right)^2 \\ p_I^{SA} = p_I^S + O\left(\dfrac{p_I^{SA}}{p_I^S} - 1\right)^2 \end{cases} \quad (13)$$

Using Equation (11), one has

$$u_I^{SA} - u_I^e = O\left(\frac{p_I^{SA}}{p_I^S} - 1\right)^2 + O\left(\frac{p_I^e}{p_L} - 1\right)^3 \quad (14)$$

$$p_I^{SA} - p_I^e = O\left(\frac{p_I^{SA}}{p_I^S} - 1\right)^2 + O\left(\frac{p_I^e}{p_L} - 1\right)^3 \quad (15)$$

For the GFM Riemann problem (6), in a similar way, one can show that error estimates (C) and (D) in Theorem 1 are true.

The states of the ghost cell for RGFM are in accord with those for SGFM except the density. So Theorem 1 is also true for RGFM. It is shown that the solution of the SGFM and RGFM approximate the exact solution to at least second-order accuracy in the sense of comparing to the exact solution of a Riemann problem at the material interface.

6. The Conservation Errors

Both overall conservation and conservation for each medium will be evaluated using Equation (1) over the computational domain $[x_A, x_B]$. We denote *RHSL(n)*, *RHSR(n)* and *RHST(n)* as the conservation errors for the medium on the left, right side of the interface and over the whole computational domain, respectively, as follows:

$$\begin{aligned} RHSL(n+1) = \frac{1}{\Delta x}\Bigg[&\int_{x_A}^{x_I^{n+1}} U^{n+1}\mathrm{d}x - \int_{x_A}^{x_I^n} U^n \mathrm{d}x \\ &+ \int_{t^n}^{t^{n+1}} \left(F_{IL} - u_I U_{IL} - F_A\right)\mathrm{d}t\Bigg] \end{aligned} \quad (16)$$

$$\begin{aligned} RHSR(n+1) = \frac{1}{\Delta x}\Bigg[&\int_{x_I^{n+1}}^{x_B} U^{n+1}\mathrm{d}x - \int_{x_I^n}^{x_B} U^n \mathrm{d}x \\ &+ \int_{t^n}^{t^{n+1}} \left(F_B - F_{IR} + u_I U_{IR}\right)\mathrm{d}t\Bigg] \end{aligned} \quad (17)$$

$$\begin{aligned} &RHST(n+1) \\ &= \frac{1}{\Delta x}\left[\int_{x_A}^{x_B} U^{n+1}\mathrm{d}x - \int_{x_A}^{x_B} U^n \mathrm{d}x + \int_{t^n}^{t^{n+1}} \left(F_B - F_A\right)\mathrm{d}t\right] \end{aligned} \quad (18)$$

Here, x_I denotes the interface position; F_A and F_B are fluxes at x_A and x_B, respectively. F_{IL} and F_{IR} are fluxes at the respective left and right sides of the interface.

The overall average conservation error can be taken as

$$\operatorname{error}(K) = \frac{1}{60}\sum_{l=60K+1}^{60(K+1)} \left|RHST(l)\right| \quad (19)$$

We shall use two cases to study the conservation errors of the RGFM and SGFM.

Case 1: a strong shock impacting on an air-air interface (*i.e.*, Problem 1 in Section 7)

From **Table 1**, it is clear that very large errors incur for the MUSCL-based MGFM-FT in the first 60 steps for this specific case, while these are suppressed very well by the MUSCL-based SGFM-FT and RGFM-FT. The conservative errors of the RGFM-FT and SGFM-FT are tend to zero faster than the MGFM-FT.

Case 2: This is a problem of a water column separated by the air and oscillating in the middle of a closed tube [8]. The problem is sketched in **Figure 3**. The initial status of the air and water are defined as

$$(\rho, u, p, \gamma, B) = \begin{cases} (0.001, 1, 1, 1.4, 0) & -1 < x < -0.1 \\ (1, 1, 1, 7, 3000) & -0.1 \le x \le 0.1 \\ (0.001, 1, 1, 1.4, 0) & 0.1 < x < 1 \end{cases}$$

With the fluids moving to the right, the air on the left of the water column expands while the air on the right of the water column is compressed. The pressure decreases at the former and increases at the latter; this leads to the deceleration and eventually stagnation of the fluids in the tube. Then the fluids move in the reverse direction due to the pressure gradient, and so on with the water column oscillating. Pressure waves are evident in the tube due to the impact of the air on the end boundaries and interacttion with the water column. The conservation errors of the air mass $M(t)$ are given as

$$M(t) = \frac{m_a(t) - m_a(0)}{m_a(0)} \quad (20)$$

Table 1. The conservation errors of the RGFM-FT, SGFM-FT and MGFM-FT.

K	RGFM-FT	SGFM-FT	MGFM-FT
0	2.5679419E−004	1.0022954E−003	0.9505743
1	5.8678547E−004	7.6730609E−004	3.4564280E−002
2	1.8233535E−004	2.2801129E−004	1.9138482E−003
3	1.4738074E−004	1.9033586E−004	2.1914005E−002
4	8.1406301E−005	9.2554928E−005	2.5471797E−003
5	7.0434890E−005	7.9106984E−005	2.3205920E−003

Figure 3. Initial status of the oscillating water column problem.

where $m_a(t)$ is the total mass of the air in the tube at time t. It can indeed be observed that the conservation errors associated with the air mass appear to be oscillating similar to that observed in [8]. **Figure 4(a)** depicts that MGFM has an oscillatory $M(t)$ with a much larger trend of net conversion of the air mass into water when compared to RGFM. **Figures 4(b)** and **(c)** show that RGFM and SGFM have the same time evolution of the conservation errors.

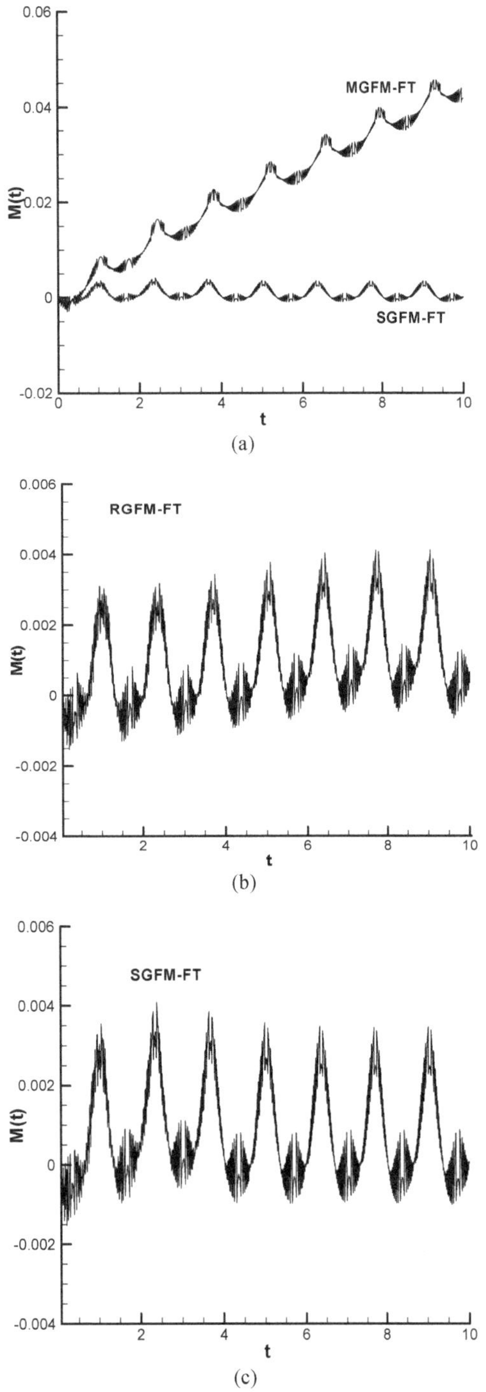

Figure 4. Comparison of conservation errors among the MGFM, the RGFM and the SGFM for Case 2 in Section 6. (a) the MGFM-FT and the SGFM-FT; (b) the RGFM-FT; (c) the SGFM-FT.

The wall pressure coefficients are given as

$$P(x=-1,t)=\frac{p(x=-1,t)-p_0}{p_0} \tag{21}$$

$$P(x=1,t)=\frac{p(x=1,t)-p_0}{p_0} \quad (p_0=1) \tag{22}$$

Figure 5 shows the time evolution of the pressure coef-

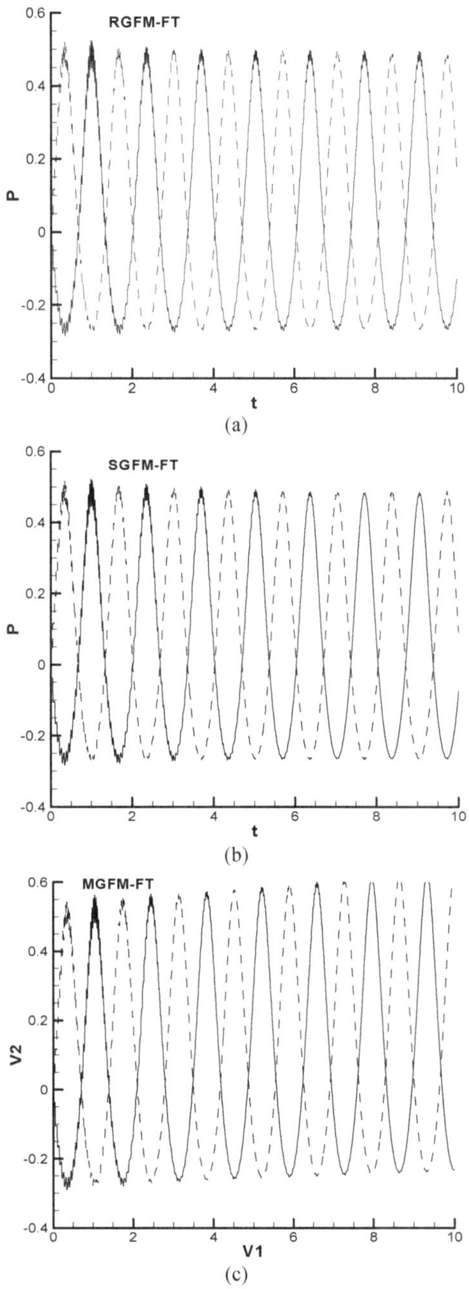

Figure 5. Time evolution of the pressure coefficients at boundaries for Case 2 in Section 6. Solid line: at the left boundary; dashed line: at the right boundary.

ficients at the wall boundaries calculated using MGFM, RGFM and SGFM. **Figure 5(c)** shows that there is a increase of the pressure amplitude with time for MGFM, while the pressure amplitude with RGFM and SGFM is changeless on the whole as show in **Figures 5(a)** and **(b)**.

7. Applications

For all the one-dimensional problems the computational domain is taken as [0,1] with 201 nodes uniformly distributed unless stated otherwise.

Problem 1: *Gaseous shock in impedance-matching medium* [8]. In this case a strong shock on the left side of the interface impacts on a gas-gas interface; The shock strength is 100 and the initial position of the shock is same as the interface, which is located at $x_0 = 0.2$. The states on the left and right sides of the interface are defined respectively as

$$(\rho, u, p, \gamma, B) = \begin{cases} (0.82369077, 0.0, 1.0, 1.67, 0.0) & x < 0.2 \\ (1.0, 0.0, 1.0, 1.2, 0.0) & x \geq 0.2 \end{cases}$$

This problem is solved at time $t = 0.06$. The shock limiter is used [11]. **Figures 6(a)-(c)** show that the density, velocity and pressure obtained by the MUSCL-based SGFM-FT method compare to the analytical solution at time 0.06.

Problem 2: *Strong shock impacting on a gas-water interface.* A shock is located at the same position as the interface, which is initially located at $x_0 = 0.5$. The strength of the incident gas shock is 1000. The states on the left and right sides of the interface are defined respectively as

$$(\rho, u, p, \gamma, B) = \begin{cases} (0.001, 0.0, 1.0, 1.4, 0.0) & x < 0.5 \\ (1.0, 0.0, 1.0, 7.15, 3309.0) & x \geq 0.5 \end{cases}$$

For this case, a very strong shock is physically reflected back into the air. **Figures 7(a)-(c)** show that the density, velocity and pressure obtained by the MUSCL-based SGFM-FT method compare to the analytical solution at time 0.0007.

Problem 3: we solve the numerical example

$$(\rho, u, p, \gamma) = \begin{cases} (2.5, 0.0, 8.0\times 10^4, 1.2) & x < 0.3 \\ (1.5, 0.0, 1.0\times 10^5, 1.4) & 0.3 \leq x < 0.6 \\ (0.5, 0.0, 1.2\times 10^5, 1.67) & x \geq 0.6 \end{cases}$$

This test has solution consisting of a rarefaction wave, two contact discontinuities and three shock waves. The MUSCL-based SGFM-FT method is used. **Figures 8(a)-(c)** describe the surprisingly good solution profiles at time 0.00028.

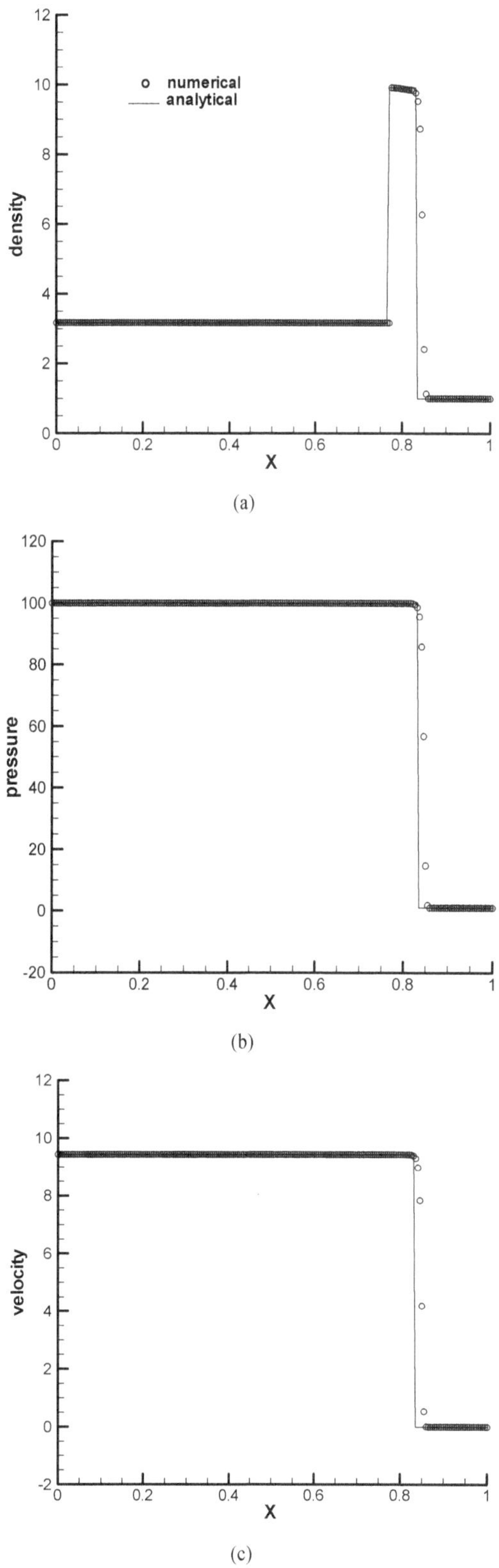

Figure 6. Problem 1.

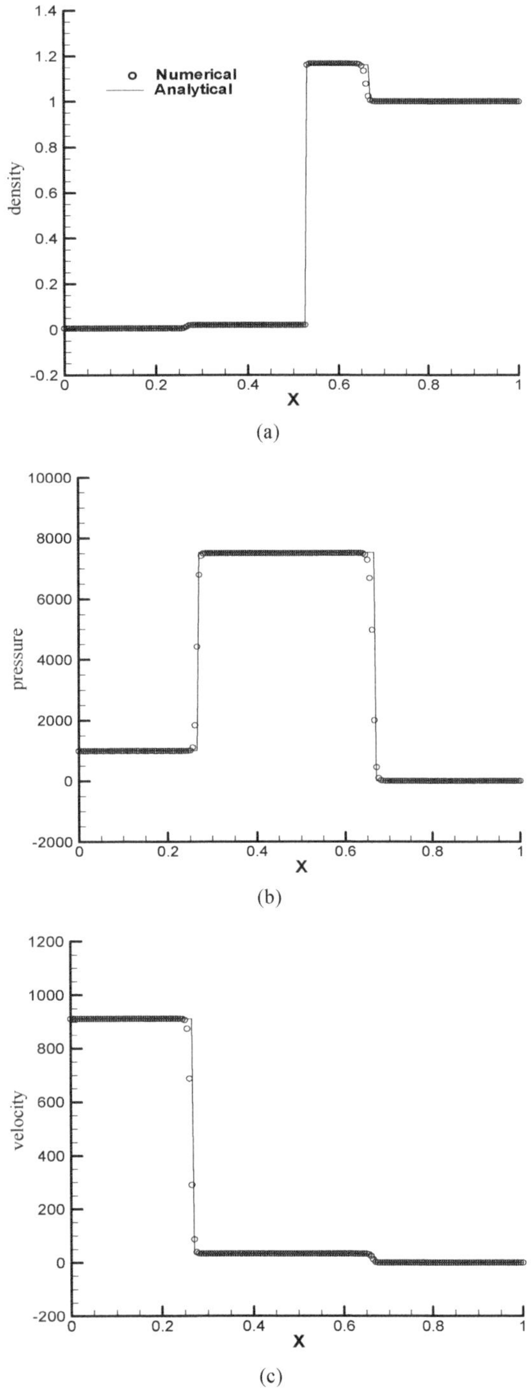

Figure 7. Problem 2.

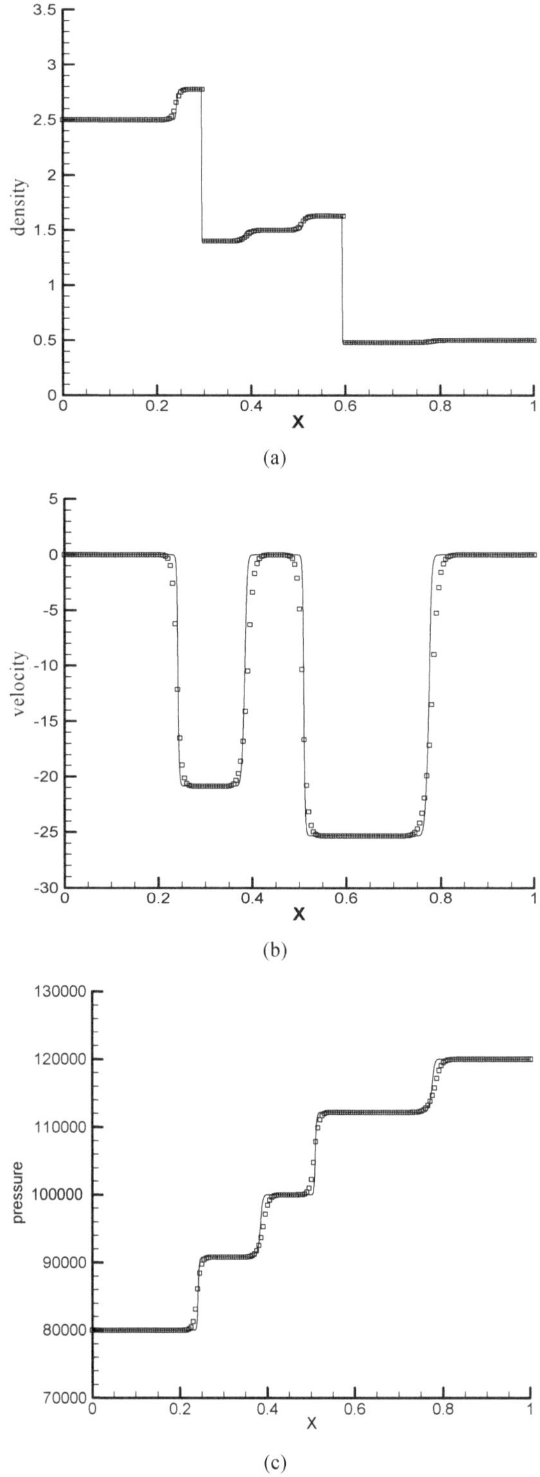

Figure 8. Problem 3: The "o" denotes the numerical results. The solid line denotes the exact solution.

8. Conclusion

In this paper, we developed a kind of GFM based on Riemann solutions. The RGFM is one of this type of GFM. A simplified front tracking algorithm had been used to track contact discontinuity. It is noted that different isentropic entropy can produce little effect on the conservation error and this kind of GFM has the property of reduced conservation error. A rigorous analysis is carried out on the accuracy of the GFM when applied to the gas-gas Riemann problem. It is shown that at the material interface this type of GFM solution approximates the exact solution to at least second-order accuracy in the sense of comparing to the exact solution of a Riemann problem. When the shock monitor is used, discernible non-physical hump and trough which can not be avoided in shock impedance matching (-like) problems [7,8] do not appear for our proposed method. The ghost cell technique can be used for simulating compressible multimedium flow.

9. Acknowledgements

The researched work was supported by the National Natural Science Foundation of China (NO.91130030 and NO.11271188) and NUAA Research Funding (NO. NP2011033).

REFERENCES

[1] C. Hirt and B. Nichols, "Volume of fluid (VOF) Method for the Dynamics of Free Boundaries," *Journal of Computational Physics*, Vol. 39, No. 1, 1981, pp. 201-225.

[2] D. Adalsteinsson and J. A. Sethian, "A Fast Level Set Method for Propagating Interfaces," *Journal of Computational Physics*, Vol. 118, 1995, pp. 269-277.

[3] H. Terashima and G. Tryggvason, "A Front-Tracking/Ghost-Fluid Method for Fluid Interfaces in Compressible Flows," *Journal of Computational Physics*, Vol. 228, No. 11, 2009, pp. 4012- 4037.

[4] R. P. Fedkiw, T. Aslam, B. Merriman and S. Osher, "A Non-Oscillatory Eulerian Approach to Interfaces in Multimaterial Flows (the Ghost Fluid Method)," *Journal of Computational Physics*, Vol. 152, No. 2, 1999, pp. 457-492.

[5] T. G. Liu, B. C. Khoo and K. S. Yeo, "Ghost Fluid Method for Strong Shock Impacting on Material Interface," *Journal of Computational Physics*, Vol. 190, No. 2, 2003, pp. 651-681.

[6] X. Y. Hu and B. C. Khoo, "An Interface Interaction Method for Compressible Multifluids," *Journal of Computational Physics*, Vol. 198, No. 1, 2004, pp. 35-64.

[7] T. G. Liu, B. C. Khoo and C. W. Wang, "The Ghost Fluid Method for Compressible Gas-Water Simulation," *Journal of Computational Physics*, Vol. 204, No. 1, 2005, pp. 193-221.

[8] C. W. Wang, T. G. Liu and B. C. Khoo, "A Real Ghost Fluid Method for the Simulation of Multimedium Compressible Flow," *Journal on Scientific Computing*, Vol. 28, No. 1, 2006, pp. 278-302.

[9] Y. Hao and A. Prosperetti, "A Numerical Method for Three-Dimensional Gas-Liquid Flow Computations," *Journal of Computational Physics*, Vol. 196, No. 1, 2004, pp. 126-144.

[10] G. Tryggvason, B. Bunner, A. Esmaeeli, D. Juric, N. A. Rawahi, W. Tauber, J. Han, S. Nas and Y. J. Jan, "A Front Tracking Method for the Computations of Multiphase Flow," *Journal of Computational Physics*, Vol. 169, No. 2, 2001, pp. 708-759.

[11] D. H.Wang, N. Zhao, O. Hu and J. M. Lui, "A Ghost Fluid Based Front Tracking Method for Multimedium Compressible Flows," *Acta Mathematica Sinica*, Vol. 29B, No. 6, 2009, pp. 1629-1646.

[12] J. Glimm, J. Grove, X. L. Li and N. Zhao, "Simple Front Tracking," *Contemporary Mathematics*, Vol. 238, 1999, pp. 133-149.

[13] T. G. Liu and B. C. Khoo, "The Accuracy of the Modified Ghost Fluid Method for Gas-Gas Riemann Problem," *Applied Numerical Mathematics*, Vol. 57, No. 5-7, 2007, pp. 721-733.

Numerical Hydrodynamic of Multihull Tunnel Vessel

Morteza Ghassabzadeh, Hassan Ghassemi
Department of Ocean Engineering, Amirkabir University of Technology, Tehran, Iran

ABSTRACT

The research described in this paper was carried out to predict the numerical hydrodynamic of multihull tunnel vessel at various speeds. The hull form of vessel is fairly generated by the tunnel hull form generator (THFG) code using the Non Uniform Rational B-Spline (NURBS) method. Then, the hydrodynamic simulation of high speed vessel is carried out based on finite volume discretization method using volume of fluid (VOF) model to consider free surface between water and air phases around the vessel. A dynamic mesh restructuring method is applied for grid generation regarding to the heave and pitch motions of vessel in each time step. Calculated drag and trim angles at various speeds are in good agreement with experimental data. More results are carried out at the speed 15 knots to understand the convergency of the pitch and heave motions.

Keywords: Multi-Hull; Tunnel Vessel; Hydrodynamic Simulations; Finite Volume Method

1. Introduction

Tunnel vessels are especial types of the advanced planing vessels that have complexity hull forms. Tunnel form, the main part of this hull form, significantly affects hydrodynamic behavior. Access to an automated quickly precise method for generating this hull form is an essential task based on the least input control parameters, but till now a few researches have been studied on these hull form design.

Some researchers have worked on the generation of different simple hull forms and most of them are based on Spline curves. For instance, Hinatsu [1] and Sariouz [2] developed a method to generate the surface of vessel hull using the Fourier series and non-uniform. The boundary curves such as centerline, chins and sheer lines of a vessel were modeled by B-Spline curves [3]. The parametric geometry generation in curved bilge fishing vessels was proposed based on the non-uniform B-Spline surfaces by Perez & Clemente [4]. Ventura *et al.* [5] developed a methodology for the computation of the intersect on of parametric surfaces using NURBS. Calkins *et al.* [6] focused on the mono-hull and catamaran planing vessels hull form design. They developed a computational automated method to create the hull form of a planing vessel using a mathematical procedure in the conceptual design step. Recently, authors developed a new software code, namely THFG, to generate automatically the hull form of planing tunnel vessel by various mathematical functions [7].

The existing techniques for analysis of high-speed planing hulls are classified into two major categories, *i.e.* analytical-experimental and numerical techniques. The numerical techniques are further divided into methods for boundary value problems such as boundary element method (BEM) and domain-dependent problems such as Finite volume method (FVM).

Hydrodynamic behavior of tunnel vessels is similar to general planing vessel and most changes are happened due to the increasing speed. The physical conditions (draft and trim) of the vessel which are significantly changing from displacement condition to planing condition could be significantly high.

So far, some researchers carried out both experimental and numerical work in this field. Most works have been done for the planing hull by Savitsky. His studies included the prismatic high speed vessel via definition of systematic model tests focusing mainly on the effects of design parameters, such as location of the centre of gravity, forward speed, and other geometric parameters of the planing hull (Savitsky & Brown [8], Savitsky *et al.* [9], Savitsky & Morabito [10]).

Numerical methods such as boundary element method (BEM) and finite volume applying a volume of fluid (VOF) model have been employed to determine the planing vessel hydrodynamic behavior. Ghassemi and Ghiasi [11] developed a combined method consisting of a potential-based boundary element method (BEM) for the induced pressure resistance, the boundary layer theory for the frictional resistance and practical method for the spray resistance to determine the hydrodynamic characteristics of planing vessels in calm water. As a complementary research, Ghassemi & Yu-min [12] extended this method to determine hydrodynamic forces on a planing hull in steady motion.

The VOF model is a surface-tracking technique applied to a fixed Eulerian mesh to study a multiphase problem. It is designed for two or more immiscible fluids where the position of the interface between the fluids is of interest [13]. Since the analysis of air-water interface changes is essential to investigate the hydrodynamics behavior of planing vessels, the VOF model seems to be an adequate method to simulate planing vessels. Yumin *et al.* [14] presented a new CFD method based on Reynolds-Averaged Navier-Stokes (RANS) equations and volume of fluid (VOF) method and the six-degrees-of-freedom equations were applied to calculate hydrodynamic forces of planing vessels.

The hydrodynamic effect of propeller tunnels in high speed vessel was studied by way of modified resistance and pressure distribution, using both numerically CFD method and experimentally geometric scaled models [15]. Width and depth parameters of channel for numerical or experimental analyses of high speed hull and hydrofoil vessels have most effect on resistance test results studied in a Froude number range of 0.3 - 1.0 using experimental results and theoretical work [16,17].

Most of researches have been studied on the simple hull form, not complex multi-hull tunnel vessels because of their complicated nature. Complementary to our previous work, an innovative quick precise method is proposed to generate the tunnel hull form automatically using the least control and hull form adjustment input parameters. Then, hydrodynamic behavior of tunnel vessel analyzed via finite volume method by using VOF model. A four-tunnel vessels model is tested experimentally so that the numerical results are compared by towing test resistance results. Finally, a full scale tunnel vessel is studied completely and various numerical results are presented.

2. Modeling and Simulation Methods

2.1. Geometry Modeling Method

The hull form geometry is generated using our developed new mathematical procedure to automatically generation of the planing tunnel vessels hull form. The THFG code is an efficient and applicable method in complex bodies like multihull shape to produce the hull geometrical models [7].

In the THFG code, minimum number of input control parameters has been used to design the geometry. At first, four longitudinal guidelines functions are defined by polynomial functions and NURBS curves. Next, using specified key points, the sections' curves are generated. Then, the fair transverse sections, the fair surfaces and the solid models could be generated consecutively. In **Figure 1**, the steps of hull form generation are shown.

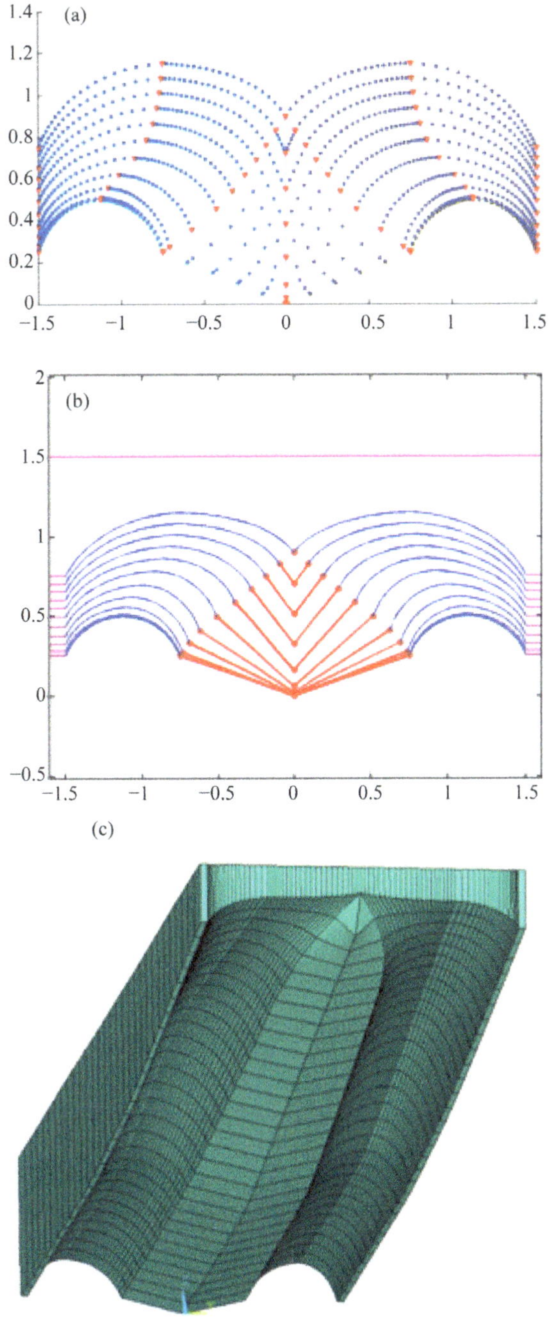

Figure 1. Generation of the hull form by THFG code. (a) Generate point with guidelines; (b) Generate curve of sections; (c) Generate hull form.

In THFG code, to produce guidelines and tunnel sections, the third order NURBS curves are used. The relations applied for this method are expressed as follows [18]:

$$C(k)=\frac{\sum_{i=0}^{n} N_{i,q}(k)\cdot W_i\cdot P_i}{\sum_{i=0}^{n} N_{i,q}(k)\cdot W_i},\quad K=\{k_0,k_1,\cdots,k_{r-1},k_r\} \quad (1)$$

In Equation (1), q is assumed to 3; $n + 1$ is 6 for guidelines and 3 for sections in each part of tunnel; $r + 1$ is fixed to 10 for guidelines and 8 for sections in each part of the tunnel. The knot vector values (k_i), the control points (P_i) and their weights (W_i) are assumed as optional values for each guideline or section curve of the tunnel. The shape of the curve can be properly adjusted by controlling these parameters.

To control the shape, the number of control points is fixed to 6 and also uniform knot vectors are used. The knot vectors values of guidelines and sections are shown in Equation (2).

$$U_{\text{Guidelines}}=\left\{0,0,0,0,\frac{1}{3},\frac{2}{3},1,1,1,1\right\}$$
$$U_{\text{Sections}}=\{0,0,0,0,1,1,1,1\} \quad (2)$$

2.2. Calculation Domain

In order to reduce computations and because of the symmetry existence in the analyses domain, half of the body vessel is considered in the multiphase (air and water) domain. The dimensions of computational domain and under studied vessel position inside it are shown in **Figure 2**. To prevent from reaching the wave motion of vessel to outflow boundary, the outflow boundary is adjusted too far (72.5 m) from behind the vessel. Furthermore, a distance of 40 meters between the vessel and bottom of domain is because of observance of deep water condition. The created sub domain containing vessel inside it at initial position is divided to two parts by water surface. For suitable simulation to make a precise calculation of free surface and around the vessel, the calculation domain is divided by air-water interface to two separate parts and also the sub domain is divided to two boxes; a small box around vessel and another out of that.

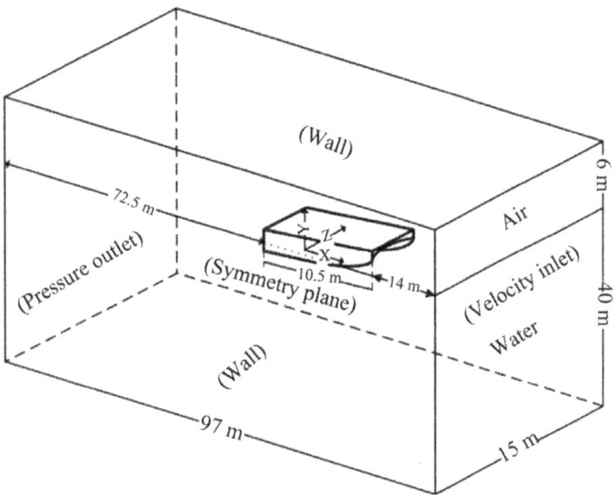

Figure 2. Computational domain of the vessel in real data.

The constructed vessel by THFG code and divided domain are applied to the commercial grid generation tool, GAMBIT 2.3 (Fluent Inc.), to generate the desirable calculation control volumes structure. To divide the calculation space into adequate discrete control volumes, three-dimensional tetrahedral computational cells were used. In addition, triangular elements were used for surface boundaries, including: the vessel, walls, interface, inlet and outlet of the domain.

During solving procedure, a dynamic mesh restructuring method is applied regarding to the heave and pitch motion of vessel. The grid section around body near the free surface between air and water is refined to achieve a desirable meshing structure for interface zone to remain similarly in all time steps. This is carried out while the outer section of grid away from body is unchanged all the time.

2.3. Hydrodynamic Numerical Method

In present work, to solve the hydrodynamic behavior of high speed vessel, the longitudinal position of vessel is fixed then air and water pass through the vessel body with constant speed. This process is similar to cavitation or wind tunnel test processes.

The local and global coordinate systems are fixed at the vessel gravity center and in the first position of gravity center before motion, respectively. The degrees of freedom of vessel in this method are pitch and heave motions as shown in **Figure 3**. In this method, draft and trim angle of vessels at a constant speed are adjusted while balancing the lift force with weight.

The procedure algorithm of numerical method for hydrodynamic analyses of high speed vessel is shown in **Figure 4**. The termination criteria of method are determined as when the variation of vessel movement diminishes to zero or velocity vector approaches zero in consecutive steps. So, the vessel is achieved a balanced state in which sum of forces and moments become zero.

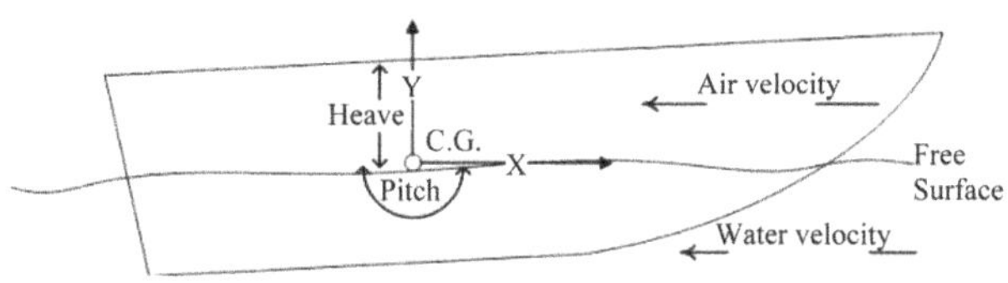

Figure 3. Degrees of freedom of vessel.

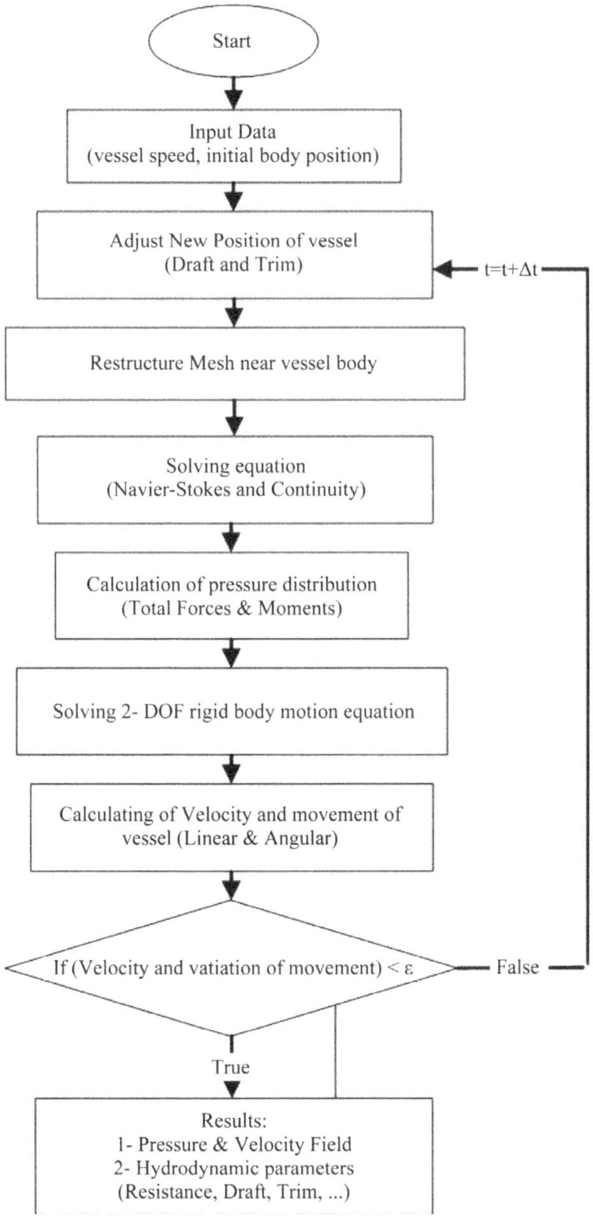

Figure 4. Solution algorithm of present methods.

At the first time step, the Navier-Stokes and Continuity equations in all nodes of domain are solved, and then by integration of pressure field around the vessel surface, the forces and moments acted on body are calculated. Then, by solving the motion of body, velocity vector of vessel gravity center is obtained. When the difference of displacement vectors (heave and pitch) and velocity vectors for previous and present time steps diminishes to zero (less than ε) and lift force equals to weight, the calculation process is terminated and the vessel becomes stable as described by Equation (3). Achieving this condition, the trim, draft and drag of vessel is fixed. If it does not happen, the vessel is moved to the new draft and trim angle based on results of solution of equilibrium equation. Then, the grid near the free surface around the vessel (small box) is restructures and more steps are iterated consecutively.

$$\text{End cond} = \begin{cases} \left(\text{heave}_t - \text{heave}_{t-1}\right) \le \varepsilon & \text{draft is fixed.} \\ \left(\text{pitch}_t - \text{pitch}_{t-1}\right) \le \varepsilon & \text{trim is fixed.} \\ v_t \le \varepsilon & \text{heave velocity is zero.} \\ \theta_t \le \varepsilon & \text{pitch velocity is zero.} \\ \left|L - m \cdot g\right| \le \varepsilon & \text{lift force equal weigth.} \end{cases} \quad (3)$$

As described earlier, the VOF approach is used for free surface modeling. In this case, the volume fraction, density and dynamic viscosity of two existent phases (air and water) are solved using following equations in each control volume for all time steps:

$$\begin{aligned} &\frac{\partial \alpha}{\partial t} + \nabla \cdot \left(\alpha U\right) = 0, \\ &\rho = \alpha \rho_2 + \left(1-\alpha\right)\rho_1, \\ &\mu = \alpha \mu_2 + \left(1-\alpha\right)\mu_1 \end{aligned} \quad (4)$$

And the corresponding equations for conservation of momentum (Navier-Stokes) are given as:

$$\begin{aligned} &\frac{\partial(\rho u)}{\partial t} + \nabla \cdot \left(\rho u U\right) = -\frac{\partial p}{\partial x} + \nabla \cdot \left(\mu \cdot \nabla(u)\right) \\ &\frac{\partial(\rho v)}{\partial t} + \nabla \cdot \left(\rho v U\right) = -\frac{\partial p}{\partial y} + \nabla \cdot \left(\mu \cdot \nabla(v)\right) \\ &\frac{\partial(\rho w)}{\partial t} + \nabla \cdot \left(\rho w U\right) = -\frac{\partial p}{\partial z} + \nabla \cdot \left(\mu \cdot \nabla(w)\right) - \rho \cdot g \end{aligned} \quad (5)$$

For movement estimation of the vessel based on velocity and pressure fields in the flow domain, the equations of linear and angular momentums at each time step are solved as following:

To reduce calculation time, the initial draft and trim angle of vessel is assumed near to the final dynamic draft and trim angle of vessel at stable condition.

In order to turbulence modeling, additional conservation equations for kinetic turbulence energy (κ) and turbulence energy dissipation (ε) are solved. In this study, Standard k-ε model has been used.

To solve hydrodynamic equations as described above, a commercial software (FLUENT 6.3) assisted by an external User Defined Function (UDF) is used for numerical calculations at each time step less than of 0.001 s. The simple discretization method is used for the Navier-Stokes system (velocity-pressure coupling) applying the first order implicit method for solving unsteady formulation of the problem.

3. Result and Discussion

3.1. Grid Independency and Validation Results

Applying the first validation approach, the comparison

between experimental and CFD simulation results for vessel drag is shown in **Figure 5**. It should be noticed that dimensions of CFD model and tested model are the same, so there is no scale effect in results. It is observed that the obtained curves are in good agreement and accordance so that the maximum difference between resulted data through experiment and CFD methods is about 7% therewith validity of CFD code is approved.

As the second approach, the resulted trim angles in comparison with experimental data reported by Subramanian and Subramanyam [15] are presented in **Figure 6** based on volume Froude numbers. It can be seen that the CFD code applies in accordance with the experiments. It is confirmed that the maximum error of CFD code is smaller than 6%.

To ensure the grid independency, different grids numbers have been chosen. Finally, it is found that with the number 2.4 million grids all parameters such as resistance and trim angle are converged. The important values y^+ is less than 50 for all speed when the resistance is converged. Gird number, resistance and its y^+ values are shown in **Table 1**.

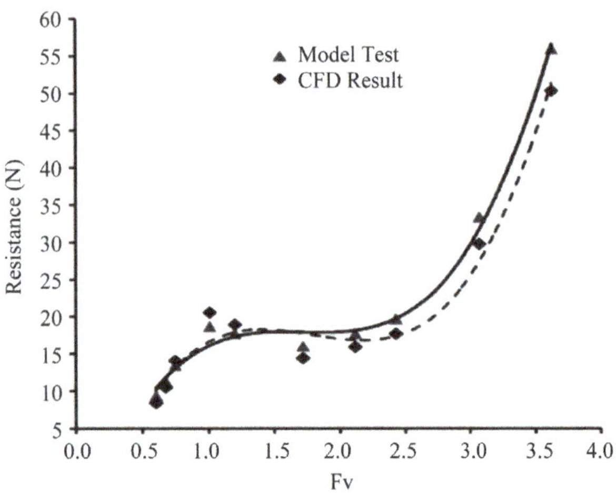

Figure 5. Comparison of drag.

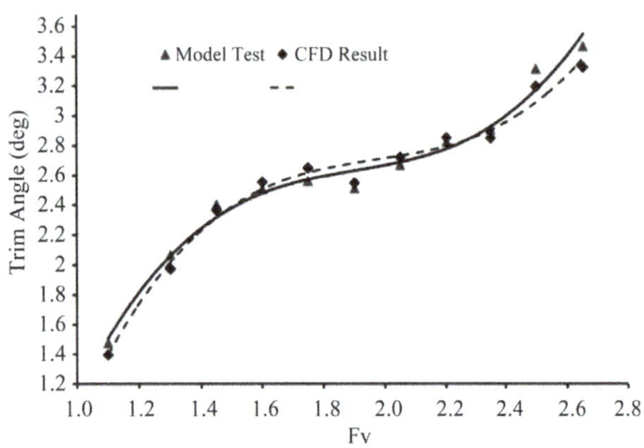

Figure 6. Comparison of trim angle.

3.2. Hydrodynamics Studies

In this study, a two-tunnel vessel with specifications according to data summarized in **Table 2** is generated which is illustrated in **Figure 7**. Masses distribution and equipments arrangement are considered according to data of **Table 3**.

Table 1. Grid number, resistance and its y^+ values (F_V = 2.5).

Number of grids	Resistance [N]	y^+
1800000	12	500
2000000	15	190
2200000	17.5	85
2400000	17.7	45

Table 2. Main specifications of the vessel.

Topic	unit	value
Length	m	10.5
Breadth	m	2.5
Depth	m	1.5
Displacement	kg	3500
V_{max}	knot	40
Deadrise at midship	deg	13
Max breadth of tunnel	m	0.75
Max height of tunnel	m	0.60
Min height of tunnel	cm	0.30

Table 3. Weight specifications of the vessel.

Parameters	Unit	Value
Weight of hull	kg	1600
Weight of 2 outboard engine	kg	500
Weight of fuel	kg	400
Weight of 4 crews	kg	400
Weight of equipment	kg	200
Total weight	kg	3300
LCG From transom	cm	315
VCG From keel	cm	65
inertia mass matrix about center of gravity	$m^2 \cdot kg$	$I_G = \begin{bmatrix} 2390 & 0 & 0 \\ 0 & 26000 & 0 \\ 0 & 0 & 24300 \end{bmatrix}$

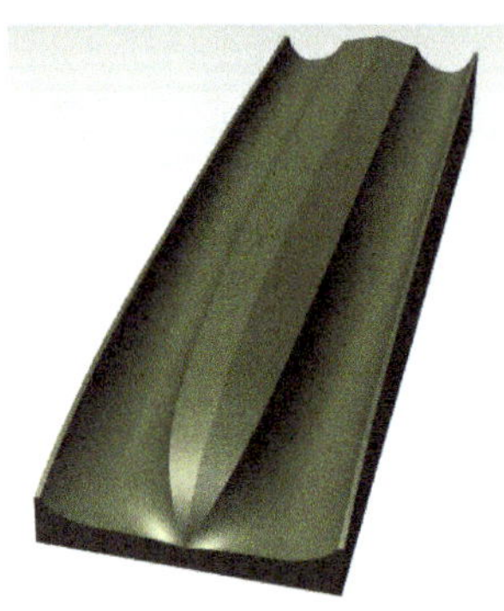

Figure 7. Bottom view of tunnel vessel hull form.

3.2.1. Variations of Heave and Pitch

The vessel is running at various speeds. Here, we focused at the speed 15 knots. The results of spray, pressure, pitch and heave motions distinguished. **Figure 8** shows the spray area and the pressure area at 15 knots. The angle between keel and stagnation line and the angle between keel and whisker spray are named as α and θ having values of 10° and 21.5°, respectively. Savitsky *et al.* [9] presented following equation for calculation of α and θ in prismatic body:

$$\tan\alpha = \frac{\pi}{2}\frac{\tan\tau}{\tan\beta}, \quad \theta = 2\alpha \tag{6}$$

where β and τ are deadrise and trim angles, respectively. At 25 knots, τ is 2.7° and in this longitudinal position, β equals to 21°. Assuming the middle V-shape body independent of tunnel body and similar to prismatic body, α and θ can be calculated based on Equation (6) as presented in **Table 4**.

Comparing these values for α and θ with numerical results, the hypothesis of similarity of middle part of tunnel vessel with simple prismatic is approved.

As previously mentioned, the pressure field around the vessel, hydrodynamic forces and momentums acted on it, are obtained for each time step. The vessel is moving up and down and rolling about transverse axis until the body would be stable and the lift force equals to weight so that the resultant momentum of them diminishes to zero. In this situation, the pitch angle, draft and drag of vessel become fixed and the vertical and angular velocities are approached to zero. In the procedure of stabilizing the vessel motions, the numerical hydrodynamic analysis of vessel is repeated for more than 8000 iterations. **Figures 9-12** show the convergency of heave, pitch, pitch and heave velocities in hydrodynamic analyses of tunnel vessel at 15 knots. It can be seen from **Figure 9** that variations of heave decrease rapidly to reach a small constant value of 0.05. It is observed from **Figure 10** that pitch angle is increased during time passing to achieve a constant level after 9.45 seconds. A damping trend also can be observed for pitch and heave velocities from **Figures 11** and **12**, respectively. The pitch velocity diminishes to −0.002 rad/s and heave velocity becomes about 0.004 m/s. Hence, considering the equality of weight and lift forces, all five termination conditions are satisfied.

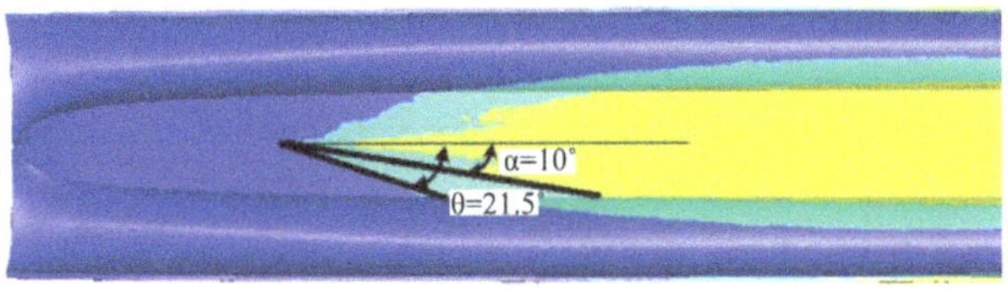

Figure 8. Spray (in green color) and pressure (in yellow color) area in prismatic body at 15 knots.

Table 4. Comparison of the spray angle between Savitsky's formula and presented calculated.

Angle (degree)	Calculated	Savitsky's formula	% error
α	10.9	10	9
θ	21.8	21.5	1.4

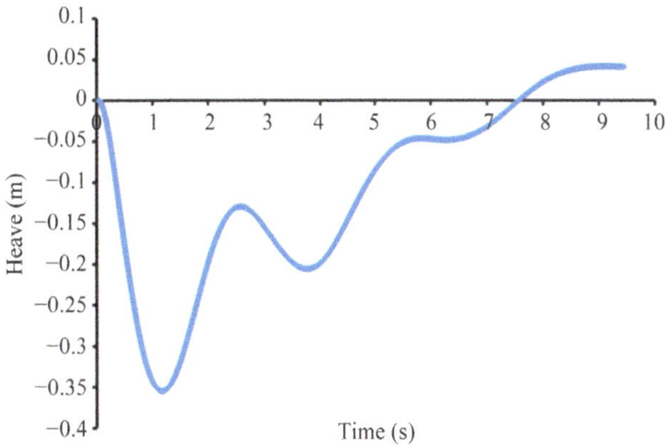

Figure 9. Variation of heave based on time steps.

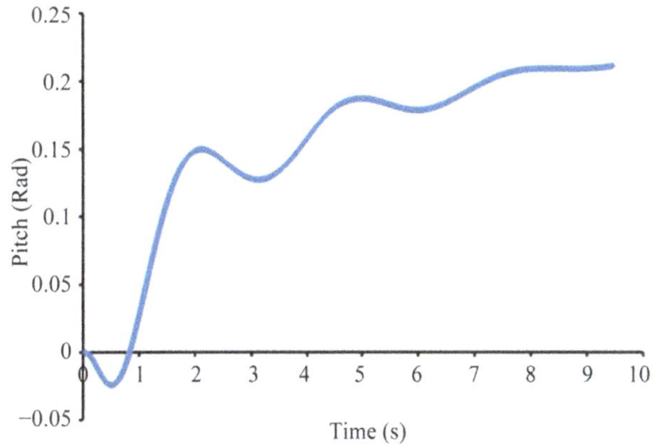

Figure 10. Variation of pitch angle based on time steps.

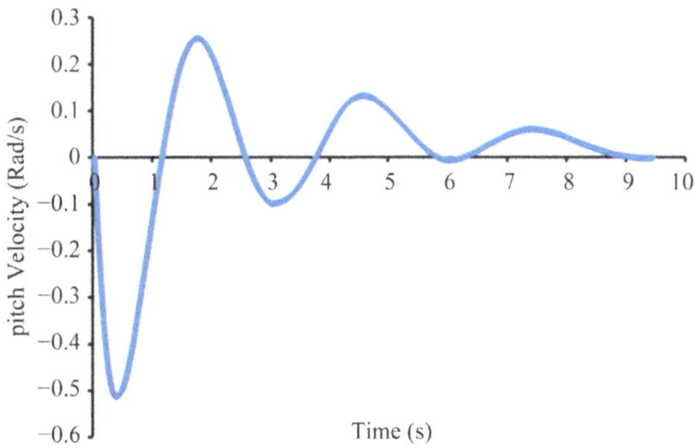

Figure 11. Variation of pitch angle velocity based on time steps.

4. Concluding Remarks

In this paper, a multihull tunnel vessel hull form was

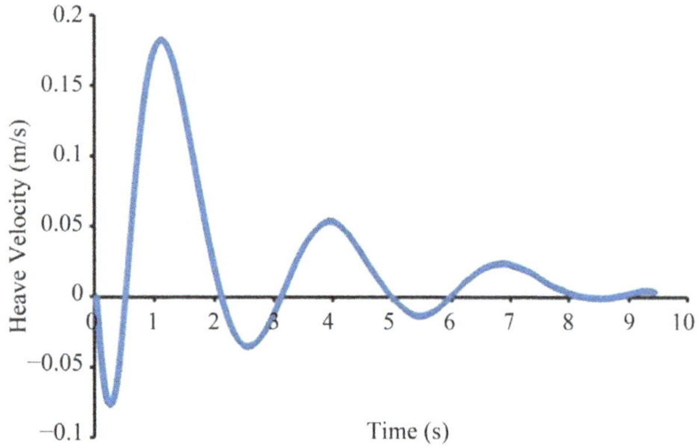

Figure 12. Variation of heave velocity based on time steps.

successfully generated by our developed THFG code. Then, a numerical approach was used to hydrodynamic study of the designed vessel. The THFG code is a suitable tool for hull form design of tunnel vessels that can be used for experimental model building and numerical hydrodynamic analyses and predictions. Its fast and precise procedure suggests its utilization for parametric study and using as an optimization approach for hull form of tunnel vessel types. Entering twenty geometry control parameters in THFG code by using NURBS curve can generate fair tunnel vessel hull form at the minimum time.

Continuously, a case study of 10.5 m tunnel vessel was designed by THFG code and studied via hydrodynamic analyses method. At 15 knots speed, angle of spray line and stagnation line is in a good accordance with Savitsky's reported experimental data. Also, at this speed, convergency of motion parameters, including pitch angle, heave, pitch and heave velocity, during iterations of solving procedure are presented.

REFERENCES

[1] M. Hinatsu, "Fourier NUBS Method to Express Ship Hull Form," *Journal of Marine Science and Technology*, Vol. 9, 2004, pp. 43-49.

[2] E. Sarioz, "An Optimization Approach for Fairing of Ship Hull Forms," *Ocean Engineering*, Vol. 33, No. 13, 2006, pp. 2105-2118.

[3] F. Perez and J. A. Suarez, "Quasi-Developable B-Spline Surfaces in Ship Hull Design," *Computer-Aided Design*, Vol. 39, No. 10, 2007, pp. 853-862.

[4] F. L. Perez and J. A. Clemente and J. A. González, "Parametric Generation, Modeling, and Fairing of Simple Hull Lines with the Use of Non-Uniform Rational B-Spline Surfaces," *Journal of Ship Research*, Vol. 52, No. 1, 2008, pp. 1-15.

[5] M. Ventura and C. Guedes Soares, "Surface Intersection in Geometric Modeling of Ship Hulls," *Journal of Marine Science and Technology*, Vol. 17, No. 1, 2012, pp. 114-124.

[6] D. Calkins, R. D. Schachter and L. T. Oliveira, "An Automated Computational Method for Planing Hull Form Definition in Concept Design," *Ocean Engineering*, Vol. 28, No. 3, 2001, pp. 297-327.

[7] M. Ghassabzadeh and H. Ghassemi, "An Innovative Method for Parametric Design of Planing Tunnel Vessel Hull Form," *Ocean Engineering*, Vol. 60, No. 4, 2013, pp. 14-30.

[8] D. Savitsky and P. W. Brown, "Procedures for Hydrodynamic Evaluation of Planing Hulls in Smooth and Rough Water," *Marine Technology*, Vol. 13, No. 4, 1976, pp. 381-400.

[9] D. Savitsky, M. F. Delorme and R. Datla, "Inclusion of Whisker Spray Drag in Performance Prediction Method for High-Speed Planing Hulls," *Maine Technology*, Vol. 44, No. 1, 2007, pp. 35-56.

[10] D. Savitsky and M. Morabito, "Surface Wave Contours Associated with the Forebody Wake of Stepped Planing Hulls," *Marine Technology*, Vol. 47, No. 1, 2010, pp. 1-16.

[11] H. Ghassemi and M. Ghiasi, "A Combined Method for the Hydrodynamic Characteristics of Planing Crafts," *Ocean Engineering*, Vol. 35, No. 3-4, 2008, pp. 310-322.

[12] H. Ghassemi and S. Yumin, "Determining the Hydrodynamic Forces on a Planing Hull in Steady Motion," *China Ocean Engineering*, Vol. 7, 2008, pp. 147-156.

[13] FLUENT Inc., "Fluent 6.3 User's Guide," 2006.

[14] S. Yumin, C. Qingtong, S. Hailong and L. Wei, "Numerical Simulation of a Planing Vessel at High Speed," *Journal of Marine Science and Application*, Vol. 11, No. 2, 2012, pp. 178-183.

[15] A. V. Subramanian, P. V. V. Subramanyam and A. N. Sulficker, "Pressure and Drag Influences Due to Tunnels in High-Speed Planing Craft," *International Shipbuilding Progress*, Vol. 54, No. 1, 2007, pp. 25-44.

[16] A. Millward, "The Effect of Channel Width and Depth on the Resistance of a High Speed Hull," *International Shipbuilding Progress*, Vol. 51, No. 4, 2004, pp. 377-394.

[17] M. Reichel and A. Bednarek, "The Experimental Studies on Hydrofoil Resistance," *Archives of Civil and Mechanical Engineering*, Vol. 7, No. 3, 2007, pp. 167-175.

[18] L. Piegl and W. Tiller, "The NURBS," Springer, Berlin, 1997.

A Study of Thermal Response and Flow Field Coupling Simulation around Hayabusa Capsule Loaded with Light-Weight Ablator

Hisashi Kihara, Naoya Hirata, Ken-ichi Abe
Department of Aeronautics and Astronautics, Kyushu University, Fukuoka, Japan

ABSTRACT

The numerical simulation of flow field around Hayabusa capsule loaded with light-weight ablator thermal response coupled with pyrolysis gas flow inside the ablator was carried out. In addition, the radiation from high temperature gas around the capsule was coupled with flow field. Hayabusa capsule reentered the atmosphere about 12 km/sec in velocity and Mach number about 30. During such an atmospheric entry, space vehicle is exposed to very savior aerodynamic heating due to convection and radiation. In this study, Hayabusa capsule was treated as a typical model of the atmospheric entry spacecraft. The light-weight ablator had porous structure, and permeability was an important parameter to analyze flow inside ablator. In this study, permeability was a variable parameter dependent on density of ablator. It is found that the effect of permeability of light-weight ablator was important with this analysis.

Keywords: Light-Weight Ablator; Coupling Calculation; Pyrolysis Gas Flow; Hayabusa Capsule

1. Introduction

About the space mission in near future, it is expected that the sample return mission like Stardust and Hayabusa increases. In those missions, the space vehicle reenters to atmosphere at very high speed. In an atmospheric entry condition, a strong shock wave and a high enthalpy flow field are generated around the flight vehicle. For example, Hayabusa reentry capsule, which returned to Earth in 2010, had the velocity of about 12 km/sec and experienced the heating rate of about 20 MW/m^2 [1]. To protect an entry vehicle from such a severe aerodynamic heating, thermal protection system (TPS) is one of the key components in its design. Especially, an ablator has been widely used as a TPS material for a vehicle under such a high-heating condition. In addition, in such high enthalpy flow, the estimation of heat flux from flow to the vehicle is very complex and difficult. Although a carbon ablator is known as an often-used one, its heaviness had been a critical problem in considering the severe weight limitation of a space vehicle. However, a light-weight ablator named phenolic impregnated carbon ablator (PICA) was developed at NASA Ames Research Center in 1990's and used for the Stardust sample return capsule fore body heat shield. Its density is lower than 300 kg/m^3 [2], while most of the conventional carbon ablators have the density of over 1000 kg/m^3.

In Japan, a light-weight ablator (300 - 400 kg/m^3) for middle range heat flux is being developed in JAXA [3]. These light-weight ablators will become more and more important for expanding the capability of future missions and thus sophisticated numerical models are needed to predict their thermal response more correctly.

An ablator under severe heating environment involves several complex physics. When the ablator is heated, the inside resin pyrolyzes and the pyrolysis gas is generated. Then, the pyrolysis gas flow through the ablator, which is porous media, ejects from its surface to the outer flow field. This flow contributes to reduction of the heat flux to the ablator surface. Furthermore, the pyrolysis of the resin that is endothermic reaction contributes to cooling the ablator itself. From these characteristics, the ablator can be used as an effective passive thermal protection. However, the aforesaid complex phenomena are involved in its analysis and thus there still remain uncertainties in predicting the thermal response of the ablator. Several efforts have focused on identifying thermal response of an ablator. Charring Materials Ablation (CMA) code and Fully Implicit Ablation and Thermal response (FIAT) program [4] simulated the internal heat conduction and the thermal decomposition in one dimension. More sophisticated simulation code named Two-dimensional Implicit Thermal response and Ablation (TITAN) [5] enabled two-dimensional simulation.

On the other hand, Ahn H. *et al.* developed a computational code named Super Charring Materials Ablation (SCMA) [6]. In SCMA code, the motion of the pyrolysis gas inside an ablator was simulated in one dimension by solving the mass and momentum equations for the gas phase. Suzuki T. *et al.* extended SCMA code for two dimensional simulation and named the code (SCMA2) [7]. Moreover, they coupled SCMA2 code with a CFD code using the Park's two-temperature model and chemical reactions with 21 species for a thermochemical nonequilibrium flow [8].

In our research group, numerical and experimental studies have been conducted for determining the thermal response of a light-weight ablator exposed to a flow field generated by an arc-heated wind tunnel, which is often used for a heating test of an ablator [9-11]. Therefore, in the present study, the thermal response simulation code is extended so that it can treat the pyrolysis gas flow as unsteady phenomena in two-dimensional axisymmetric coordinate. Besides, the CFD code is also extended so as to include more chemical species, leading to more detailed evaluation of the ablation-gas injection consisting of the pyrolysis gas and the carbonaceous gas generated by the surface reactions. Furthermore, the code is developed for application to wider range of the flow conditions not only for a nitrogen flow but also for an air flow.

2. Computational Models

To simulate the process of heat shield using ablator, we have to treat many complex situations like the blowing flow of pyrolysis gas from ablator, chemical reaction on the surface recession of ablator, radiation heat conduction from strong shock heated gas, and interaction between blowout gas and free stream, etc. To understand such complex phenomena, the coupling simulation of outer flow-field and inside of ablator was employed.

2.1. Flow-Fields

Flow-field is assumed continuous and axisymmetric flow. Navier-Stokes equation extends to thermo-chemical nonequilibrium flow is adopted as governing equation. The present simulation treats 11 species (N_2, O_2, NO, N_2^+, O_2^+, NO^+, N, O, N^+, O^+, e^-) for free stream and 9 species (H_2, C_2, C_3, CN, CO, H, C, H^+, C^+) are related to ablation gas. These 20 Species and 153 chemical reactions are considered [12].

The four temperature model is employed to handle in detail thermal non-equilibrium. These are translation, rotation, vibration and electron respectively. The electro excited temperature treats as equilibrium to electron temperature. The energy transfers between each of the internal energy modes are considered as follows: translation-rotation [13], translation-vibration [14], translation-electron [15,16], rotation-vibration [13], rotation-electron [17] and vibration-electron [18]. The energy losses for vibration and rotation due to the heavy particle-impact dissociation [16] and the electron energy loss due to the electron-impact dissociation/ionization reactions are considered. In addition, radiated heat transfer is not ignored, so the radiation field of high temperature gas is calculated coupled with flow field around capsule. For the calculation of radiation field, the present paper employed 3-band model [19] extended to chemical nonequilibrium flow field, which is very low cost method [20].

The char layer at the ablator surface is lost due to the surface reactions and then the ablator surface recedes. The recession rate r is determined using the carbon mass loss rate:

$$r=\frac{1}{\rho_{char}}\left(\frac{M_C}{M_{CN}}J_{CN}^{nit}+\frac{M_C}{M_{CO}}J_{CO}^{oxi}+\frac{M_C}{M_C}J_C^{sub}+\frac{M_C}{M_{C2}}J_{C2}^{sub}+\frac{M_C}{M_{C3}}J_{C3}^{sub}\right) \quad (1)$$

The shape change should be accounted in the simulation because the magnitude and distribution of the heat flux are very sensitive to the surface shape [21]. Thus, in order to consider the shape change due to the surface recession, the computational domains for both the ablator and the flow field are reconstructed according to the surface recession. In this calculation, the grid system was rebuilt at the each 1 μm recession quantity.

2.2. Thermal Response of Ablator

In this simulation, the ablator is made of carbon foam material impregnated with phenolic resin. It is easy predicted that the flow of pyrolysis gas inside the light-weight ablator was different because it had much vacant spaces. The permeability of virgin layer of conventional ablator has quite little compare with the permeability of light-weight ablator's one. Therefore, it is considered the phenomena inside the ablator include the conversion from the solid to the gas phases (pyrolysis), the heat conduction in the solid and the gas phases and the gas phase flow with convective heat transfer. The governing equations inside ablator are shown as follows.

Solid mass conservation:

$$\frac{\partial \rho_s}{\partial t}=-R \quad (2)$$

Solid energy conservation:

$$\frac{\partial E_s}{\partial t}=\frac{\partial}{\partial x_j}\left(\lambda_s\frac{\partial T_s}{\partial x_j}\right)-Q_{sg}-R\cdot e_s+(1-a)R\Delta h^{pyr} \quad (3)$$

Gas mass conservation

$$\frac{\partial(\varepsilon\rho_g)}{\partial t}+\frac{\partial}{\partial x_j}(\varepsilon\rho_g u_i)=-R \quad (4)$$

Gas momentum conservation

$$\frac{\partial(\varepsilon\rho_g u_j)}{\partial t}+\frac{\partial}{\partial x_j}(\varepsilon\rho_g u_i u_j+\varepsilon\delta_{ij}p)=\frac{\partial(\varepsilon\tau_{ji})}{\partial x_j}-\varepsilon f_i \quad (5)$$

Gas energy conservation

$$\frac{\partial(\varepsilon E_g)}{\partial t}+\frac{\partial}{\partial x_j}\left[\varepsilon(E_g+p)u_j\right]=\frac{\partial}{\partial x_j}\left[\varepsilon u_i\tau_{ji}+\varepsilon\lambda_g\frac{\partial T_g}{\partial x_j}\right]+Q_{sg}+R\cdot e_g+aR\Delta h^{pyr} \quad (6)$$

where subscripts *s* and *g* represent the solid and the gas phases, respectively.

In Equation (2), R represents the pyrolysis rate. It shows as follow [22]:

$$R=-\left(\frac{\partial(\rho_s)}{\partial t}\right)=A\exp\left(-\frac{B}{T_s}\right)\rho_{\text{virgin}}\left(\frac{\rho_s-\rho_{char}}{\rho_{\text{virgin}}}\right) \quad (7)$$

In Equations (3) and (6), a represent the effectiveness of reaction heat for gas phase and $(1-a)$ is for solid phases respectively. The third term in Equation (3) and fourth term in Equation (6) represents the energy exchange between the solid and the gas phases. The fourth term in Equation (3) and fifth term in Equation (6) are the energy loss and energy gain due to the pyrolysis, respectively. The last term in Equations (3) and (6) are the source terms due to the reaction heat of the pyrolysis. The pyrolysis gas made from C, H, O and calculated by chemical equilibrium code [23].

The solid phase consists of char and resin and then the solid density is given by the sum of those densities:

$$\rho=\rho_{resin}+\rho_{char} \quad (8)$$

Note that the char density ρ_{char} is constant, while the resin density ρ_{resin} decreases by the pyrolysis. Similarly, the internal energy of the solid phase is divided into the parts of char and resin:

$$E_s=\rho_{resin}e_{resin}+\rho_{char}e_{char} \quad (9)$$

$$e_{resin}=\int C_{p,resin}(T)\mathrm{d}T+\Delta h^0_{resin} \quad (10)$$

$$e_{char}=\int C_{p,char}(T)\mathrm{d}T+\Delta h^0_{char} \quad (11)$$

The specific heat is given in the form proposed by Potts [24]. This form is based on the characteristics that the specific heat of graphite.

$$C_{p,s}=C_\infty\frac{T_s}{\sqrt{T_s^2+(C_\infty/c_1)}} \quad (12)$$

where C_∞ and c_1 are determined by curve fitting based on JANAF's data sheet of graphite [25].

The pressure of the gas phase is given by the equation of state:

$$p=\rho_g RT_g \quad (13)$$

The internal energy of the gas phase is given by

$$E_g=\rho_g e_g=\rho_g\left(C_v T_g+\Delta h_g^0+\frac{1}{2}q^2\right) \quad (14)$$

where $q^2=u_iu_i$.

The thermal conductivity is determined by a curve-fitting method with a manufacturing data sheet on carbon fiber material. Porosity is given by

$$\varepsilon=\varepsilon_{\min}-\frac{\rho_{\text{virgin}}-\rho_s}{\rho^*_{resin}} \quad (15)$$

where $\varepsilon_{\min}$ represents the porosity in case that the resin before pyrolyzes.

The friction force per unit volume $\boldsymbol{f}$ is given by

$$\boldsymbol{f}=\mu\boldsymbol{K}^{-1}\boldsymbol{u} \quad (16)$$

where $\boldsymbol{K}$ is the permeability tensor.

$$\boldsymbol{K}=\begin{bmatrix}\cos\theta & -\sin\theta\\ \sin\theta & \cos\theta\end{bmatrix}\begin{bmatrix}K_1 & 0\\ 0 & K_2\end{bmatrix}\begin{bmatrix}\cos\theta & \sin\theta\\ -\sin\theta & \cos\theta\end{bmatrix} \quad (17)$$

θ is the direction along the capsule surface for light-weight ablator or the laminated direction for conventional ablator. K_1 is the permeability of θ direction, and K_2 is the normal direction of θ. K_1 and K_2 are measured by experiment in our laboratory.

2.3. Coupling and Conditions

The analysis of Flow field and thermal response of the ablator coupling is done through the boundary condition of these two fields. The boundary conditions change time after time. The reentry trajectory determined by JAXA [26] is treated in present study. The calculation condition shows in **Table 1**. Time 0sec was altitude at 200 km. The calculation was done during 40 sec to 100 sec for ablator. As the boundary condition of inflow, pseudo steady condition is used. Inflow boundary condition changed every 5 sec. The calculation conditions are shown in **Table 1**.

Numerical domain is shown in **Figure 1**. The computation mesh are 110 × 60 for flow field and 60 × 60 for ablator.

3. Result and Discussions

Figure 2 shows temperatures distribution along center

Table 1. Free stream conditions at calculated point.

Flight time S	Altitude Km	Velocity km/s	Density kg/m^3	Temperature K
55	77.75	11.695	2.408×10^{-5}	213.7
60	68.59	11.546	9.065×10^{-5}	226.8
65	59.94	11.061	2.897×10^{-4}	242.3
70	52.16	9.868	7.785×10^{-4}	256.4
75	45.74	7.807	1.785×10^{-3}	258.1
80	40.99	5.422	3.516×10^{-3}	247.7

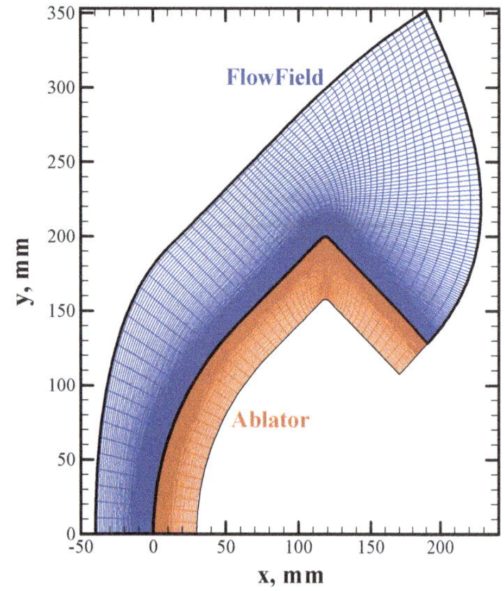

Figure 1. Numerical grid system.

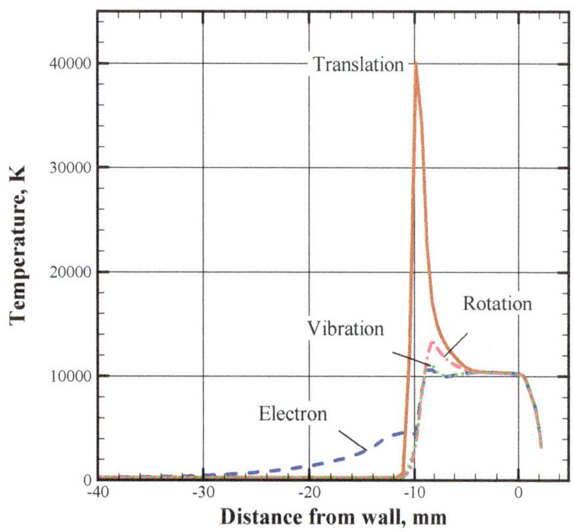

Figure 2. On axis temperatures distribution at 52.16 km.

axis in front of vehicle at 52.12 km in altitude and 70 sec after reentry as a typical condition. We discuss about condition 52.12 km which the typical data of present simulation. The zero point of x-axis shows the initial position of capsule surface. A precursor is remarkably seen because gas becomes a very high temperature with the shock wave. The numerical domain of flow field is selected that the effect of precursor could not ignore. The translational temperature becomes about 40,000 K. It can be seen that it is the strong thermal nonequilibrium upstream than 5 mm of initial surface. The temperature of the plateau part behind shock wave exceeds 10,000 K and the four temperatures are almost equilibrium.

Figure 3 shows temperatures distribution along center shows the difference in the temperature profiles along the centerline without ablator or ablator. The zero point of x-axis shows the surface of capsule. To be easy to look, only translational temperature and vibrational temperature are shown here. There seems to be no difference between two conditions. However, the peak translational temperature and the temperature gradient of approach to the surface are different and this difference indicates ablating cooling.

Figure 4 shows temperature profiles for each time along the center line using heavy ablator. The calculation started t = 40 sec cause of the heat flux is enough small to able to ignore. At t = 40 sec to 60 sec, it assumed that the heat flux and pressure interpolated by quadratic function. On the other hand, it assumed that the heat flux became 0 at t = 100 and calculation was stopped. The peak temperature of the surface is about 3100 K at 70 sec. This value has good agreement with measured value [26] when Hayabusa reentered.

Figures 5 and **6** show pressure distribution and mass flux vectors of pyrolysis gas for heavy ablator and lightweight ablator at 52.16 km, respectively. The permeability of conventional heavy ablator at virgin layer is quite large so that the pyrolysis gas could not penetrate to the direction of rear wall. So the pyrolysis gas must flow out

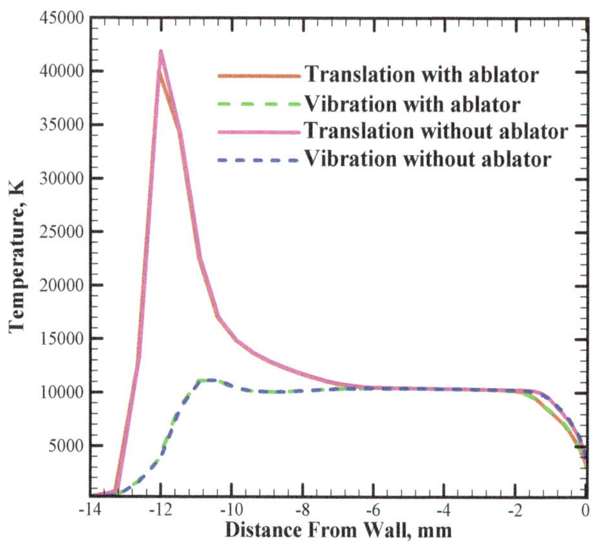

Figure 3. Difference of temperatures distribution between with and without ablated flow at 52.16 km.

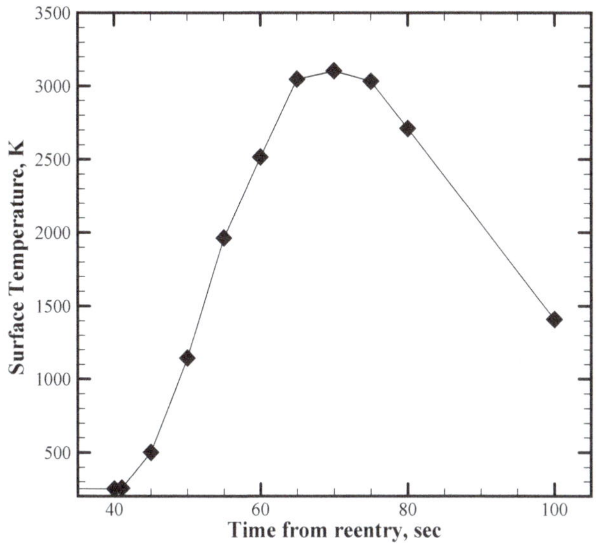

Figure 4. Surface temperature history at stagnation with heavy ablator.

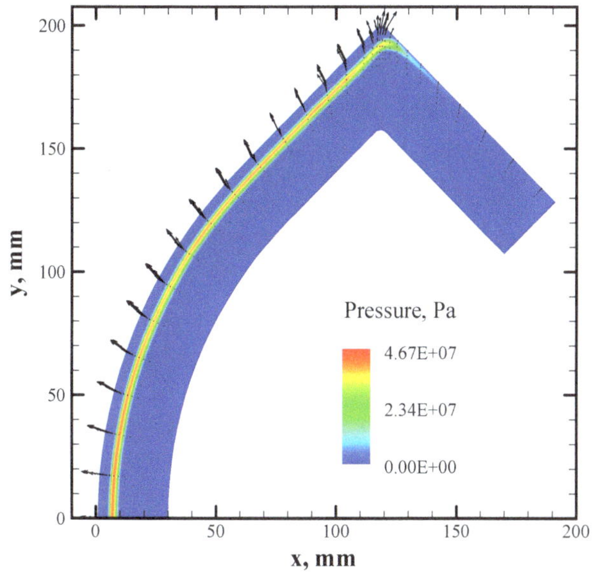

Figure 5. Pressure distribution and mass flux vectors of pyrolysis gas at 52.16 km with heavy ablator.

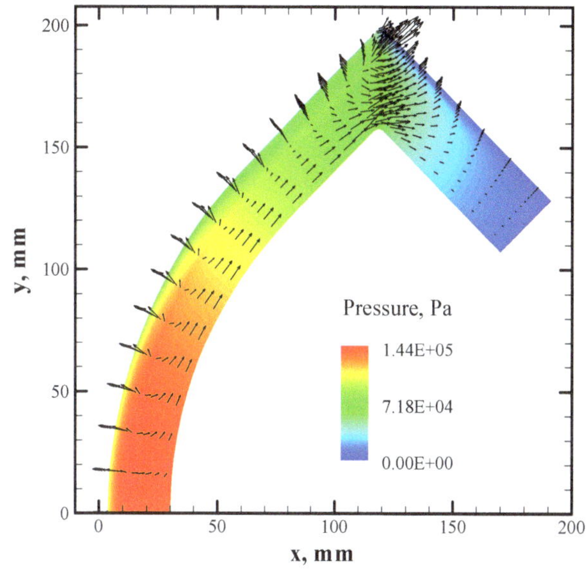

Figure 6. Pressure distribution and mass flux vectors of pyrolysis gas at 52.16 km with light-weight ablator.

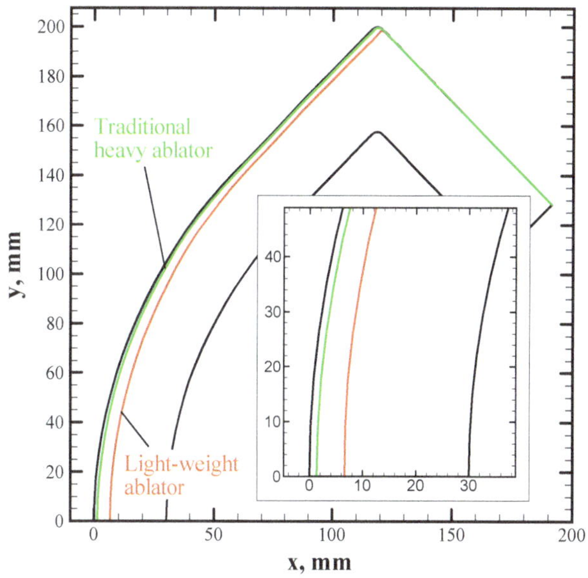

Figure 7. Shape change of ablator at 100 sec.

to front side only. On the other hand, the pyrolysis gas inside light-weight ablator flows not only front but also rear and radial direction. The pyrolysis gas in light-weight ablator is more transmissible than heavy weight ablator. Therefore, the pressure distribution of pyrolysis-gas in light-weight type became broad. In addition, cause of the anisotropy of permeability (distribution of porosity) the flow to the radial direction is occurred strongly. Particularly, the flow of the shoulder part in the ablator is influenced strong by the expansion wave from the corner.

The shapes of ablator surface at 100 sec are shown in **Figure 7**. The black line is the initial line, the green line and the red line show heavy ablator and light-weight ablator respectively.

The zooming up around stagnation region is imposed. The quantity of the recession at stagnation point is large for high heat flux. However, extreme change of shape does not occur and it keeps similarity shape mainly. The recessions of surface at stagnation point at 100 sec of heavy ablator and light-weight ablator are 1.4 mm and 6.6 mm respectively. In addition, the surface temperature at 100 sec is between 1300 K and 1450 K. The surface recession does not occur dramatically in such temperature region. On the other hand, there is little the surface recession at side wall. The temperature of side wall is not more than about 1000 K. The total surface recession at stagnation point has predicted with 2 mm from 1 mm by Suzuki *et al*. [26]. The present study has good agreement to heavy ablator. **Figure 8** shows the history of convective heat flux at stagnation point. The Solid lines show conventional heat flux and the black line shows non ablated wall condition for compared. The radiative equilibrium condition is used at the surface condition. Green and red lines are heavy and light-weight respectively. The dashed line shows net heat flux with surface chemical reaction that includes surface reactions (nitrization, oxization, sublimation). The effusion of pyrolysis gas works effectively to decrease heat flux. The dashed lines are higher than 30% of solid lines. We can see that the surface reaction play important role. In comparison with

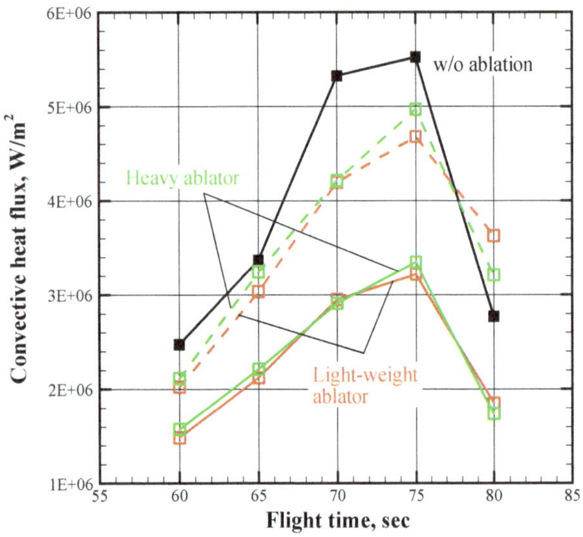

Figure 8. Time profiles of convective heat flux at stagnation point.

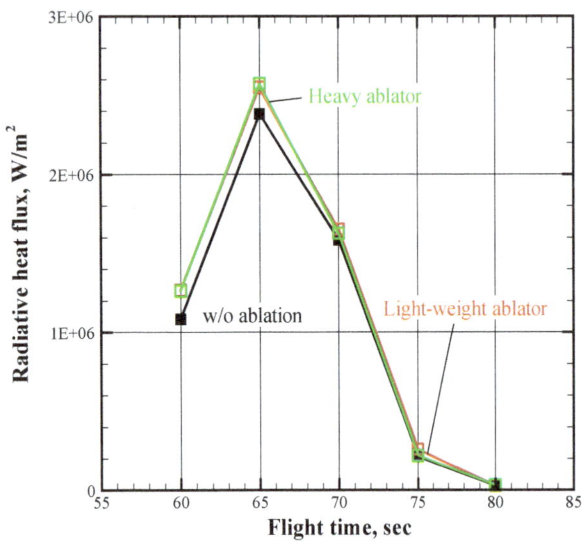

Figure 9. Time profiles of radiative heat flux at stagnation point.

it, the difference between heavy ablator and light-weight ablator is not so much. On the other hand, radiative heat flux (**Figure 9**) does not decrease using by ablating material. It rather increase slightly cause of radiation from ablating species. It can find a very slight difference between heavy ablator and light-weight ablator in radiative heat flux.

Temperature distribution along center axis at each time is shown in **Figure 10**. The peak temperature is 3100 K for heavy ablator and 3200 K for light-weight ablator at 70 sec. The rear wall temperature does not rise until 80 sec. Though there is no heating, the rear wall temperature is increasing at 100 sec. On other hand, the rear wall temperature of heavy ablator keeps initial temperature. The light-weight ablator does not have enough thickness for this condition. However, it suggests that it can prevent heat by slightly increasing the thickness of the light-weight ablator. Present study calculated until 100 sec, because the heat flux became quite small estimated in ref. [26].

4. Concluding Remarks

A trajectory-based flow field simulation around Hayabusa capsule loaded light-weight ablator coupled with thermal response was carried out as a test case of near future mission. To validate this calculation, the conventional heavy ablator condition was carried out, too. The light-weight ablator in this calculation had been made and tested in our laboratory. In addition, the radiative heat flux had computed using 3-band model extended to nonequilibrium flow.

To calculate flow field around capsule, present simulation adopted four-temperature model and it was clear that the flow field had strong thermal nonequilibrium behind shock wave and thermo-chemical equilibrium near the

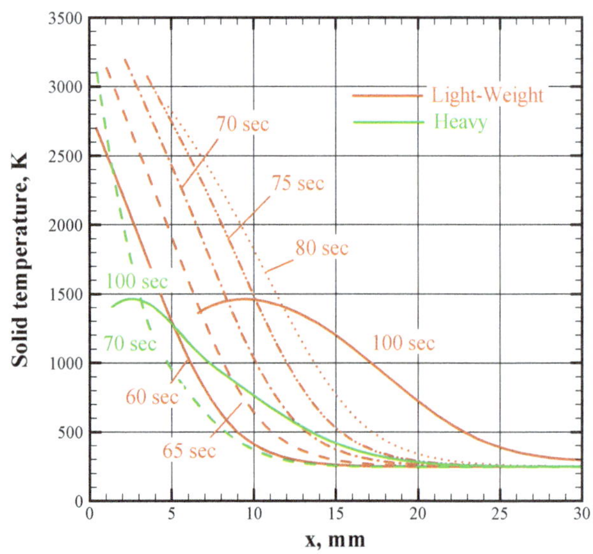

Figure 10. Temperature profiles along center axis.

wall surface. The precursor phenomenon could see using four-temperature model.

In thermal response for light-weight ablator, the state of the pyrolysis gas and its movement play important roles to the flow inside of ablator. The pyrolysis gas in heavy ablator flows towards a near boundary surface but the pyrolysis gas in light-weight ablator flows none one-dimensionally. The pyrolysis gas flow transports energies to the direction of rear wall and it makes the apparent heat conduction higher.

The difference of stagnation heat flux between heavy ablator and light-weight ablator is small. Especially radiated heat fluxes of these are almost the same. On the other hand, the stagnation heat flux taking account of the surface reaction becomes 30% larger than convectional heat flux without surface reaction.

The surface recession of light-weight ablator is few times larger than heavy ablator in addition; the heat transfer is slightly higher than heavy ablator. Therefore, this study shows light-weight ablator in this paper cannot use same configuration of heavy ablator and same thickness but it suggests that it can prevent heat by slightly increasing the thickness of the lightweight ablator.

5. Acknowledgements

The computation was mainly carried using the computer facilities at the Research Institute for Information Technology, Kyushu University. This work was partially supported by Grant-in-Aid for Scientific Research, No. 23460954, sponsored by Ministry of Education, Culture, Sports, Science and Technology, Japan.

REFERENCES

[1] T. Yamada, N. Ishii, Y. Inatani and M. Honda, "Thermal Protection System of the Reentry Capsule with Superorbital Velocity," Institute of Space and Astronautical Science Report Source Publishing No. 17, 2003, pp. 245-261.

[2] M. A. Covington, J. M. Heinemann, H. E. Goldstein, Y.-K. Chen, I. Terrazas-Salinas, J. A. Balboni, J. Olejniczak and E. R. Martinez, "Performance of a Low Density Ablative Heatshield Material," *Journal of Spacecraft and Rockets*, Vol. 45, No. 4, 2008, pp. 854-864.

[3] T. Yamada, Y. Ishida, T. Suzuki, K. Takasaki, K. Fujita, T. Ogasawara and T. Abe, "Development of High and Low Density Ablators for Dash-II and Future Reentry Missions," *Proceedings of the 27th International Symposium on Space Technology and Science*, Tsukuba, 5-12 July 2009, p. e-03.

[4] Y. K. Chen and F. S. Milos, "Ablation and Thermal Response Program for Spacecraft Heatshield Analysis," *Journal of Spacecraft and Rockets*, Vol. 36, No. 3, 1999, pp. 475-483.

[5] Y. K. Chen and F. S. Milos, "Two-Dimensional Implicit Thermal Response and Ablation Program for Charring Materials," *Journal of Spacecraft and Rockets*, Vol. 38, No. 4, 2001, pp. 473-481.

[6] H. Ahn, C. Park and K. Sawada, "Response of Heatshield Material at Stagnation Point of Pioneer-Venus Probes," *Journal of Thermophysics and Heat Transfer*, Vol. 16, No. 3, 2002, pp. 432-439.

[7] T. Suzuki, K. Sawada, T. Yamada and Y. Inatani, "Thermal Response of Ablative Test Piece in Arc-Heated Wind Tunnel," AIAA Paper No. 2004-0341, 2004.

[8] T. Suzuki, T. Sakai and T. Yamada, "Calculation of Thermal Response of Ablator under Arcjet Flow Condition," *Journal of Thermophysics and Heat Transfer*, Vol. 21, No. 2, 2007, pp. 257-266.

[9] S. Nozawa, T. Kanazaka, H. Kihara and K. Abe, "Experimental and Numerical Studies of Spallation Particles Ejected from A Light-Weight Ablator," *Proceedings of 61st International Astronautical Congress*, Prague, 27 September-1 October 2010, IAC-10-C2.7.7.

[10] Y. Takahashi, H. Kihara and K. Abe, "Effect of Radiative Heat Transfer in Arc-Heated Nonequilibrium Flow Simulation," *Journal of Physics D: Applied Physics*, Vol. 43, 2010, Article ID: 185201.

[11] N. Hirata, S. Nozawa, Y. Takahashi, H. Kihara and K. Abe, "Numerical Study of Pyrolysis Gas Flow and Heat Transfer inside an Ablator," *Computational Thermal Sciences*, Vol. 4, No. 3, 2012, pp. 225-242.

[12] C. Park, R. L. Jaffe and H. Partridge, "Chemical-Kinetic Parameters of Hyperbolic Earth Entry," *Journal of Thermophysics and Heat Transfer*, Vol. 15, No. 1, 2001, pp. 76-90.

[13] C. Park, "Rotational Relaxation of N_2 behind a Strong Shock Wave," *Journal of Thermophysics and Heat Transfer*, Vol. 18, No. 4, 2004, pp. 527-533.

[14] C. Park, "Problems of Rate Chemistry in the Flight Regimes of Aeroassisted Orbital Transfer Vehicles," AIAA Paper 84-1730, 1985.

[15] J. P. Appleton and K. N. C. Bray, "The Conservation Equations for a Nonequilibrium Plasma," *Journal of Fluid Mechanics*, Vol. 20, No. 4, 1964, pp. 659-672.

[16] P. A. Gnoffo, R. N. Gupta and J. L. Shinn, "Conservation Equations and Physical Models for Hypersonic Air Flows in Thermal and Chemical Nonequilibrium," National Aeronautics and Space Administration, Washington DC, 1989.

[17] S. S. Lazdinis and S. L. Petrie, "Free Electron and Vibrational Temperature Nonequilibrium in High Temperature Nitrogen," *Physics of Fluids*, Vol. 17, No. 8, 1974, pp. 1539-1546.

[18] J.-H. Lee, "Electron-Impact Vibrational Relaxation in High-Temperature Nitrogen," *Journal of Thermophysics and Heat Transfer*, Vol. 7, No. 3, 1993, pp. 399-405.

[19] T. Sakai, "Computational Simulation of High-Enthalpy Arc Heater Flows," *Journal of Thermophysics and Heat Transfer*, Vol. 21, No. 1, 2007, pp. 77-85.

[20] M. Yamada, Y. Matsuda, H. Hirata, Y. Takahashi, H.

Kihara and K. Abe, "Numerical Simulation of Flow Field and Heat Transfer around HAYABUSA Reentry Capsule," *Proceedings of the 28th International Symposium on Space Technology and Science*, Okinawa, 5-12 June 2011, p. e-38.

[21] Y. K. Chen and F. S. Milos, "Loosely Coupled Simulation for Two-Dimensional Ablation and Shape Change," *Journal of Spacecraft and Rockets*, Vol. 47, No. 5, 2010, pp. 775-785.

[22] T. Kanzaka, S. Nozawa, Y. Takahashi, H. Kihara and K. Abe, "Numerical Investigation of Thermal Response of Ablator Exposed to Thermochemical Nonequilibrium Flow," *Proceediongs of the 6th Asian-Pacific Conference on Aerospace Technology and Science*, Huangshan, 15-19 November 2009, B2-2.

[23] S. Gordon and B. J. McBridge, "Computer Program for Calculation of Complex Chemical Equilibrium Computation and Application," NASA Reference Publication 1311, 1994.

[24] R. L. Potts, "Application of Integral Methods to Ablation Charring Erosion: A Review," *Journal of Spacecraft and Rockets*, Vol. 32, No. 2, 1995, pp. 200-209.

[25] M. W. Chase Jr., C. A. Davies Jr., J. R. Downey, D. J. Frurip, R. A. McDonald and A. N. Syverud, "JANAF Thermochemical Tables," *Journal of Physical and Chemical Reference Data*, Vol. 14, 1985, pp. 535.

[26] T. Suzuki, K. Fujita, T. Yamada, Y. Inatani and N. Ishii, "Post-Flight TPS Analysis fo Hayabusa Reentry Capsul," AIAA Paper 2011-3759, 2011.

22

MHD Effect on an Oscillatory Mixed Convective Flow Past an Infinite Vertical Porous Plate with Radiation

Nazibuddin Ahmed, Sujan Sinha
Department of Mathematics, Gauhati University, Guwahati, India

ABSTRACT

A parametric study to investigate the effects of magnetic field and radiation on a mixed convective flow past an infinite vertical porous plate with constant suction velocity when the free stream varies periodically with time about a steady mean is presented. A uniform magnetic field is applied normal to the plate. The Rosseland approximation is used to describe the radiative heat flux in the energy equation and the resultant set of non-dimensional equations are solved analytically by adopting regular perturbation technique. The profiles of velocity, temperature, skin-friction and nusselt number are demonstrated graphically for various values of the parameters involved in the problem and the results are physically interpreted.

Keywords: Mixed Convection; Porous Medium; Radiation

1. Introduction

The problems of interaction of free and forced convection with thermal radiation of viscous incompressible MHD unsteady flow past infinite vertical plates with mass transfer are being studied nowadays due to many applications of such problems in Astrophysics, Geophysical and different Engineering fields. The heating of rooms and buildings by use of radiators is an example of heat transfer by free convection. The study of MHD is quite important in the field of aeronautics, especially in missile aerodynamics, since the temperature that occurs in such flight speeds are sufficient to dissociate or even ionize the air appreciably. For example, when a high speed missile re-enters the earth's atmosphere, a very large amount of heat is generated due to the friction of air molecules and this viscous heating may some times be so considerable as to ionize the air near the stagnation point. This ionized gas or plasma interacts with the magnetic field and alter the heat transfer and friction characteristics. In addition to this, some fluids can also emit and absorb thermal radiation. In such a situation, the study of the effect of the magnetic field on the flow characteristics becomes more interesting as the fluid is not only electrical conductor but also capable of emitting and absorbing thermal radiation. This type of investigation is carried out because of its importance in space and temperature related problems.

MHD is the science of motion of electrically conducting fluid in presence of magnetic field. There are numerous examples of application of MHD principles, including MHD generators, MHD pumps and MHD flow meters etc. The dynamo and motor is a classical example of MHD principle. MHD principles also find its application in medicine and biology. The present form of MHD is due to the pioneer contribution of several authors like Alfven [1], Cowling [2], Crammer and Pai [3], Shercliff [4] and Ferraro and Pulmption [5]. Model studies on MHD free and forced convection with heat and mass transfer problems have been carried out by many of the authors due to their application in many branches of science and technology. Some of them are Ahmed [6], Elbashbeshy [7] and Singh and Singh [8]. Gregantopoulos *et al.* [9] studied two-dimensional unsteady free convection and mass transfer flow of an incompressible viscous dissipative and electrically conducting fluid past an infinite vertical porous plate.

Radiation is a process of heat transfer through electromagnetic waves. Radiative convective flows are encountered in countless industrial and environment process. For e.g. heating and cooling chambers, fossil fuel combustion energy processes, evaporation from large open water reservoirs, astrophysical flows etc. Radiative heat and mass transfer play an important role in manufacturing industries for the design of reliable equiponent. If the temperature of the surrounding field is rather high, radiation effect plays an important role in space related technology. The effect of radiation on various convective flows under different conditions has been studied by many researchers including Hussain and Takhar [10], Ahmed and Sarmah [11], Rajesh and Varma [12] and Ahmed *et al.* [13]. As the present authors are aware till now no attempt has been made to study the magnetic field effect on an oscillatory mixed convective flow past an infinite vertical porous plate with radiation following the technique adopted by Mansour [14]. Such an attempt has been made in the present paper.

The main objective of the present investigation is to study the MHD effects on the unsteady mixed convective mass transfer flow past an infinite vertical porous plate with constant suction, where the plate temperature oscillates with the same frequency as that of variable suction velocity. It is assumed that the plate is embedded in a uniform porous medium and moves with a constant velocity in the direction of the flow in presence of a transverse magnetic field. This work is an extension to the work done by Mansour [14] to consider the effect of MHD.

2. Mathematical Analysis

We now consider an MHD conducting flow of a viscous incompressible electrically conducting fluid past an infinite vertical porous plate with constant suction under the influence of a uniform transverse magnetic field. Our investigation is restricted to the following assumptions:

- The polarization effects are assumed to be negligible and hence the electric field is also negligible.
- The suction velocity is constant.
- The variations of all fluid properties other than the variations of density except in so far as they give rise to a body force are ignored completely.
- All the physical variables are functions of y' and t' only as the plate are infinite.
- It is assumed that the variation of expansion co-efficient is negligibly small and the pressure and influence of the pressure on the density are negligible.
- The free stream oscillates about a steady mean.

We introduce a co-ordinate system (x', y', z') with *X*-axis vertically upwards along the plate, *Y*-axis perpendicular to the plate and directed into the fluid region and *Z*-axis along the width of the plate as shown in **Figure 1**. Let the components of velocity along with *X* and *Y* axes should be u' and v'. Let these velocity components are chosen in the upward direction along the plate and normal to the plate respectively.

Under these assumptions, the equations that describe the physical situation are given by

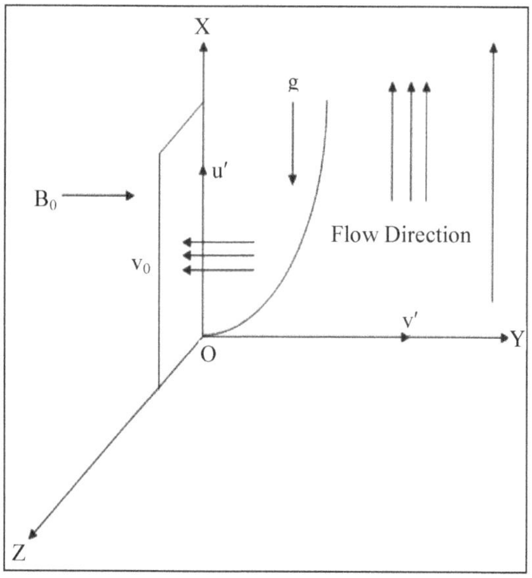

Figure 1. Physical model of the problem.

$$\frac{\partial v'}{\partial y'} = 0 \tag{1}$$

$$\frac{\partial u'}{\partial t'} + v'\frac{\partial u'}{\partial y'} = -\frac{1}{\rho'}\frac{\partial p'}{\partial x'} + \upsilon\frac{\partial^2 u'}{\partial y'^2} + g\beta\left(T' - T'_\infty\right) - \upsilon\frac{u'}{K'} - \frac{\sigma B_0^2 u'}{\rho'} \tag{2}$$

$$\frac{\partial v'}{\partial t'} = -\frac{1}{\rho}\frac{\partial p'}{\partial y'} \tag{3}$$

$$\frac{\partial T'}{\partial t'} + v'\frac{\partial T'}{\partial y'} = \alpha\left(\frac{\partial^2 T'}{\partial y'^2} - \frac{1}{K_T}\frac{\partial q_r}{\partial y'}\right) \tag{4}$$

All the physical quantities involved in the above equations are defined in the nomenclature.

The boundary conditions are:

$$\begin{aligned} &u' = 0, T' = T'_w \text{ at } y' = 0 \\ &u' \to U'(t) = U_0\left(1 + \varepsilon e^{i\omega' t'}\right), T' \to T'_\infty \text{ as } y' \to \infty \end{aligned} \tag{5}$$

As there is a constant suction velocity v_0 at the plate, Equation (1) yields

$$v' = -v_0 \tag{6}$$

where v_0 is a scale of suction velocity which is non-zero positive constant. The negative sign indicates that the suction is towards plate.

Outside the boundary layer, Equation (2) gives

$$-\frac{1}{\rho'}\frac{\partial p'}{\partial x'} = \frac{\mathrm{d}U'}{\mathrm{d}t'} + \frac{\upsilon}{K'}U' + \frac{\sigma B_0^2 U'}{\rho'} \tag{7}$$

In order to write the governing equations and the boundary conditions in dimensional form, the following non-dimensional quantities are introduced:

$$u=\frac{u'}{U_0},y=\frac{y'v_0}{\upsilon},t=\frac{t'v_0^2}{4\upsilon},Gr=\frac{\upsilon g\beta\left(T_w'-T_\infty'\right)}{U_0v_0^2},$$
$$\theta=\frac{T'-T_\infty'}{T_w'-T_\infty'},\Pr=\frac{\upsilon}{\alpha},M=\frac{\sigma B_0^2\upsilon}{\rho v_0^2},$$
$$\omega=\frac{4\upsilon\omega'}{v_0^2},U=\frac{U'}{U_0},C_T=\frac{T_w'}{T_w'-T_\infty'},\tag{8}$$
$$R=\frac{4\sigma'\left(T_w'-T_\infty'\right)^3}{kK_T},K=\frac{K'v_0^2}{\upsilon^2}$$

In view of the Equations (6)-(8), the Equations (2)-(4) reduce to the following dimensional form:

$$\frac{1}{4}\frac{\partial u}{\partial t}-\frac{\partial u}{\partial y}=\frac{1}{4}\frac{\mathrm{d}U}{\mathrm{d}t}+\frac{\partial^2 u}{\partial y^2}+Gr\theta+N\left(U-u\right)\tag{9}$$

$$\frac{1}{4}\frac{\partial\theta}{\partial t}-\frac{\partial\theta}{\partial y}=\left[1+\frac{4}{3}R\left(\theta+C_T\right)^3\right]\frac{\partial^2\theta}{\partial y^2}+4R\left(\theta+C_T\right)^2\left(\frac{\partial\theta}{\partial y}\right)^2\tag{10}$$

The corresponding boundary conditions are:

$$\begin{aligned}&u=0,\theta=1\ \text{ at }\ y=0\\&u=U\left(t\right),\theta=0\ \text{ as }\ y\to\infty\end{aligned}\tag{11}$$

3. Method of Solution

Equations (9) and (10) are coupled non-linear partial differential and these cannot be solved in closed-form. However, these equations can be reduced to a set of ordinary differential equations, which can be solved analytically. This can be done by representing the velocity, temperature and concentration of the fluid in the neighbourhood of the plate as

$$\begin{aligned}&u=u_0\left(y\right)+\varepsilon e^{i\omega t}u_1\left(y\right)+o\left(\varepsilon^2\right)+\cdots\\&\theta=\theta_0\left(y\right)+\varepsilon e^{i\omega t}\theta_1\left(y\right)+o\left(\varepsilon^2\right)+\cdots\\&U=1+\varepsilon e^{i\omega t}\end{aligned}\tag{12}$$

Substituting (12) in Equations (9) and (10), equating the harmonic terms and neglecting the higher order terms of $0\left(\varepsilon^2\right)$, we obtain

$$u_0''+u_0'-Nu_0=-N-Gr\theta_0\tag{13}$$

$$u_1''+u_1'-\left(N+\frac{i\omega}{4}\right)u_1=-N-Gr\theta_1-\frac{i\omega}{4}\tag{14}$$

$$\left[1+\frac{4}{3}R\left(\theta_0+C_T\right)^3\right]\theta_0''+4R\left(\theta_0+C_T\right)^2\theta_0'+\Pr\theta_0=0\tag{15}$$

$$\left[1+\frac{4}{3}R\left(\theta_0+C_T\right)^3\right]\theta_1''+8R\left(\theta_0+C_T\right)^2\left(\frac{\partial\theta_0}{\partial y}\right)\left(\frac{\partial\theta_1}{\partial y}\right)+8R\left(\theta_0+C_T\right)^2\left(\frac{\partial\theta_0}{\partial y}\right)^2\theta_1+4R\left(\theta_0+C_T\right)^2\theta_0''\theta_1+\Pr\theta_1'-\frac{i\omega}{4}\Pr\theta_1=0\tag{16}$$

where primes denotes ordinary differentiation with respect to y.

The corresponding boundary conditions can be written as

$$\begin{aligned}&u_0=0,u_1=0,\theta_0=1,\theta_1=0\ \text{ at }\ y=0\\&u_0\to 1,u_1\to 1,\theta_0\to 0,\theta_1\to 0\ \text{ as }\ y\to\infty\end{aligned}\tag{17}$$

If we assume the radiation parameter R to be small, we expand the velocity and temperature as

$$\begin{aligned}&u_0\left(y\right)=u_{01}\left(y\right)+Ru_{02}\left(y\right),u_1\left(y\right)=u_{11}\left(y\right)+Ru_{12}\left(y\right)\\&\theta_0\left(y\right)=\theta_{01}\left(y\right)+R\theta_{02}\left(y\right),\theta_1\left(y\right)=\theta_{11}\left(y\right)+R\theta_{12}\left(y\right)\end{aligned}\tag{18}$$

Substituting (18) in Equations (13)-(16), we obtain the following set of equations for the mean flow:

$$u_{01}''+u_{01}'-Nu_{01}=-N-Gr\theta_{01}\tag{19}$$

$$u_{02}''+u_{02}'-Nu_{02}=-N-Gr\theta_{02}\tag{20}$$

$$u_{11}''+u_{11}'-\left(\frac{i\omega}{4}+N\right)u_{11}=-Gr\theta_{11}-\frac{i\omega}{4}-N\tag{21}$$

$$u_{12}''+u_{12}'-\left(\frac{i\omega}{4}+N\right)u_{12}=-Gr\theta_{12}\tag{22}$$

$$\theta_{01}''+\Pr\theta_{01}'=0\tag{23}$$

$$\theta_{02}''+\frac{4}{3}\left(\theta_{01}+C_T\right)^3\theta_{01}''+4\left(\theta_{01}+C_T\right)^2\theta_{01}'^2+\Pr\theta_{02}'=0\tag{24}$$

$$\theta_{11}''+\Pr\theta_{11}'-\frac{i\omega}{4}\Pr\theta_{11}=0\tag{25}$$

$$\theta_{12}''+\Pr\theta_{12}'-\frac{i\omega}{4}\Pr\theta_{12}=-\frac{4}{3}\left(\theta_{01}+C_T\right)^3\theta_{11}''-8\left(\theta_{01}+C_T\right)^2\theta_{01}'\theta_{11}'-8\left(\theta_{01}+C_T\right)^2\theta_{01}'^2-4\left(\theta_{01}+C_T\right)^2\theta_{01}''\theta_{11}\tag{26}$$

with the boundary conditions

$$\begin{aligned}&u_{01}=0,u_{02}=0,u_{11}=0,u_{12}=0,\theta_{01}=1,\\&\theta_{02}=0,\theta_{11}=0,\theta_{12}=0\ \text{ at }\ y=0\\&u_{01}\to 1,u_{02}\to 0,u_{11}\to 1,u_{12}\to 0,\\&\theta_{01}\to 0,\theta_{02}\to 0,\theta_{11}\to 0,\theta_{12}\to 0\ \text{ as }\ y\to\infty\end{aligned}\tag{27}$$

The solutions of zeroth and first order of velocity and temperture are:

$$u_0(y)=1+A_6 e^{-A_3 y}-A_5 e^{-\Pr y}$$
$$+R\left(A_{15}e^{-A_3 y}+A_{14}e^{-\Pr y}+A_9 e^{-4\Pr y}\right.$$
$$\left.+A_{10}e^{-3\Pr y}+A_{11}e^{-2\Pr y}-A_{12}ye^{-\Pr y}\right)$$
$$u_1(y)=1-e^{-m_2 y}$$
$$+R\left(A'_{11}e^{-m_2 y}+A'_8 e^{-m_1 y}+A'_9 e^{-3\Pr y}+A'_{10}e^{-2\Pr y}\right)$$
$$\theta_0(y)=e^{-\Pr y}$$
$$+R\left(A_2 e^{-\Pr y}-\frac{4}{9}e^{-4\Pr y}-2C_T e^{-3\Pr y}\right.$$
$$\left.-4C_T^2 e^{-2\Pr y}+\frac{4}{3}\Pr yC_T^3 e^{-\Pr y}\right)$$
$$\theta_1(y)=R\left(A'_5 e^{-m_1 y}+A'_3 e^{-3\Pr y}+A'_4 e^{-2\Pr y}\right)$$

The solutions for velocity and temperature distributions are:

$$u(y,t)=1+A_6 e^{-A_3 y}-A_5 e^{-\Pr y}+R\left(A_{15}e^{-A_3 y}+A_{14}e^{-\Pr y}\right.$$
$$\left.+A_9 e^{-4\Pr y}+A_{10}e^{-3\Pr y}+A_{11}e^{-2\Pr y}-A_{12}ye^{-\Pr y}\right)$$
$$+\varepsilon e^{i\omega t}\left[1-e^{-m_2 y}\right.$$
$$\left.+R\left\{A'_{11}e^{-m_2 y}+A'_8 e^{-m_1 y}+A'_9 e^{-3\Pr y}+A'_{10}e^{-2\Pr y}\right\}\right]$$
$$\theta(y,t)=e^{-\Pr y}+R\left(A_2 e^{-\Pr y}-\frac{4}{9}e^{-4\Pr y}-2C_T e^{-3\Pr y}\right.$$
$$\left.-4C_T^2 e^{-2\Pr y}+\frac{4}{3}\Pr yC_T^3 e^{-\Pr y}\right)$$
$$+\varepsilon e^{i\omega t}R\left(A'_5 e^{-m_1 y}+A'_3 e^{-3\Pr y}+A'_4 e^{-2\Pr y}\right)$$

Skin-friction:

The non-dimensional form of skin-friction at the plate are given by:

$$Cf=\tau=\left[\frac{\partial u}{\partial y}\right]_{y=0}$$
$$=-A_3A_6+\Pr A_5+R$$
$$\left[-A_3A_{15}+\Pr A_{14}-4\Pr A_9-3\Pr A_{10}-2\Pr A_{11}-A_{12}\right]$$
$$+\varepsilon e^{i\omega t}\left[m_2+R\left\{-m_2A'_{11}-m_1A'_8-3\Pr A'_9-2\Pr A'_{10}\right\}\right]$$

Nusselt number:

The non dimensional form of the rate of heat transfer in terms of Nusselt number at the plate are given by:

$$Nu=-\left[\frac{\partial \theta}{\partial y}\right]_{y=0}$$
$$=\Pr-R\left[-A_2\Pr+\frac{16}{9}\Pr+6\Pr C_T+8C_T^2\Pr+\frac{4}{3}\Pr C_T^3\right]$$
$$-\varepsilon e^{i\omega t}R\left(-m_1A'_5-3\Pr A'_3-2\Pr A'_4\right)$$

4. Results and Discussion

In order to get the physical insight in to the problem, we have carried out numerical calculations for non-dimensional velocity field, temperature field, nusselt number and skin-friction have been carried out by assigning specific values to the different parameters involved in the problem, viz, Hartmann number *M*, Grashof number *Gr* , Prandtl number Pr, Porosity parameter *K*, Radiation conduction parameter R and Temperature difference parameter C_T. The effects of these values are demonstrated through different graphs and the results are interpreated physically.

The **Figures 2-4** depict the variation of velocity field u against *y* under the effect of Hartmann number *M*, Porosity parameter *K* and radiation conduction parameter *R*. **Figure 2** illustrates that the fluid velocity decreases with the increase in magnetic intensity. In other words the imposition of the transverse magnetic field tends to retard the fluid flow. This phenomenon has an excellent agreement with the physical fact that the Lorentz force generated in the present flow model due to interaction of the transverse magnetic field and the fluid velocity acts as a resistive force to the fluid flow which serves to decelerate the flow. **Figures 3** and **4** demonstrate the effect of Porosity number *K* and radiation conduction parameter *R* on velocity field. It is simulated from the figures that an increase in Porosity parameter *K* and radiation conduction parameter *R* leads to a decrease in fluid velocity. *i.e.* the fluid motion is decelerated for increasing the permeability of the medium and for low thermal conductivity.

The **Figures 5-7** exhibit the variation of temperature field θ against *y* under the influence of temperature difference parameter C_T, radiation-conduction parameter *R* and Prandtl number Pr. **Figures 5** and **6** present the variation of Temperature difference parameter C_T and Radiation conduction parameter *R* on temperature field. It is

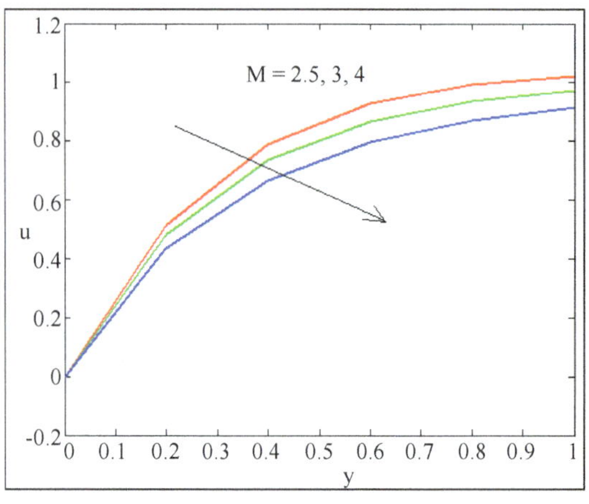

Figure 2. Velocity profile versus y for *Gr* = 5, Pr = 0.71, *R* = 0.1, *Ct* = 0.01, *K* = 1, ε = 0.01, ωt = π/2.

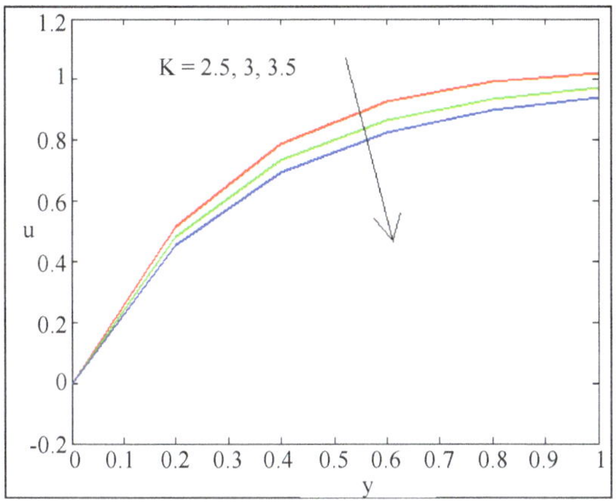

Figure 3. Velocity profile versus *y* for *Gr* = 5, Pr = 0.71, *R* = 0.1, *Ct* = 0.01, *M* = 1, *ε* = 0.01, *ωt* = *π*/2.

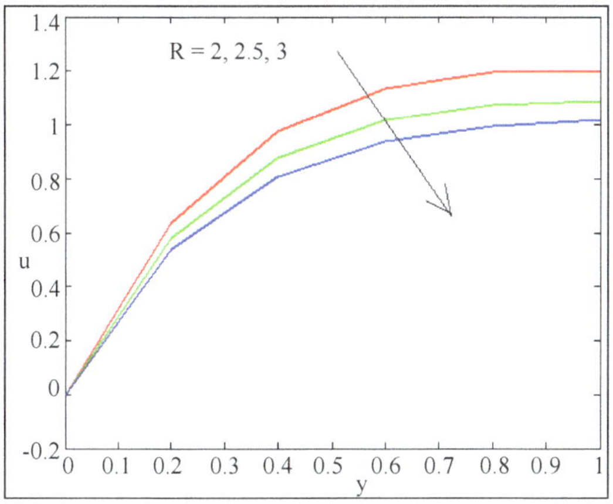

Figure 4. Velocity profile versus y for *Gr* = 5, Pr = 0.71, *K* = 1, *Ct* = 0.01, *M* = 1, *ε* = 0.01, *ωt* = *π*/2.

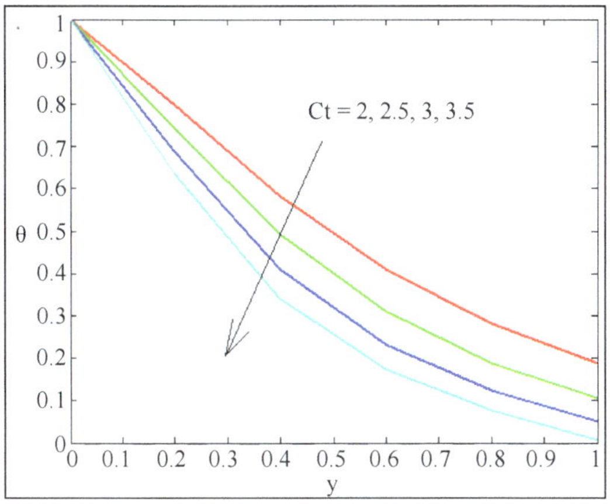

Figure 5. Temperature profile versus *y* for *R* = 0.1, Pr = 0.71, *ε* = 0.01, *ωt* = *π*/2.

observed from both the figures that there is a steady drop in temperature due to small temperature difference and low thermal conductivity. **Figure 7** indicates that the fluid temperature decreases with the increase in Prandtl number. *i.e.* thermal diffusivity reduces the fluid temperature.

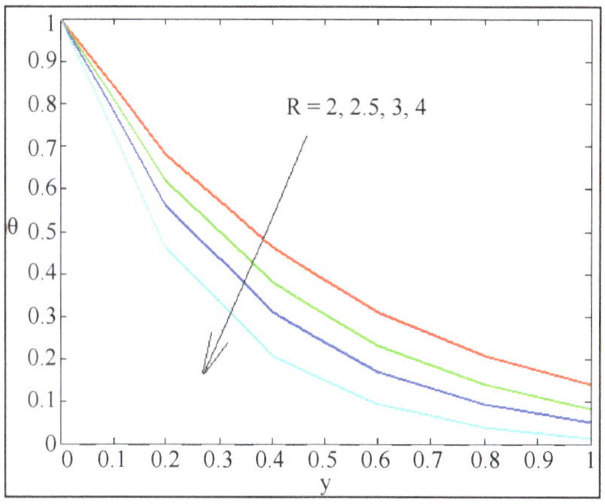

Figure 6. Temperature profile versus *y* for *Ct* = 0.01, Pr = 0.71, *ε* = 0.01, *ωt* = *π*/2.

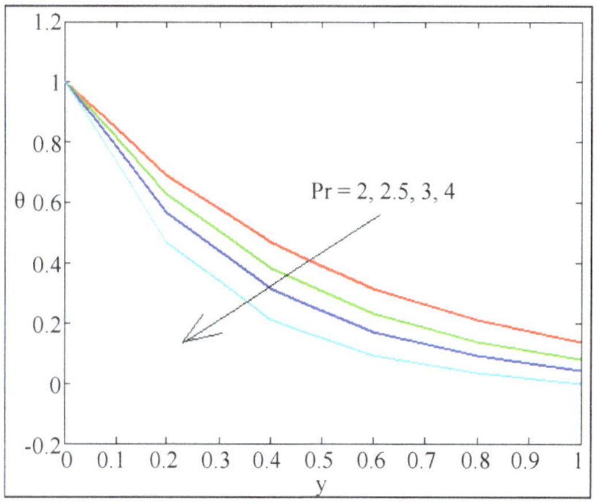

Figure 7. Temperature profile versus *y* for *R* = 0.1, *Ct* = 0.01, *ε* = 0.01, *ωt* = *π*/2.

This result is in a good agreement to the fact that the thickness of the thermal boundary layer may be prevented with the increasing Prandtl number.

The effects of Prandtl number Pr, radiation-conduction parameter *R* and temperature difference parameter C_T on the co-efficient of rate of heat transfer in terms of Nusselt number have been displayed in **Figures 8-10**. It is found from the **Figures 8** and **9** that the magnitude of the rate of heat transfer is considerably increased with the increase in Prandtl number Pr and Radiation-conduction parameter *R*. *i.e.* the energy flux is rised due to increasing the permeability of the medium as well as low thermal conductivity. **Figure 10** predicts that the energy flux from the plate to the fluid in terms of Nusselt number get increased with the increase in Temperature difference parameter C_T.

The **Figures 11-14** correspond to the co-efficient of skin-friction τ against radiation conduction parameter *R* under the influence of magnetic parameter *M*, Porosity parameter *K*, temperature difference parameter C_T and Grashof number *Gr*. In the **Figures 11, 12** and **14** it is

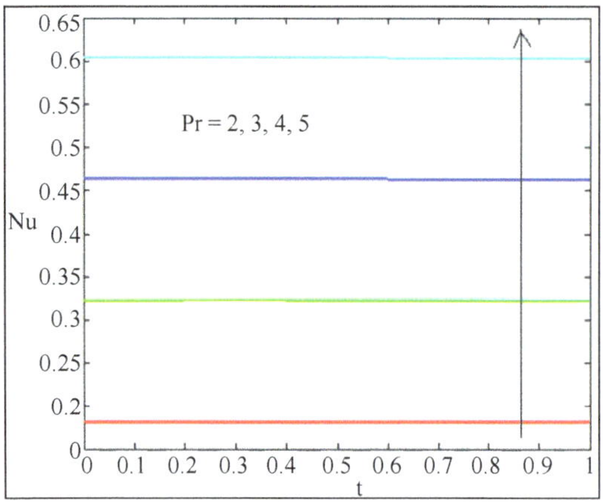

Figure 8. Nusslet number versus *t* for *R* = 0.1, *Ct* = 0.01, *ε* = 0.01, *ω* = 1.

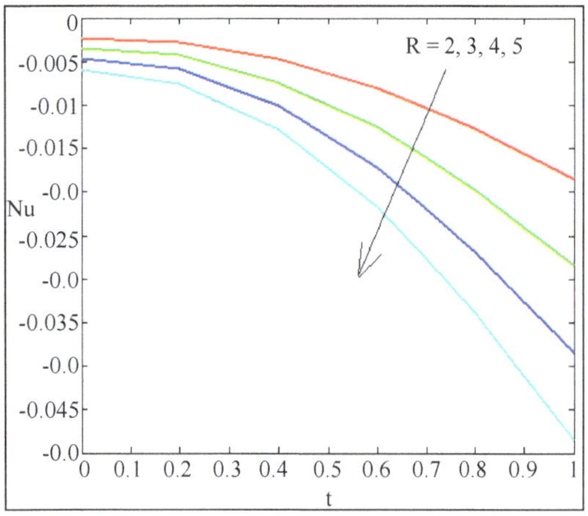

Figure 9. Nusslet number versus *t* for Pr = 0.71, *Ct* = 0.01, *ε* = 0.01, *ω* = 1.

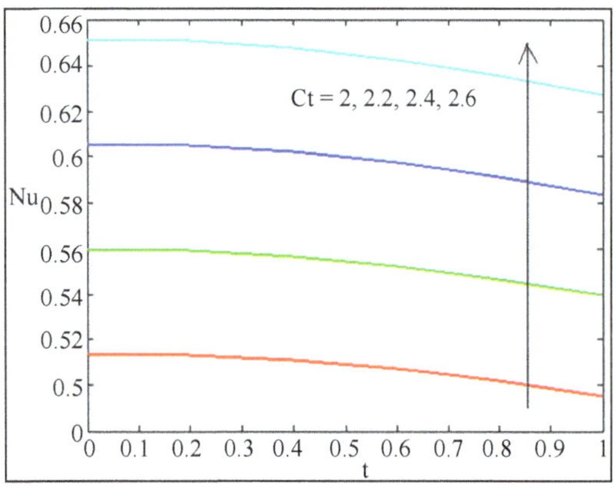

Figure 10. Nusslet number versus *t* for Pr = 0.71, *R* = 0.1, *ε* = 0.01, *ω* = 1.

observed that an increase in Magnetic parameter *M*, Porosity parameter *K* and Grashof number *Gr* tends to raise the viscous drag whereas viscous drag falls with the increase in Temperature difference parameter C_T in **Figure 13**.

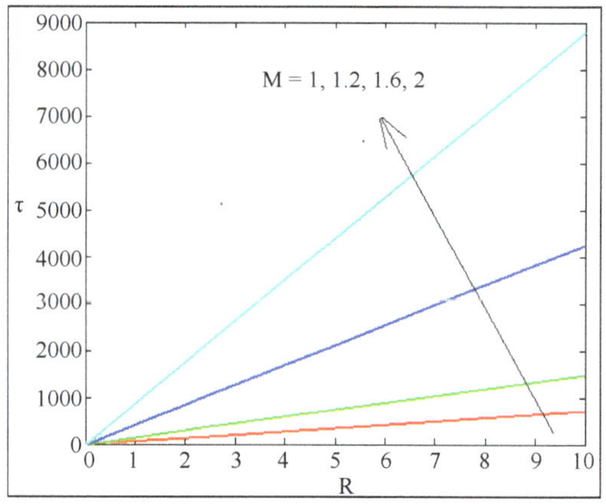

Figure 11. Co-efficient of skin friction versus *R* for *Ct* = 0.01, Pr = 0.71, *K* = 1, *Gr* = 5, *ε* = 0.01, *ω* = 1, *t* = 11/7.

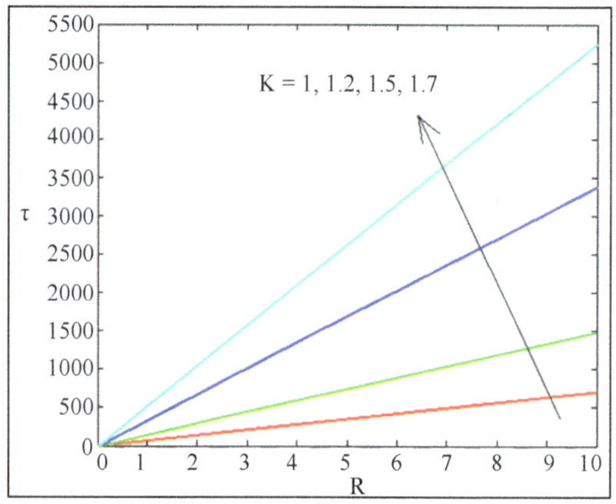

Figure 12. Co-efficient of skin friction versus *R* for *Ct* = 0.01, Pr = 0.71, *M* = 1, *Gr* = 5, *ε* = 0.01, *ω* = 1, *t* = 11/7.

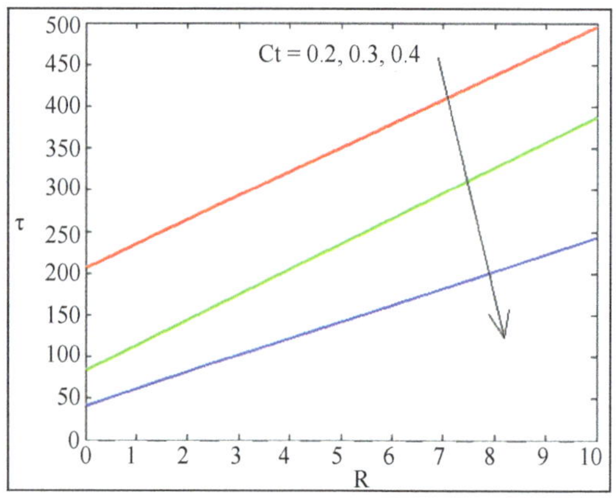

Figure 13. Co-efficient of skin friction versus *R* for *K* = 1, Pr = 0.71, *M* = 1, *Gr* = 5, *ε* = 0.01, *ω* = 1, *t* = 11/7.

4. Conclusions

Our investigation of the problem setup leads to the following conclusions:

- The fluid motion is decelerated under the action of tranverse magnetic field, Porosity parameter and radiation conduction parameter.

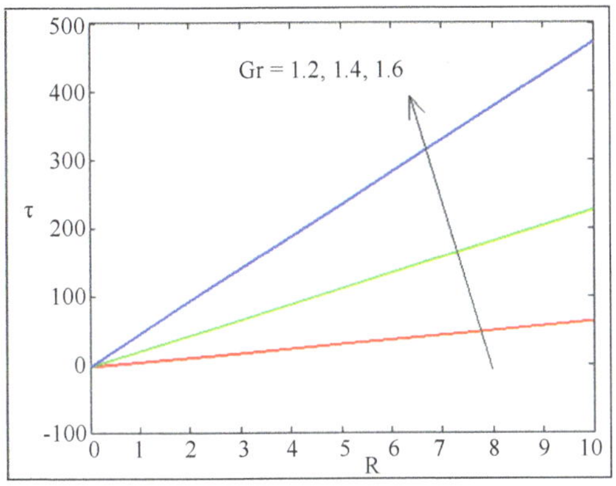

Figure 14. Co-efficient of skin friction versus R for $K = 0.01$, Pr = 0.71, $M = 1$, $Ct = 5$, $\varepsilon = 0.01$, $\omega = 1$, $t = 11/7$.

- Temperature falls due to small temperature difference and low thermal conductivity.
- The energy flux rises due to increasing the permeability of the medium as well as low thermal conductivity.
- Magnetic parameter raises the viscous drag on the plate but it falls with the increase in temperature difference parameter.

5. Acknowledgements

The authors are grateful to the CSIR-HRDG (India) for sponsoring this work through their Research-Grant-In-Aid. No. 25 (0209)/12/EMR-I.

REFERENCES

[1] H. Alfven, "Discovery of Alfven Waves," *Nature*, Vol. 150, No. 3805, 1942, pp. 405-406.

[2] T. G. Cowling, "Magneto Hydrodynamics," Wiley Inter Science, New York, 1957.

[3] K. P. Crammer and S. L. Pai, "Magneto-Fluid Dynamics for Engineers and Applied Physics," McGraw Hill book Co., New York, 1978.

[4] J. A. Shercliff, "A Text Book of Magneto-Hydrodynamics," Pergamon Press, London, 1965.

[5] V. C. A. Ferraro and C. Plumpton, "An Introduction to Magneto-Fluid Mechanics," Clarandon Press, Oxford, 1966.

[6] N. Ahmed, "MHD Free and Forced Convection with Mass Transfer from an Infinite Vertical Porous Plate," *Journal of Energy, Heat and Mass Transfer*, Vol. 32, No. 1, 2010, pp. 55-70.

[7] E. M. A. Elbashbeshy, "The Mixed Convection along a Vertical Plate Embedded in Non-Darcian Porous Medium with Suction and Injection," *Applied Mathematics and Computation*, Vol. 136, No. 1, 2003, pp. 139-149.

[8] N. P. Singh and A. K. Singh, "MHD Effects on Heat and Mass Transfer in a Flow of Viscous Fluid with Induced Magnetic Field," *Indian Journal of Pure and Applied Physics*, Vol. 38, No. 3, 2000, pp. 182-189.

[9] G. A. Gregantopoulos, J. Koullias, C. L. Goudas and C. Courogenis, "Free Convection and Mass Transfer Effects on the Hydromagnetic Oscillatory Flow Past an Infinite Vertical Porous Plate," *Astrophysics and Space Science*, Vol. 74, No. 2, 1981, pp. 357-389.

[10] M. A. Hussain and H. S. Takhar, "Radiation Effect on Mixed Convection along a Vertical Plate with Uniform Surface Temperature," *Heat Mass Transfer*, Vol. 31, No. 4, 1996, pp. 243-248.

[11] N. Ahmed and H. K. Sarmah, "The Radiation Effect on a Transient MHD Flow with Mass Transfer Past an Impulsively Fixed Infinite Vertical Plate," *International Journal of Applied Mathematics and Mechanics*, Vol. 5, No. 5, 2009, pp. 87-98.

[12] V. Rajesh and S. V. K. Varma, "Radiation Effects on MHD Flow through a Porous Medium with Variable Temperature or Variable Mass Diffusion," *International Journal of Applied Mathematics and Mechanics*, Vol. 6, No. 11, 2010, pp. 39-57.

[13] N. Ahmed, S. Sinha and S. Talukdar, "Dufour Effect on a Transient MHD Flow Past a Uniformly Moving Porous Plate with Heat Sink," *Advances and Applications in Fluid Mechanics*, Vol. 13, No. 1, 2013, pp. 1-24.

[14] M. A. Mansour, "Radiative and Free-Convection on the Oscillatory Flow Past a Vertical Plate," *Astrophysics and Space Science*, Vol. 166, No. 2, 1990, pp. 269-275.

Nomenclature

B_0 = Strength of the applied magnetic field; C_T = Temperature difference parameter; G_r = Thermal Grashof number; g = Acceleration due to gravit; K_T = Thermal conductivity; k = Mean absorption co-efficient; K = Porosity parameter; M = Local Hartmann number; Nu = Nusselt number; Pr = Prandtl number; p' = Pressure; q_r = Radiative heat flux term; R = Radiation-conduction parameter; T' = Dimensional temperature; T'_w = Dimensional temperature at the plat; T'_∞ = Dimensional temperature in the free stream; U' = Dimensional free stream velocity; U_0 = Mean stream velocity; U = Non-Dimensional fluid velocity; (u', v') = Components of the fluid velocity; t' = time; (x', y', z') = Cartesian coordinates

Greek Symbols

ρ = Density of the fluid; υ = Kinematic viscosity; α = Thermal diffusivity; ω = Frequency parameter; θ = Dimensionless temperature; σ' =; tephan-Boltzmann constant; σ = Electrical conductivity; β = Co-efficient of volume expansion for heat transfer; ε = Small reference parameter.

Unsteady Cavitation Analysis Using Phase Averaging and Conditional Approaches in a 2D Venturi Flow

Vincent Aeschlimann[1], **Stéphane Barre**[2], **Henda Djeridi**[3]
[1]LAMH, Université Laval, Quebec, Canada
[2]LEGI, UMR 5519, Grenoble, France
[3]LPO, UMR 6523, Brest, France

ABSTRACT

The present study refers to a cavitating Venturi type section geometry characterized by a convergent angle of 18° and a divergent angle of about 8° where the sheet cavity presents typical self-oscillation behavior with quasi-periodic vapor clouds shedding. This work is an extension of previous works concerning void ratio measurements and velocity fields using double optical probe and constitutes a complete analysis of the two-phase structure of unsteady cavitating flow. This paper provides a new method based on conditional and phase averaging technique with wall pressure signal to treat experimental data in order to evaluate more precisely time-averaged and rms values of the void ratio and instantaneous velocity fields. Conditional analysis shows a different behavior of the two-phase flow dynamics leading to highlight high void ratio events linked to the break-off cycle. Unsteady phase averaging of the optical probe signal gives the evolution of the void ratio at each studied location in the venturi and shows that the fluctuations close to the wall (where the re-entrant jet is predominant) are in phase with the upper part of the cavity instead of the thickness of the cavity which is unchanged.

Keywords: Unsteady Cavitation; Sheet Cavity; Optical Probes; Venturi-Type Flow; Turbo Pump

1. Introduction

In cryogenic rocket engines, an inducer stage is used to protect the main turbopump from cavitation. Because of the low pressure used in cryogenic fuels tanks, cavitation occurs at the inducer's blade suction side. In this situation, cavitation has two main effects. At first, it affects the inducer performance and then the global pump performance. In a second step, it can induce several instabilities which may affect the global behavior of the entire engine fueling system (tanks, pipes, valves, pumps and combustion chamber).

A lot of experimental works (Kamijo *et al.* [1], De Bernardi *et al.* [2] and Hassan *et al.* [3] among others) have been performed on real inducers geometry in order to describe and classify all these cavitation instabilities. However, due to the excessive complexity of real inducer flows, it is difficult to develop a clever physical understanding of the physics of such cavitation instabilities from real inducers measurements. Then, in order to obtain a more accurate description of the cavitation sheets existing on the suction side of the inducer blade, a Venturi type geometry has been proposed. The main idea is to obtain a 2D flow which can be easily instrumented to perform more accurate diagnostics than in the real inducer geometry. It is decided to design this Venturi in such a way that the pressure field in these devices must be the same as the one existing over an inducer blade suction side (where the cavitation mainly develops). Two kinds of Venturi flows are then designed.

The first one which has a 4° divergent angle (Venturi 4°) corresponds to the nominal operating point of the inducer. It features a quasi stable cavitation sheet. It has been extensively described in Stutz and Reboud [4], Stutz [5,6] and Barre *et al.* [7]. The main interest of these studies as compared to the one performed on real inducers relies in the fact that it permits the investigation of the cavitation sheet itself. It is then shown that the sheet, most of the time, is not composed of pure vapor. At the opposite, the majority of the sheet flow consists in a two-phase mix between the vapor and the liquid phase. It has been also pointed out that it exists, even in stable cavity case such as the Venturi 4°, a re-entrant jet in the rear part of the cavity. A lot of information concerning the void ratio distribution and some characteristics of the internal velocity field has been obtained in Venturi 4° flows thanks to these efforts [7].

The main consequence of these results was a great change in the physical perception of such flows which were only considered in the past as pure vapor sheets with almost no internal dynamics and no interaction with the external flow. This new approach leads to subsequent progress in numerical simulations of such flow configuration making possible a global homogeneous modelisation in RANS code (Coutier *et al.* [8], Reboud *et al.* [9], Goncalves *et al.* [10-12] and Decaix *et al.* [13]) compared to the very simplified approach used in the past.

The second Venturi flow developed was the Venturi 8°. It corresponds to a partial flow-rate operating point for the inducer. In this case, the flow incidence on the inducer blades is quite high leading to a partial stall of these blades. In this situation, the cavitation sheet mainly consists in a detached vortical zone in which a two-phase fluid is embedded and convected downstream in a very unsteady manner. It was pointed out that this unsteadiness maybe responsible for the observed cavitation instabilities on real inducers geometry. Then, these flow configurations, and particularly, these unsteadiness characteristics, were extensively studied [14,15]. It was found that the cavity dynamics was quasi-periodic and that the void ratio values were quite low and never exceeded 25% for unsteady situations. Nevertheless, the unsteady behavior of the sheet cavity leads a major difficulty for simulations taking into account the interplay between cavitation and turbulence regarding the unsteadiness and the specific break-off cycle of the vapor phase. Indeed, the time averaged velocity profiles showed discrepancies in the cavity closure region due to the RANS approach where the turbulence-cavitation interactions were not fully taken into account. To get the unsteady behavior of quasi-peridodic cavitating structures, Large Eddy Simulations may be used to simulate the sheet/cloud cavitation phenomena. In this way, Dittakavi *et al.* [16] performed a LES of cavitation in a venturi nozzle in order to capture at the venture throat the specific irregular shedding of small scale vapor structures near the turbulent cavity closure region. They showed that these structures were responsible of vorticty production resulting in the formation of hair-pin vortices. De Lange *et al.* [17] observed that the re-entrant jet had a spanwise component which resulted in the formation of hair-pin vortices. This is the reason why it is obvious to get more insight about the physical mechanism of the shedding process using new experimental data-processing based on unsteadiness study. The main topic of this paper is to give a new insight about unsteady cavitation process in the venturi 8° flow type. The idea is to join phase averaging techniques and conditional measurements of the void ratio and vapor phase velocity in the sheet cavity. This technique allows a refined analysis of the cyclic phenomenon. Phase averaging process is performed using wall pressure signal as time reference.

The experimental facilities and operating conditions are presented in Section 2. The measuring techniques—dual fiber-optic probe for the dynamic characterization of the vapor phase are also described. An overview of experimental results concerning void ratio and vapor phase velocity is given in Sections 3 and 4.

2. Experimental Set-Up and Measurements

2.1. Experimental Set-Up

Experiments have been performed in a closed loop at CREMHyg lab in Grenoble and the test section is 520 mm long, 44 mm wide and 50 mm high in the inlet. The tested flow configuration corresponds to flat upper and side walls, whereas the height of the lower wall varied in order to form a convergent-divergent nozzle. In the present considered experiments, "venturi 8°", the convergent part of the lower wall (between inlet and throat) exhibits a 18° angle and the maximum angle of the divergent part was 8°. The inlet pressure P_{inlet} can be tuned and was lowered until the desired cavitation number allowing a specific unsteady behavior of the sheet cavity is obtained. The cavitation number at the inlet section was defined as:

$$\sigma_{inlet} = \frac{P_{inlet} - P_{vap}}{\frac{1}{2}\rho V_{inlet}^2} \quad (1)$$

where P_{vap} was the vapor pressure and P_{inlet}, V_{inlet}, were the pressure and velocity respectively at the reference section upstream of the venturi.

The divergent part was equipped with eight probe holes to take various measurements such as the instantaneous pressure, local void ratio and velocity in the sheet cavity. Specific probe locations are presented in **Figure 1** for the tested case.

By fine tuning both the velocity and the pressure (V_{inlet} and P_{inlet}) it was possible to obtain several sheet cavity behaviors. The present choice leads us to tune the flow in order to obtain a sufficiently large and unstable sheet but

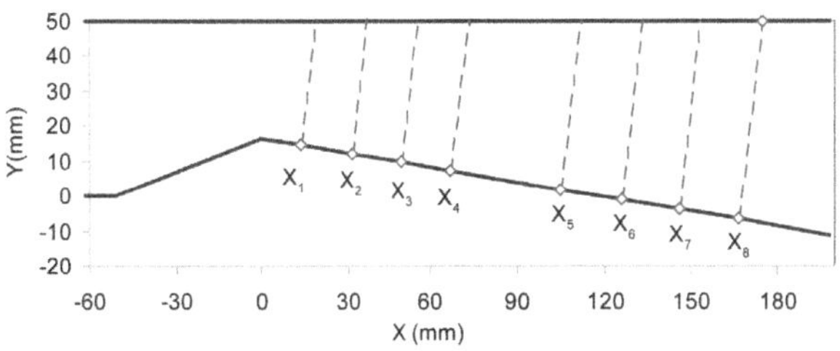

Figure 1. Venturi design and probe locations. X_1 **= 13.7 mm;** X_2 **= 31.5 mm;** X_3 **= 49.9 mm;** X_4 **= 67.7 mm;** X_5 **= 106 mm;** X_6 **= 126.9 mm;** X_7 **= 147.1 mm;** X_8 **= 168.3 mm.**

without very huge break-off cycles which may lead to unmanageable global flow-rate fluctuations of the entire testing loop. An operating point was then chosen to match all the previous constraints. Concerning the geometrical data, $L_{ref} = 224$ mm is the reference length and corresponds to the distance between the inlet section and the last probed station (X_8). The Reynolds number Re = 3.1×10^5 is based on V_{inlet} and on the test section width (44 mm). X_{ref} is the averaged cavity length (45 mm).

With these previous parameters and according to experimental observations, cavitation sheets developed from the venturi throat. A typical self-oscillation behavior was observed with quasi-periodic vapor clouds shedding and the mean attached cavity length value was 45 ± 5 mm (≈0.2 L_{ref}). Some instantaneous pictures of the cloud shedding are presented in **Figure 2**.

The stream goes from right to left and the throat is located on the right side of the pictures. At the first step (**Figure 2(a)**) an attached cavity is observed at the throat and slightly downstream a horseshoe shaped detached cloud is observed. At the second step (**Figure 2(b)**), the sheet reaches its maximum length and the cavity breaks in two parts at the third step (**Figure 2(c)**). The detachment of the vapor phase is due to the onset of reverse flow between the cavity and the venturi wall. This reverse flow is called "re-entrant jet". At the fourth step, the downstream part is swept along within the stream and starts to collapse while the attached part starts another cycle (**Figure 2(d)**). This particular behavior has been observed within various experimental studies (De lange *et al.* [17] and Foeth *et al.* [18] and). In the present case the corresponding frequency is close to 45 Hz (which corresponds roughly to $\approx 1.4 V_{ref}/L_{ref}$) leading to a classical Strouhal number of about 0.3 when based on the cavity length.

2.2. Measurements

2.2.1. Pressure Measurements

Pressure measurements were carried out using a piezoelectric sensor DRUCK model PMP4070, with a maximum range of 70000 Pa and absolute precision of ±19 Pa

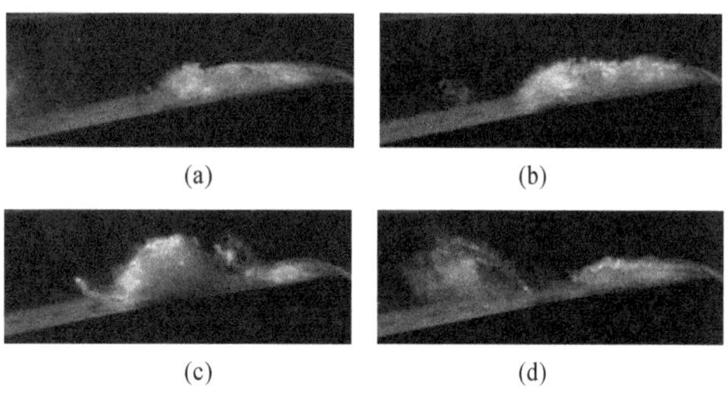

Figure 2. Instantaneous pictures of a break off cycle. (Flow is from right to left).

and the relative precision is ±0.027%. The mean and fluctuating values of the wall pressure have been measured upstream to characterize the cavitation number and at any stations in the divergent area especially inside the vapor cavity and in its wake. In order to perform a phase averaging of the data, the pressure signal was recorded at station 5 simultaneously with the optical probe one; this signal, being periodic, was then used as a time reference for the phase averaging process. Pressure signal was recorded at a sampling rate of 50 kHz. 6 millions of samples were taken for each measurements leading to an observation time of 120 s which is sufficient to perform the mean and root mean square (*rms*) value computation. The cavitation number was σ_{inlet} = 1.998% ± 0.11%, where P_{inlet} = 0.517% ± 0.03% and V_{inlet} = 7 m/s, at the reference section upstream of the venturi and it was verified that the mean and the most probable values are almost even. **Figures 3(a)** and **(b)** show the longitudinal evolution of the mean and normalized rms value of the wall pressure along the venturi. The mean pressure is here presented via the ratio:

$$\frac{C_p}{\sigma_{ref}} = \frac{P - P_v}{P_{ref} - P_v} \tag{2}$$

where P is the measured mean pressure value. The mean observed sheet length ($X_{ref} \approx 45$ mm) is also represented (gray area) and the longitudinal coordinate X is normalized to X_{ref}. The mean pressure increases continuously when going downstream. The positive pressure gradient is maximum at the sheet rear end near $X/X_{ref} \approx 1$ corresponding to the cavity closure zone. Downstream, a quasi linear monotonic pressure increase due to the venturi geometry is observed. It is easily shown that, almost in the probed range, the mean wall pressure never reach the vapor pressure value (P_v). However, the flow is diphasic and vapor is created even if the mean pressure remains superior to P_v in almost all the sheet zone. In fact, vapor is created due to the unsteady and turbulent nature of the flow in which turbulent eddies generate unsteady low pressure zone where the vapor creation process may take place and then, this vapor is transported by those eddies in the break-off cycle process.

The maximum level of the rms value (**Figure 3(b)**) is reached at the first probed position with a value of P_{RMS}/P_{mean} close to 20%. Further downstream, this level start to decrease quite sharply to reach an almost null value downstream of the sheet rear end (near $X/X_{ref} \approx 1.5$). A "remanent" level of about 5% is obtained near $X/X_{ref} \approx$ 2.3 which may corresponds to some complex interaction between the shedding process and the wall itself. However it seems clear that the initial detachment process occurring upstream of the sheet leads to large pressure fluctuations which seems to be damped when going downstream.

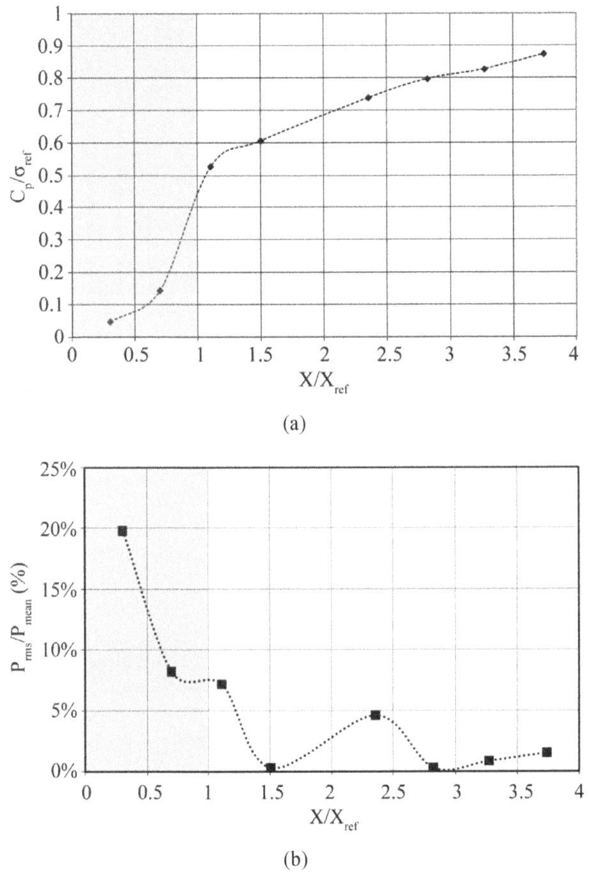

Figure 3. (a) Longitudinal evolution of the mean wall pressure across the flow; (b) Longitudinal evolution of the normalized RMS pressure fluctuations.

2.2.2. Two Phase Flow Measurements

For the two-phase flow measurements, double optical probe was used to evaluate void ratio and velocity fields. It was constituted by two sapphire optical fibers (see Barre *et al.* [7] for technical details about the probe). Due to the low void ratio in this flow configuration refined data acquisitions have been carried out in order to obtain sufficiently converged statistics and to perform an unsteady phase averaging analysis of the data. The most important difference with respect to previous work[15] concerns the geometry of the double optical probes and the data processing. Indeed, in the present study, we used an optical probe made of two sapphire optical fibers of 80 μm diameter and with frayed ends. The inter-tips distance is 1.05 mm ± 0.02. In previous work, the inter-tips distance was about 2 mm. The optical probe previously used was constituted of two glass fibers with a 50 μm tip diameter. This characteristic caused a lot of breaking probe problems due to the aggressiveness of the studied unsteady flow. The data acquisitions were carried out on a fast A/D data acquisition board NI-DAQ PCI-6110E. For each acquisition we recorded 4 million points per way (one way per optical probe), with a sampling rate of 100 kHz (giving 40 seconds of acquisition per measurement). For each *Y* position of the double optical probe in the venturi, the process was repeated 100 times in order to generate 400 million data per position in the cavity corresponding to an observation time of 4000 s per *Y* position which can be compared to 0.45 million data (6 s of acquisition) per probe position recorded and treated in previous study [15].

3. Processing Technique for Void Ratio and Vapor Phase Velocity

3.1. Estimation of the Local Void Ratio

The local void ratio α of a gas/liquid mixture is defined as the ratio between the cumulated time of gas phase at the tip of the probe (T_{vap_i}) and a given time of observation. It is given by:

$$\alpha = \frac{\sum_{i=1}^{n} T_{vap\_i}}{T_{tot\_obs}} \quad (3)$$

Figure 4 shows an idealized optical probe signal, the maximum value $V_{\max}$ corresponds to the vapor state, and the minimal value $V_{\min}$ to the liquid state. The used post-processing algorithm enables us to exploit the signal from the optical probe, by distinguishing the gas phase from the liquid phase according to the value of the measured tension. In order to estimate the local void ratio, a threshold in tension $V_{threshold}$ was chosen carefully and it allows us to determine the phase of the fluid around the probe: liquid state if $V < V_{threshold}$ and gas state if $V > V_{threshold}$. This threshold was fixed according to the parameter:

$$\beta = \frac{V_{threshold} - V_{\min}}{V_{\max} - V_{\min}} \quad (4)$$

The value of this parameter was gauged in great detail for this type of flow [4-6,15]. They retained a value of $\beta = 0.1$ to give correct void ratios for this typical flow pattern. We used this value for the determination of the void ratio. For more details about the void ratio determination, see Barre *et al.* [7].

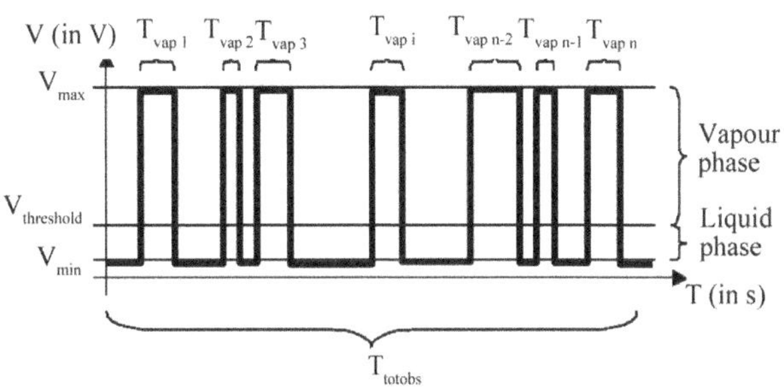

Figure 4. Schematic probe signal—Phase detection.

3.2. Estimation of the Instantaneous Local Velocity

The determination of the local velocity was based on the method detailed in Barre *et al.* [7]. The used double optical probe, aligned in the direction of the flow, produces two distinct signals (S1 upstream and S2 downstream). Those electric signals are thus analyzed and broken up by introducing two tension thresholds as shown in **Figure 5**. The low threshold V_{low}, taken on the rising part of the electric signal, leads us to detect the beginning of the crossing of a vapor bubble. The high threshold V_{high} is considered to detect the end of the crossing of the bubble on the downward part of the signal (see **Figure 5**). A bubble is taken into account in the analysis only when these two thresholds are crossed consecutively.

The values of the low and high thresholds are defined from the imposed values of β_{low} and β_{high} with:

$$\beta_{low} = \frac{V_{low} - V_{\min}}{V_{\max} - V_{\min}} \text{ and } \beta_{high} = \frac{V_{high} - V_{\min}}{V_{\max} - V_{\min}} \quad (5)$$

In the present study we used the value of $\beta_{low} = 0.2$ proposed by Stutz *et al.* [4-6,14,15].

Concerning β_{high}, its value has been reduced to 0.54 instead of 0.8 used by the same authors in order to obtain a substantial increase of the detected bubbles population. Some validations were carried out and confirm that this data rate increase is obtained without any degradation of the quality of the velocity field determination.

From these measurements, it was possible to deduce the tangential component (parallel to the profile) of the instantaneous local velocity by considering the temporal shift between the two traces provided by a bubble crossing successively the two tips of the probe. This optimal shift, called interval I_{tauopt}, was recorded and corresponded to the number of sampling data between the considered traces as shown in **Figure 6**.

Then I_{tauopt} is converted into a time interval using T_{ech}, the sampling period ($T_{ech} = 10^{-5}$ s)

$$\Delta t = I_{tauopt} \cdot T_{ech} \quad (6)$$

The velocity is then deduced by:

$$V = \frac{\Delta x}{\Delta t} = \frac{\Delta x}{I_{tauopt} \cdot T_{ech}} \quad (7)$$

where Δx is the distance between the two probe tips (Δx = 1.05 mm).

This method which aims to obtain instantaneous velocities values was based on the determination of the convective velocity of some selected bubbles being respectively detected by the two probes. As a matter of fact, the velocity calculation algorithm rejects a lot of bubbles and then only about 10% of the detected bubbles are validated to give a useable velocity value.

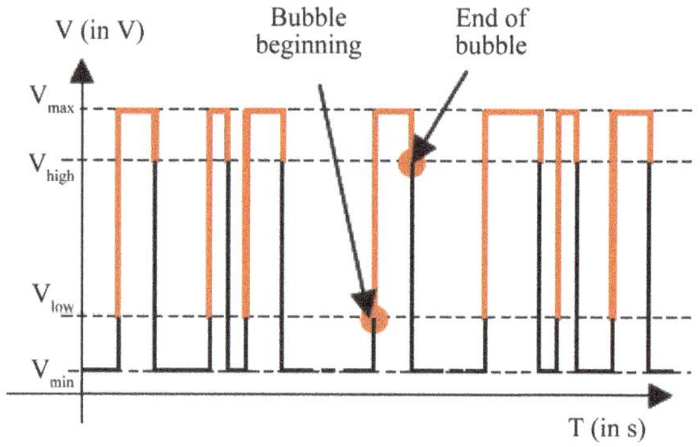

Figure 5. Bubble localization using threshold method.

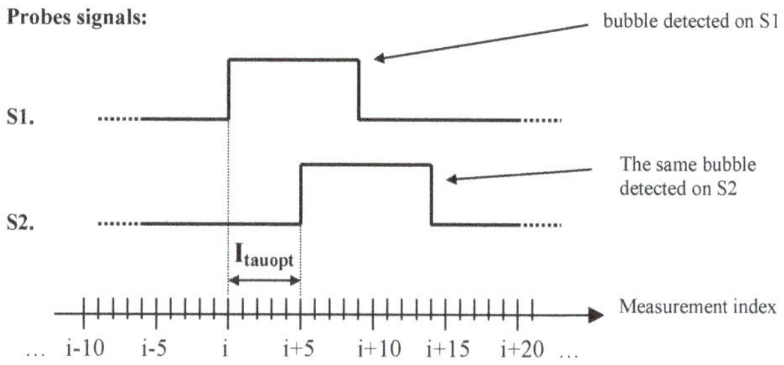

Figure 6. I_{tauopt} definition.

4. Unsteady Void Ratio and Velocity Fields Description with Phase Averaging and Conditional Analysis

As described in Stutz *et al.* [14], the periodic fluctuations of the cavity can be characterized by a "break off cycle" where the length of the cavity grows with a mean velocity of 0.4 V_{ref}, then the cavity is divided in two, the first part stays attached to the throat and a vapor cloud is convected downstream and collapses. In order to characterize a typical vapor cloud shedding, the choice of computing a phase averaged vapor shedding by dividing the period into five parts has been made. The pressure signal recorded downstream (at position X_4 = 67.7 mm) is used as time reference because of its quasi-periodicity at 45 Hz like the "break-off cycle". The phase averaging algorithm does not use a constant value for the pulsating period but determines, for each pulsation of the flow, the corresponding period. With this precaution, phase averaging can be done even in the present situation where the flow pulsation frequency is not perfectly constant but depends on the considered period. The main objective was then to compare the void ratio and velocity field distribution between the conditional and phase averaged methods and classical method in order to highlight the difficulties relative to the measurements of the unsteady behavior of the vapor phase. In this section, results concerning the void ratio are presented firstly followed by the velocity distributions and discussion about the physical mechanism of the break-off cycle.

4.1. Void Ratio

To describe the void ratio field dynamics, unsteady phase averaging has been performed and the evolution of the mean and standard deviation of the void ratio during a typical (phase-averaged) break-off cycle are presented on **Figures 7** and **8** for each probed station. The typical period has been divided in 5 parts, each step corresponds to roughly 4.4 ms and then a phase averaging process has been performed. This phase averaging process consists, for each break-off cycle, in cutting the pressure signal in five equal parts. After that, for each time part of the cycle, corresponding bubbles are detected on the optical probe signal. Then, these bubbles are processed in order to calculate their void ratio and a phase-averaged mean void-rate.

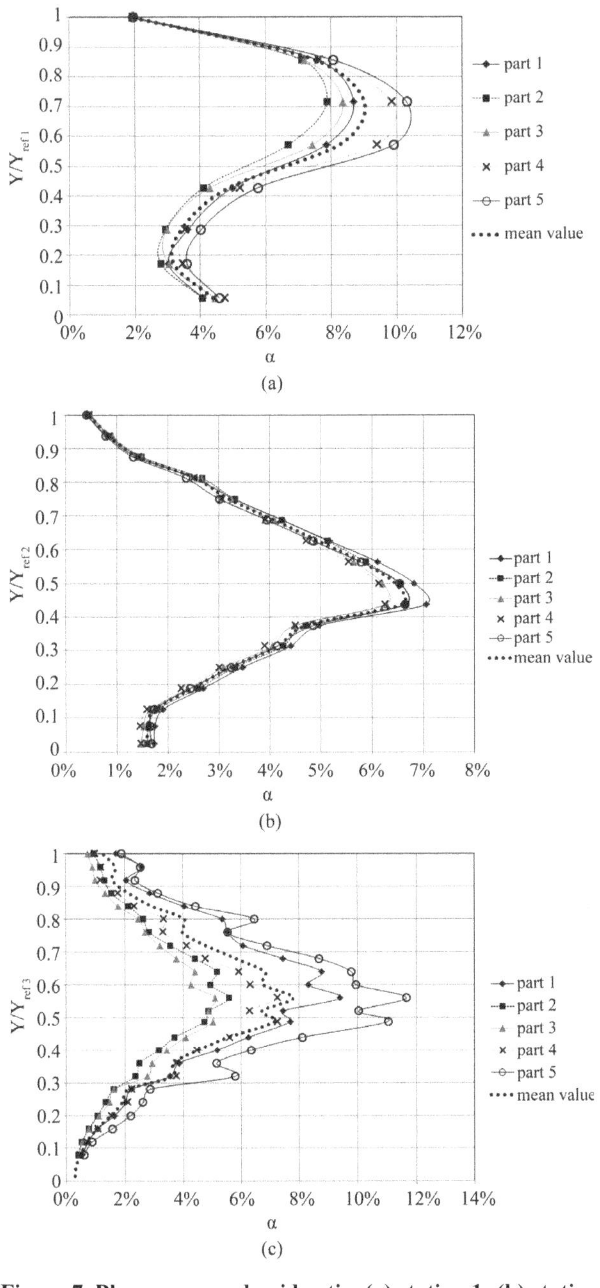

Figure 7. Phase averaged void ratio. (a) station 1; (b) station 2 and (c) station 3.

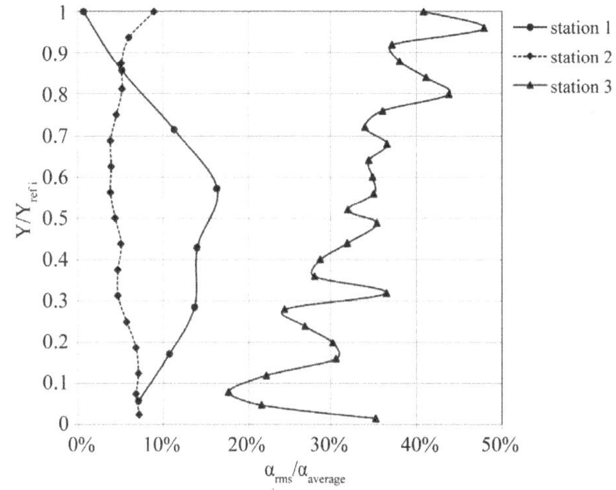

Figure 8. Standard deviation of the void ratio.

The standard deviation was defined as follow:

$$\sqrt{\frac{\sum_{i=1}^{5}\left(\alpha_i-\left[\sum_{j=1}^{5}\frac{\alpha_i}{5}\right]\right)^2}{5}} \qquad (8)$$

The mean phase averaged void ratio distribution inside the cavity is presented in **Figure 7** for each probed station; X_1 = 13.7 mm; X_2 = 31.5 mm and X_3 = 49.9 mm versus the transversal position. The transversal and longitudinal coordinate are reduced using respectively the local cavitating sheet thickness ($Y_{ref,i}$) and the mean sheet length X_{ref}. Here $Y_{ref\,i}$ are the cavity thicknesses for the considered stations: Y_{ref1} = 3.5 mm; Y_{ref2} = 8 mm; Y_{ref3} = 12.5 mm and X_{ref} is the mean attached cavity length which is about 45 mm. The method for the estimation of the local sheet thickness was based on the zone determination where the void ratio was greater than 0% in the context of the spatial resolution of the void ratio measurement (0.5 mm) [7]. The classical mean values of the void ratio are superimposed to the phase averaged ones.

For stations 1 and 3, the mean phase averaged void ratio spatially oscillates in-phase and it can be seen a maximum peak value at the fifth step and a minimum value between the second and the third one. The evolution of void ratio in a given section is either growing or reducing. The fluctuations of the void ratio close to the wall (at the re-entrant jet region) and in the upper part of the cavity, are in-phase. The thickness of the cavity at each station does not change during a "break-off cycle". We notice that, at station 2, fluctuations are low and the void ratio profile is quite steady. The station 3 is located just after the extremity of the attached part of the cavity where vapor cloud breaks away. It is important to note that the void ratio is never null over the whole section at any step. There are always few bubbles of vapor even between two clouds of vapor.

Figure 8 shows the normalized RMS evolution of the void ratio for each plotted station. Maximum relative fluctuations are obtained at station 3 because of the very low mean void ratio value inducing a strong effect on the RMS value for almost all the travelling bubbles localized in this flow region. At station No.1 the spatial repartition of the void ratio RMS fluctuations is not homogeneous and it exhibits a maximum value located near the sheet half height. This place may correspond to the maximum vapor generation area. At the opposite, for station No.2, it can be observed a very low and homogeneous fluctuation level. It then can be concluded that an almost steady situation is experienced at this station.

In order to provide a better visualization of the phase averaged fluctuations dynamics, the spatially integrated dimensionless value α^* (centered and normalized) of the phase averaged void ratio has been computed using the formula:

$$\alpha^* = \frac{\alpha - \frac{\alpha_{\max} + \alpha_{\min}}{2}}{\frac{\alpha_{\max} - \alpha_{\min}}{2}} \quad (9)$$

with

$$\alpha = \frac{\int_0^{y_{ref}} \alpha(y)\,\mathrm{d}y}{y_{ref}} \quad (10)$$

In which α is the space and phase averaged void ratio and $\alpha_{\max}$ and $\alpha_{\min}$ are the spaced and phase averaged void ratio at the steps where there is respectively the maximum and the minimum amount of vapor (as an example, at station 1 and 3, the space and phase averaged void ratio is maximum at the 5th step, this is where $\alpha_{\max}$ is defined). **Figure 9** presents the space and phase averaged void ratio evolution for each step during the break-off cycle. For clarity purpose, this evolution is plotted over three periods.

It can be observed that whatever the station, the void ratio evolution is pulsated corresponding to the break-off cycle. To illustrate the relative intensity of such fluctuations in a typical cycle we compute the ratio between their root mean square and mean values over a phase averaged period. **Figure 10** displays this longitudinal evolution versus X/X_{ref}. The maximum relative void ratio fluctuations are obtained at station X_3 where, at the same time, the mean void ratio value is the lowest. This illustrates the fact that the cavitation sheet closure zone is highly unsteady leading to large relative void ratio fluctuations of up to 30% of the mean value. This behavior is opposed to the one observed for the pressure fluctuations

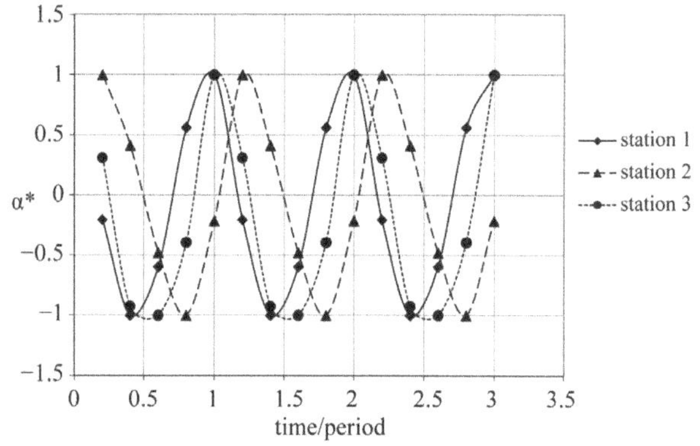

Figure 9. Spatially and phase averaged void ratio evolution over 3 periods.

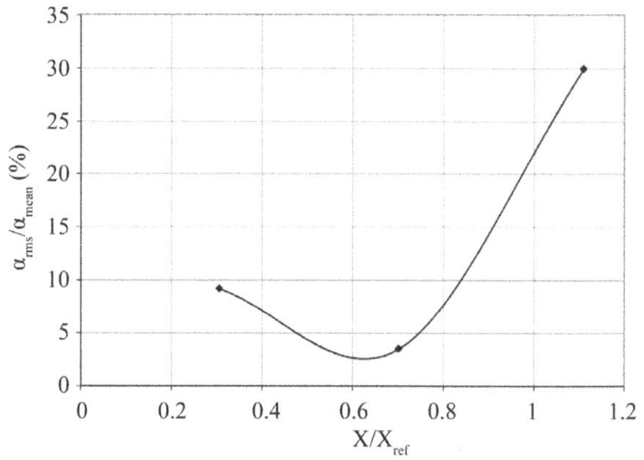

Figure 10. Normalized void ratio fluctuations over a break-off cycle.

(see **Figure 3(b)**) where the maximum relative fluctuation level is obtained at X_1 and then decreases sharply when going downstream. This experimental evidence may lead to some important consequence in term of cavitation modeling. It shows that a classical barotropic approach used in some cavitating flow numerical simulations (Fortes Patella *et al.* [19]) in which the pressure and the void ratio are linked by a direct functional dependence may not be able to reproduce the present experimental results.

4.2. Velocity Fields

In this section we firstly present the probability density function (PDF) of the velocity for each station at different vertical position in the sheet cavity. Specific transversal locations have been chosen for each station in order to take into account the longitudinal evolution of the thickness of the cavity: close to the wall, in the middle of the sheet cavity and in the cavity upper boundary. PDF are plotted on **Figures 11(a)**, **(b)** and **(c)** and exhibit an oscillating shape between positive and negative value of the velocity except at station 3 where the velocity is mainly positive.

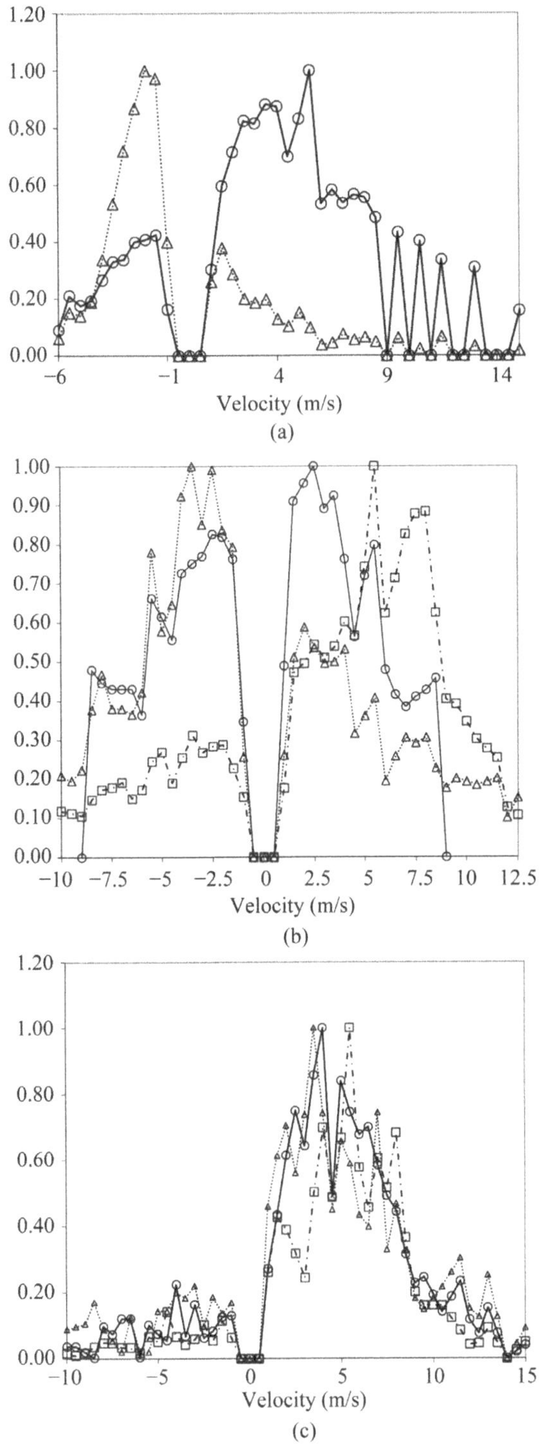

Figure 11. (a) Velocity PDF at station 1: △: Y/Y_{ref1} = 0.28, V_{mean} = −1.3 m/s and $V_{most\text{-}probable}$ = −2 m/s; O: Y/Y_{ref1} = 0.86, V_{mean} = 2.72 m/s and $V_{most\text{-}probable}$ = 5.5 m/s; (b) Velocity PDF at station 2: △: Y/Y_{ref2} = 0.125, V_{mean} = −1.25 m/s and $V_{most\text{-}probable}$ = −3.5 m/s; O: Y/Y_{ref2} = 0.375, V_{mean} = −0.49 m/s and $V_{most\text{-}probable}$ = 2.5 m/s; □: Y/Y_{ref2} = 0.75, V_{mean} = 2.65 m/s and $V_{most\text{-}probable}$ = 5.5 m/s; (c) Velocity PDF at station 3: △: Y/Y_{ref3} = 0.16, V_{mean} = 3.38 m/s and $V_{most\text{-}probable}$ = 3.5 m/s; O: Y/Y_{ref3} = 0.32, V_{mean} = 3.9 m/s and $V_{most\text{-}probable}$ = 4 m/s; □: Y/Y_{ref3} = 0.64, V_{mean} = 4.5 m/s and $V_{most\text{-}probable}$ = 5.5 m/s.

At station 2 for example, when the location measurement is just up to the border of the re-entrant jet (Y/Y_{ref2} = 0.375), it can be observed two main peaks with similar occurrences with respective most probable values of about −2.5 m/s and 2.5 m/s. In this region, the mean velocity only represents the average of those two values pondered by their probabilities of occurrence. The boundary of the re-entrant jet may then be defined by the location where occurrences of these peaks are equal. For stations 1 and 2, close to the wall, the peak with negative velocities values outweighs the one corresponding to positive ones. This behavior is characteristic of the re-entrant jet. Getting away from the wall, the negative peak disappears and the positive one increases. The two peaks are even at the re-entrant jet border and then the positive peak outweighs the negative one where the fluid goes downstream. Station 3 is located in the region corresponding to the cavity closure and the velocity is always positive. As shown in the previous figures, the obtained velocity probability distributions are not Gaussian, so the mean velocity is not representative of the dynamic behavior of the fluid. This fact constitutes the main characteristic of this unsteady velocity field. From these data, we will be able to describe the velocity field inside the cavity.

Taking into account the previous results let us now consider the transversal evolutions of the longitudinal velocity for the three considered measurement stations. A particular attention is paid on the comparison between the classical mean values and the most probable values. **Figure 12** exhibits the transversal evolution of both the classical mean velocity value and the most probable one. This representation is completed with the phase averaged values of the velocity obtained for both the five studied shedding cycle steps. In order to better describe the internal cavity dynamics, a phase averaging analysis has been performed on the velocity field. The data were obtained by cutting each signal period in five parts as was previously done for the void ratio. With this decomposition we can describe the typical cavity break-off cycle. Results are presented on the same figures which show the velocity profiles inside the cavity for the five steps observed during the break off cycle.

From a qualitative point of view, it can be seen for station 1 and 2 that a recirculating flow is present. It corresponds to the re-entrant jet associated with the negative velocity values shown on **Figure 12**. At station 3, the mean and most probable values exhibit a classical near wall turbulent boundary layer velocity profile. By comparisons, the mean value approach underestimates the extent and the intensity of the re-entrant jet, especially at station 2 where the unsteadiness is more pronounced.

Furthermore, we can observe that the velocity at the upper border of the cavity does not fluctuate in the con-

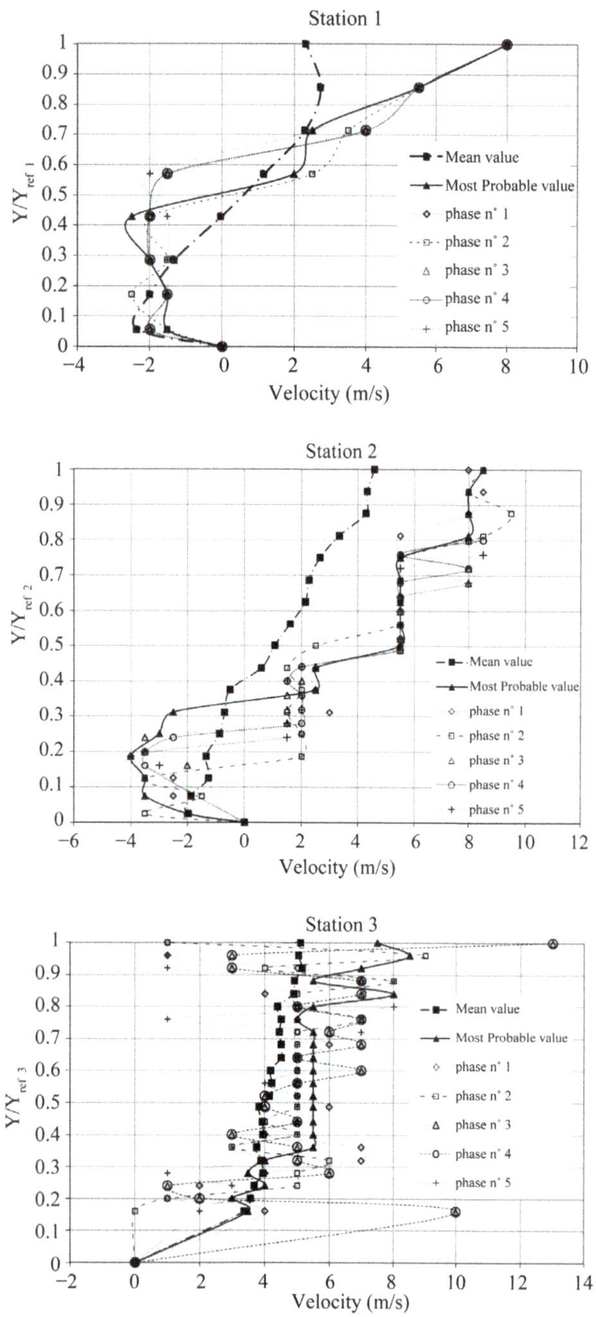

Figure 12. Velocity field inside the cavity. Comparisons with phase averaging analysis of the velocity field.

text of the present phase averaging analysis. At stations 1 and 2, the re-entrant jet is present during the whole cycle. We can also observe that the overall main feature of the cycling process relies on the fact that the re-entrant jet apparent thickness seems to evolve from one step to another. Velocity profiles lead us to describe the re-entrant jet dynamics. Indeed, the liquid gets inside the cavity from the downstream part between station 2 and 3. Then it vaporizes on the way to the throat. Finally, it changes direction and goes back swept along by the liquid above the cavity. As presented in **Figure 13**, the cavity can be dissected within three parts: the re-entrant jet, the upside part and convected structures.

4.3. Vapor Phase Dynamic inside the Cavity

The flow inside the cavity is turbulent, unsteady and quasi-periodic at a frequency close to 45 Hz. Fluctuations of the shape of the cavity have been observed showing that vapor clouds are convected by the liquid flow. The pressure has been measured at each station at the wall (station X_1 to X_8, and $y = 0$ mm). Pressure fluctuations are oscillating at a dominant frequency of 45 Hz for each location in the downstream part. Thanks to the pressure signal time reference it was possible to individually calculate the mean void ratio of each break-off cycle. With this data, an analysis of the void ratio field dynamics has been performed. The optical probe signals were then cut up using the pressure signal recorded in the downstream part as time reference to locate the beginning and the end of each break-off period. **Figure 14** shows an example of the obtained void ratio for each period during a 40 s measurement which corresponds to about 2000 periods. In this case, the average void ratio is $\alpha = 0.64\%$ (dotted line in **Figure 15**). The probability density function of the void ratio is plotted in **Figure 16** for a cumulative data set of 10 measurements of 40 s each, giving a 400 s observation time.

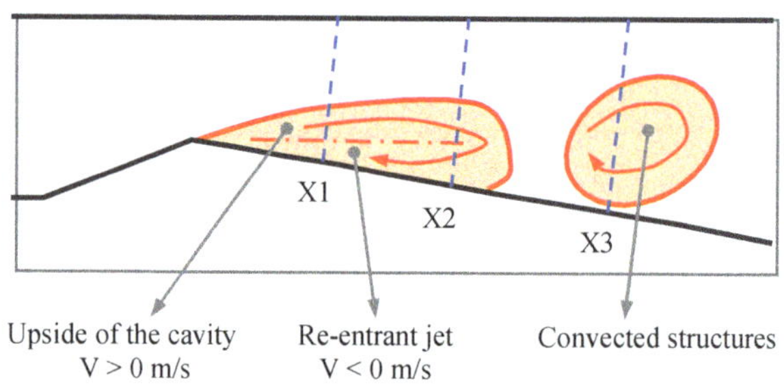

Figure 13. Sketch of the cavity structure.

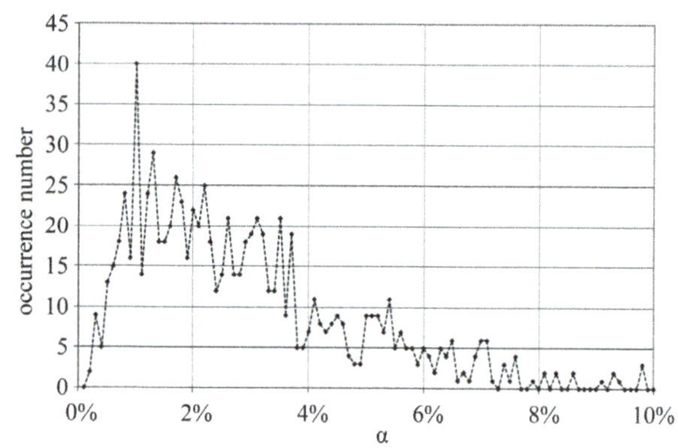

Figure 14. Void ratio PDF of the period where bubbles have provided velocity measurement at location. $X_3 = 49.9$ mm; $Y = 8$ mm (400 s measurement).

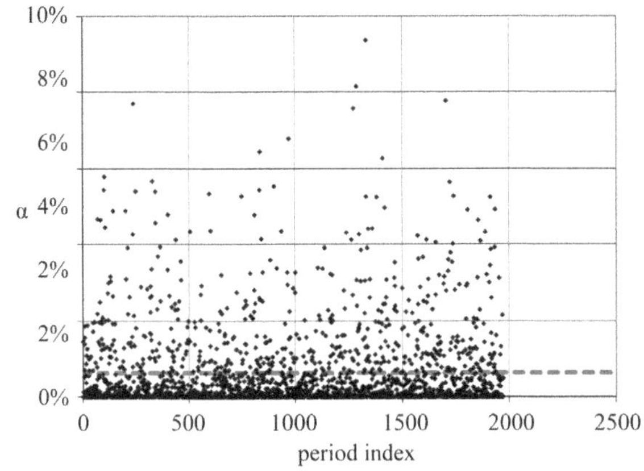

Figure 15. Period by period void ratio at location X_3 = 49.9 mm; Y = 8 mm (40s measurement). - - - mean void ratio: 0.64%.

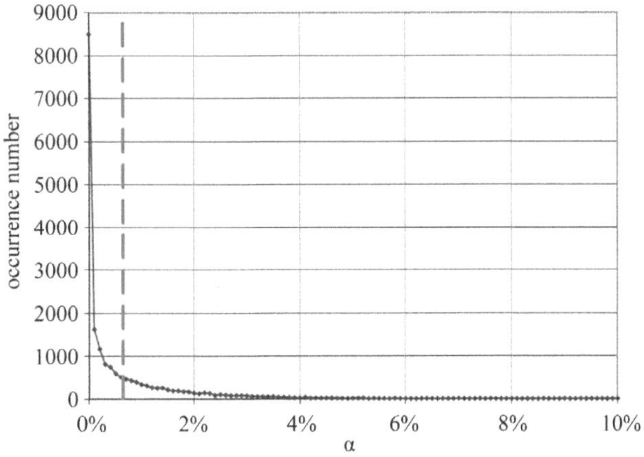

Figure 16. Period by period void ratio PDF at location X_3 = 49.9 mm; Y = 8 mm (400 s measurement).

As shown in the **Figures 15** and **16**, period by period void ratio values cover a wide range from zero to about 10%. A sizeable number among all these periods exhibit a void ratio higher than five times the average which is 0.64%. The PDF shows that 43% of the periods have a void ratio lower than 0.1% whereas the averaged void ratio is 0.64% (dotted line in **Figure 15**). This conditional analysis shows that even though the pressure signal is periodic, the vapor formation is quite irregular. Nevertheless, it can be noticed that the velocity measurement method used in the present study may leads to some limitation especially during a period with a very low void fraction. Indeed, the present study was carried out with a homogeneous approach for the two phase flow. This means that the vapor and the liquid flow are supposed to run in the same direction at the same velocity (no-slip hypothesis). From that hypothesis we are able to describe the fluid velocity by measuring only the velocity of the vapor bubbles. So, more bubbles there are, more instantaneous velocity measurements are obtained. Therefore, by using our velocity measurement method, the high void ratio events provide more contribution to the velocity values determination than the others leading to a kind of bias. In order to quantitatively describe this effect we perform a conditional analysis of the void ratio field. Following this analysis, only the events (bubbles) leading to a validated velocity value (according to the method described in section 3.2 of the present paper) were used to calculate the conditional void ratio PDF. **Figure 14** shows an example of this probability density function (PDF). The y-axis shows the number of bubbles which have provided a velocity value, the x-axis is the void ratio of the period where those specific bubbles have been observed. The averaged void ratio pondered by the occurrence number of velocity measurements obtained is then called the conditional void ratio.

The statistics show that no velocity measurements were obtained for period with void ratio lower than 0.1% which represents 43% of the events. Results have then to be observed carefully: velocity measurements for the studied location (X_3, Y = 8 mm) seems to represent a category of events that is qualitatively different from the one obtained in the first analysis performed to compute the mean void ratio distribution. It then became questionable to characterize the behavior of the fluid inside the cavity from these data. To illustrate this, the conditional analysis has been carried out through the whole cavity.

For this purpose, the conditional void ratio has been computed for all the available measurements positions of the present experiment. Results are presented on **Figures 17(a)**, **(b)** and **(c)** which display the conditional void ratio evolution for the three probed stations of the present study. The conditional void ratio is here compared to the continuous value obtained by a classical global averaging method as described in section 3.1 of the present paper.

This comparison shows that the estimated void ratio distributions are quite different and depend on the localization inside the cavity. It can be seen, for stations 1 and 3, a sizeable difference especially where the break-off cycle is predominant.

To sum up, the measured velocity is not representative of all events: it highlights high void ratio events. At the two first stations: velocity measurements come from events at a slightly higher void ratio than the average one. The corresponding velocity measurements are considered representative of the global dynamic behavior inside the cavity. In the third station case, a sizeable difference between the average and the conditional void ratio was observed. Therefore, the velocity field is characteristic of the events at high void ratio. It must be noticed that this study points out some limits of velocity measurements by optical probe technique in such unsteady two phases turbulent flow where void ratio is low (α < 1% at station 3).

The present measurements lead us to consider the

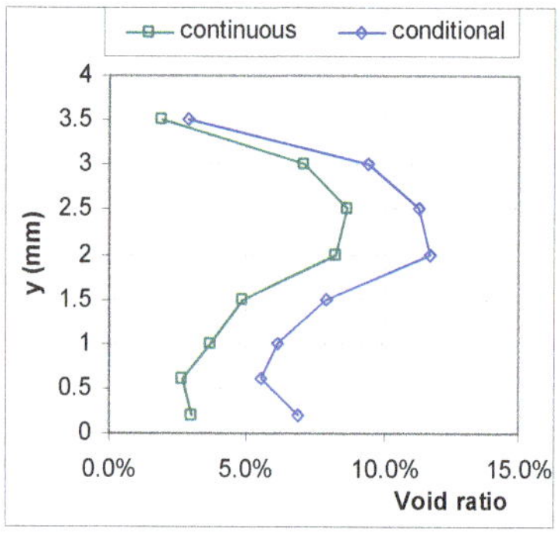

(a)

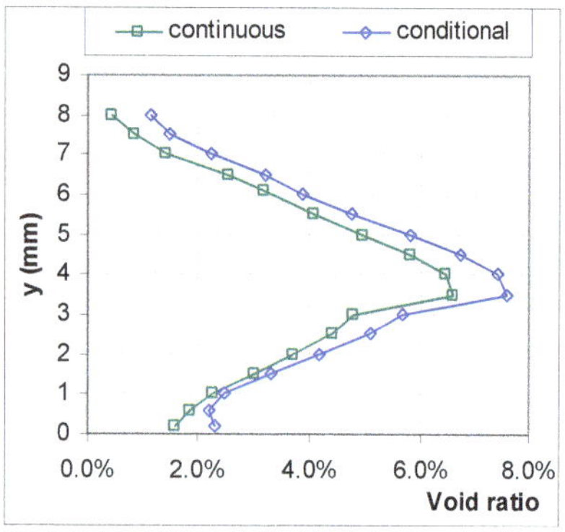

(b)

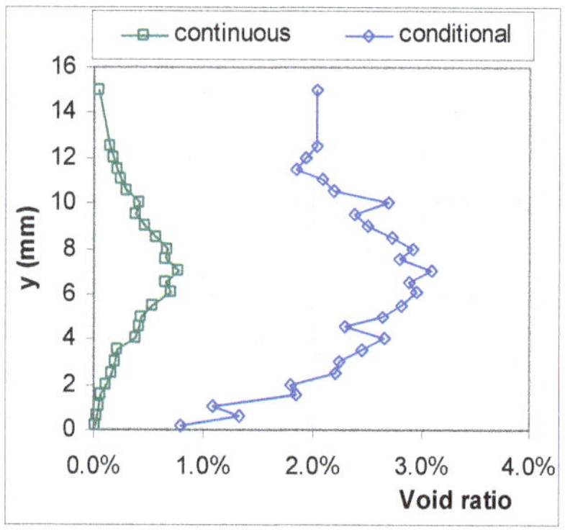

(c)

Figure 17. Comparison between mean and conditional values of the void ratio distribution.

mass and vapor flow rate inside the cavity. Indeed, from the integration of both the density and mean velocity field at a given section, we obtain the mass flow rate. For the two first sections studied, station 1 and 2, the respective values of 0.1 kg/s and 0.4 kg/s are obtained whereas at station 3 it is 2.2 kg/s (see **Figure 18**). When those values are compared to the mass flow rate through the whole duct, set at 15.5 kg/s, it appears that at station 1 and 2 they are negligible. It means that the re-entrant jet brings back as much fluid as the upper part of the cavity carries downstream. This behavior allows us to consider the border of the cavity as approximately a flow-line from the throat to the upper border of the cavity located at station 2.

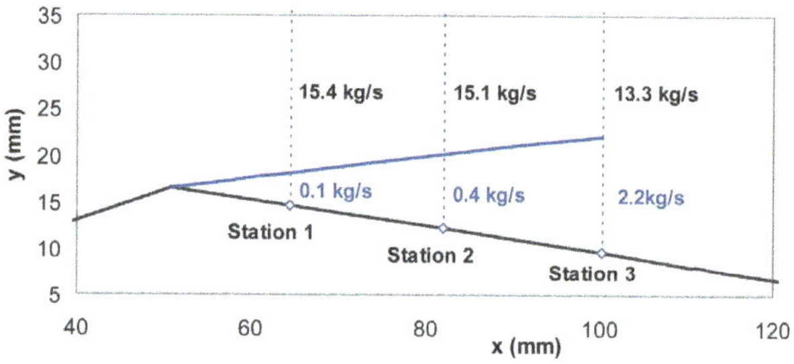

Figure 18. Illustration of the mass flow rate distribution.

The **Figure 19** presents the vapor flow rate distribution inside the cavity which has been calculated by integration of the experimental mean velocity and the void ratio over each section. For the station 1 and 2 there are two values, one corresponding to the section in the re-entrant jet and the other one corresponds to the section where the velocity is positive.

The volume of vapor is not preserved from station 2 (4.2 cm^3/s) to station 1 (2.4 cm^3/s) within the re-entrant jet nor from station 1 (11.7 cm^3/s) to station 2 (15.1 cm^3/s) within the upper part of the cavity. So the bubbles are not carried from a station to the next one without changing size, collapsing or being created. It is also plausible that some bubbles cross the border of the re-entrant jet between station 1 and 2.

Contrary to the negligible mass flow rate in the 1st and 2nd section, the vapor flow rate is quite important: 9.3 cm^3/s cumulated at station 1 and 10.9 cm^3/s at station 2. The attached part of the cavity is producing vapor. From station 2 to station 3, the vapor flow rate decreases from 10.9 cm^3/s to 8.1 cm^3/s. This means that vapor clouds have already started to collapse when they cross the third section and probably some bubbles are re-injected inside the attached cavity by the re-entrant jet.

5. Conclusions

A quite complete experimental study of an unstable cavitation sheet in a 2D Venturi configuration has been performed by conditional and phase averaging approaches. Compared to previous works done in the same geometry, some important improvements both on the experimental method and data processing algorithms were carried out to obtain more precise and reliable results concerning void ratio and instantaneous velocity fields. A global analysis of the sheet pulsation cycle was performed and provided acceptable results. Indeed, both conditional and phase approaches led to characterize the dynamics of the void ratio topology during a break-off process and allowed us to highlight the re-entrant jet thickness and associated oscillating mechanism. Actually, the thickness of the cavity was almost constant during a break-off cycle instead of the void ratio fluctuations in

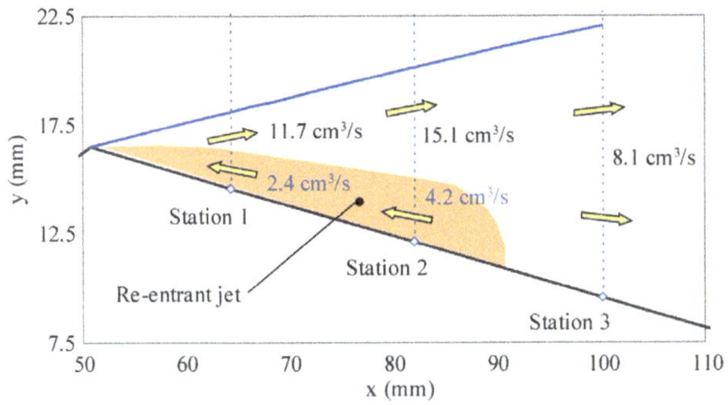

Figure 19. Illustration of the vapor flow rate repartition.

the re-entrant jet area and in the upper part of the cavity were in phase. By comparison with the classical mean estimation, the re-entrant jet is more extended where unsteadiness is more pronounced. Moreover, by performing a local and unsteady analysis of the void ratio and velocity fields, the study shows that there is no agreement between mean and conditional most probable approaches, concerning void ratio and velocity field distributions inside the attached cavitation sheet.

To conclude, each cycle of break-off mechanism of the cavity got different magnitude and the link between the void ratio and the pressure may not be direct and independent of the instantaneous transports dynamics characteristics of the flow.

As a perspective of the present work, it can be admitted that the present velocity determination method using double optical probe is found to be strongly affected by some bias when used in flows experiencing a high void ratio fluctuations level. This result leads us to completely change our measurement strategy for the next works dealing with high Reynolds number turbulent cavitating flows. PIV-LIF measurements are then developed to measure mean velocity fields and Reynolds stress tensor in this kind of flows (see Aeschlimann *et al.* [20]). We also move to X-Ray absorption techniques in order to measure mean and fluctuating void rate fields (see Aeschlimann *et al.* [21]). The future direction of this field of researches will be analyzing the spatio-temporal dynamic of the liquid and vapor phase simultaneously using optical methods. The future database will be available in order to test URANS and/or LES approaches in this class of flow where turbulence and cavitation interact.

6. Acknowledgements

The authors wish to express their gratitude to the French space agency CNES and to the SNECMA company for their continuous support.

REFERENCES

[1] K. Kamijo, T. Shimura and M. Watanabe, "An Experimental Investigation of Cavitating Inducer Instability," *ASME Winter Annual Meeting*, Atlanta, 27 November- 2 December 1977.

[2] J. De Bernardi, F. Joussellin and A. Von Kaenel, "Experimental Analysis of Instabilities Related to Cavitation in Turbopump Inducer," *Proceedings of 1st International Symposium on Pumps Noise and Vibration*, Paris, 7-9 July 1993, pp. 91-99.

[3] W. Hassan, S. Legoupil, D. Chambellan and S. Barre, "Dynamic Localization of Vapor Fraction in Turbopump Inducers by X-Ray Tomography," *IEEE-TNS* (*Transaction on Nuclear Sciences*), Vol. 55, No. 1, 2008, pp. 656-661.

[4] B. Stutz and J. L. Reboud, "Two-Phase Flow Structure of Sheet Cavitation," *Physical Fluids*, Vol. 9, No. 12, 1997, pp. 3678-3686.

[5] B. Stutz, "Analyse de la Structure Diphasique et Instationnaire de Poches de Cavitation," Ph.D. Thesis, INPG, CREMHYG Laboratory, Grenoble, 1996.

[6] B. Stutz, "Influence of Roughness One the Two-Phase Flow Structure of Sheet Cavitation," *Journal of Fluids Engeneering*, Vol. 125, No. 4, 2003, pp. 652-659.

[7] S. Barre, J. Rolland, G. Boitel, E. Goncalves and R. Fortes Patella, "Experiments and Modelling of Cavitating Flows in Venturi: Attached Sheet Cavitation," *European Journal of Mecanics B/Fluids*, Vol. 28, No. 3, 2009, pp. 444-464.

[8] O. Coutier-Delgosha, J.-L. Reboud and Y. Delannoy, "Numerical Simulation of the Unsteady Behavior of Cavitating Flows," *International Journal for Numerical Methods in Fluids*, Vol. 42, No. 5, 2003, pp. 527-548.

[9] J. L. Reboud, R. Fortes-Patella, M. Hofmann, H. Lohrberg, G. Ludwig and B. Stoffel, "Numerical and Experimental Investigations on the Self-Oscillating Behavior of Cloud Cavitation," ASME-FEDSM 99-6755/7259, San Francisco, 1999.

[10] E. Goncalves and R. Fortes-Patella, "Numerical Simulation of Cavitating Flows with Homogeneous Models," *Computers & Fluids*, Vol. 38, 2009, pp. 1682-1696.

[11] E. Goncalves and R. Fortes-Patella, "Constraints on Equation of State for Cavitating Flows with Thermodynamic Effects," *Applied Mathematics and Computation*, 2010, in Press.

[12] E. Goncalves, "Numerical Study of Expansion Tube Problems: Toward the Simulation of Cavitation," *Computers and Fluids*, Vol. 72, 2013, pp. 1-19.

[13] J. Decaix and E. Goncalves Da Silva, "Compressible Effects Modelling in Turbulent Cavitating Flows," *European Journal of Mechanics B/Fluids*, Vol. 39, 2013, pp. 11-31.

[14] B. Stutz and J. L. Reboud, "Experiments One Unsteady Cavitation," *Experiments in Fluids*, Vol. 22, 1997, pp 191-198.

[15] B. Stutz and J. L. Reboud, "Measurements within Unsteady Cavitation," *Experiments in Fluids*, Vol. 29, No. 6,

2000, pp. 545-552.

[16] N. Dittakavi, A. Chunekar and S. Frankel, "Large Eddy Simulation of Turbulent-Cavitation Interactions in a Venturi Nozzle," *Journal of Fluid Engineering*, Vol. 132, 2010, Article ID: 121301.

[17] F. De Lange and J. De Bruin, "Sheet Cavitation and Cloud Cavitation, Re-Entrant Jet and Three-Dimensionality," *Applied Scientific Research*, Vol. 58, No. 1-4, 1998, pp. 91-114.

[18] E. J. Foeth and T. V. Terwisga, "The Structure of Unsteady Cavitation. Part I: Observation of an Attached Cavity on a Three-Dimensional Hydrofoil," *6th International Symposium on Cavitation*, Wageningen, 11-15 September 2006.

[19] R. Fortes-Patella, O. Coutier-Delgosha, J. Perrin and J.-L. Reboud, "A Numerical Model to Predict Unsteady Cavitating Flow Behavior in Inducer Blade Cascades," *Journal of Fluid Engineering*, Vol. 129, No. 2, 2007, pp. 128-135.

[20] V. Aeschlimann, S. Barre and H. Djeridi, "Velocity Field Analysis in an Experimental Cavitating Mixing Layer," *Physics of Fluids*, Vol. 23, 2011, Article ID: 055105.

[21] V. Aeschlimann, S. Barre and S. Legoupil, "X-Ray Absorption Measurements in a Cavitating Mixing Layer for Instantaneous 2D Void Ratio Determination," *Physics of Fluids*, Vol. 23, 2011, Article ID: 055101.

24

Near Wake of a Horizontal Circular Cylinder in Stably Stratified Flows

Yuji Ohya, Takanori Uchida, Tomoyuki Nagai
Research Institute for Applied Mechanics, Kyushu University, Kasuga-shi, Japan

ABSTRACT

The near wake of a circular cylinder in linearly stratified flows of finite depth was experimentally investigated by means of flow visualization and measurements of vortex shedding frequencies, at Reynolds numbers 3.5×10^3 - 1.2×10^4 and stratification parameters k_d 0 - 2.0. The non-dimensional parameter k_d is defined as $k_d = Nd/U$, where N is the Brunt-Väisälä frequency, d, the diameter of the cylinder, and U, the approaching flow velocity. The study demonstrates that as k_d increases from zero, the vortex shedding from a circular cylinder progressively strengthens, while the Strouhal number gradually becomes lower than that for homogeneous flow. This phenomenon can be explained by the effect of the increasingly stable stratification which enhances the two-dimensionality of the near-wake flow of the circular cylinder; the enhanced two-dimensionality of the flow strengthens the roll-up of the separated shear layer. Above a certain value of k_d, however, vortex formation and shedding are strongly suppressed and the Strouhal number rises sharply. This observation is attributable to the development of stationary lee waves downstream of the circular cylinder because the lee waves strongly suppress vertical fluid motions.

Keywords: Stratified Flow; Circular Cylinder; Wake Pattern; Vortex Shedding; Lee Wave

1. Introduction

In stratified fluids with vertical gradients of density and/or temperature, buoyancy effects become significant. For instance, the strength of stratification is a predominant factor for fluid motion, turbulence transport, and dispersion in the atmosphere and ocean. Accordingly, a large number of investigations have been conducted on stratified flow around terrain or objects on the Earth's surface (e.g. Blumen [1]). Some of these investigations were conducted to improve understanding of meteorological phenomena such as lee waves and local winds (e.g. Snyder [2]). Other investigations have centered around issues of the atmospheric environment that require understanding of advection and dispersion of air pollutants. In contrast to the investigations on atmospheric flows, there exist few investigations on stratified flow around basic objects such as cylinders or spheres.

In stratified flow, the flow around an object is generally governed by the Reynolds number and other dimensionless stratification numbers (e.g. the Froude number). Honji [3-5] investigated two-dimensional flows around a circular cylinder placed horizontally in stratified flow for a range of Reynolds numbers less than approximately 800. With flow visualizations, Honji [3-5] determined the following properties of two-dimensional near wake flows. In stratified flow, fluid motions are suppressed in the direction of stratification. As a result, at low density stratification, Kármán-vortex-street wake-flow patterns are observed behind a horizontal circular cylinder placed perpendicular to the flow direction. However, with increasing stratification, the wake-flow pattern makes a gradual transition to that dominated by internal waves. When the stratification becomes sufficiently high, blocking occurs and upstream wakes emerge [6,7]. Furthermore, in a highly stratified fluid of finite depth, columnar disturbances emerge and propagate upstream of an object [8,9]. Boyer *et al.* [7] investigated the near-wake flows of a circular cylinder for Reynolds numbers less than 4000 and over a wide range of stratification. With a flow visualization experiment, the wake patterns of these investigated flows were classified in detail. Hwang and Lin [10] numerically computed stratified flows around a circular cylinder for Reynolds numbers less than 500. The results of their numerical computation showed suppressed vortex shedding with increasing density stratification, consistent with the experimental results above.

In more recent years, stratified wakes of tilted bluff bodies have been investigated by a number of authors (e.g. Boulanger [11,12], Meunier [13,14]). Meunier [13] reported in the study of stratified wake of a tilted cylinder, that the tendency of stabilization of the von Karman vortex street which can be seen with a moderate stratifycation is not efficient in a strongly stratified flow due to the presence of a second unstable mode characterized by tiled vortices. This leads to a discontinuity in the Strouhal number as a function of the Froude number which is related to the stratification through the Brunt-Väisälä frequency.

Besides a small number of recent studies on the tilted cylinder wakes, there have been only a limited number of results on the wakes of the cylinders in stratified flows, and almost no comprehensive publication on the near wakes of horizontal cylinder in stratified flows during last two decades.

In homogeneous flows, the effect of the Reynolds number is known to be large on the near wake flow of an object, particularly on the wake of a circular cylinder and a sphere. In contrast, the influence of stratification on the wake of an object is not yet well understood over a wide range of Reynolds numbers, particularly for high Reynolds numbers. Given this background, the present study investigates the influence of stratification on two-dimensional near wake flows around a circular cylinder with the use of a density-stratified wind tunnel. The density-stratified flow is generated with a Reynolds number in the range of 3.5×10^3 - 1.2×10^4. The near wake flow around a circular cylinder is investigated with a flow visualization technique and measurements of wake velocity fluctuations. Some of the phenomena observed in the current paper were previously reported as a letter [15]. The present paper will investigate the mechanisms of these phenomena.

2. Experimental Set-Up and Methodology

2.1. Generation and Evaluation of Density Stratification

Experiments were performed in a density-stratified wind tunnel of a closed-circuit type [16]. The test section of the wind tunnel was 0.4 m wide, 0.6 m high and 1.7 m long. Designed to generate predefined density-stratified flows, the wind tunnel, except for the contraction and test sections, was divided horizontally into six stories (**Figure 1**). A mixture of air and sulfur hexafluoride, SF_6, was injected into each story from an external source to directly produce density stratification. Sulfur hexafluoride is a high-density gas with a molecular weight approximately 5 times that of air. Various types and strengths of vertical density gradients were produced in the test section

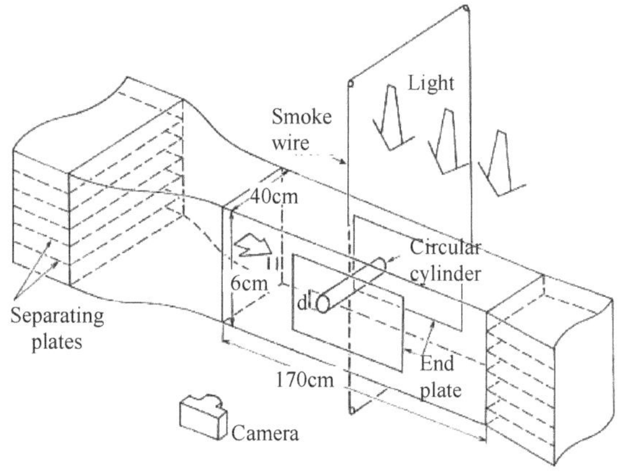

Figure 1. Test section of the density-stratified wind tunnel with a circular cylinder in place.

by appropriately adjusting the volume of SF_6 injected into each story. The density of the gas mixture in the test section was indirectly measured with the use of an oxygen meter. To evaluate the volume fraction of oxygen, the gas mixture was sampled through a gas sampling tube placed upstream of the test section and sent to the oxygen meter. The volume fraction of oxygen is approximately 1/5 of a given volume of air. Thus, the volume fraction of oxygen evaluated by the oxygen meter allowed for the calculation of the volume fraction of the air, thus that of SF_6 in the gas mixture. Finally, the molecular ratio of air to SF_6 was used to evaluate the specific density of the gas mixture with respect to the air, ρ/ρ_a, at the location of the gas sampling.

2.2. Wind Velocity Control and Measurements

The wind in the test section was generated by 18 microfans, each driven by a micromotor; three microfans were mounted side by side on each story. The generated wind velocity, U, ranged from 0 to 1.2 m/s. In homogeneous flow (air only), an I-type hot-wire anemometer was deployed to measure low wind velocities with high accuracy. As for the wind velocity measurements in the stratified flow of the air-SF_6 mixture, the smoke-tracer technique was adopted; that is, the visualized flow was video recorded and the transit time required for markers to pass between two reference points was analyzed to determine the wind velocity.

2.3. Test Models and Measurement of Velocity Fluctuations in the Wake

The models used in the present study were circular cylinders with diameters, d = 5, 7.5, 10 and 13 cm and two rectangular cylinders. The spanwise length of each of the cylinders was 30 cm. For both of the rectangular cylinders, the long side of the rectangular cross-section, d, was 10 cm. The symbol d is chosen for both the circular and rectangular cylinders because it represents the vertical height in both cases and will simplify the notation used in later analysis (see below and Section 5.1). The ratio of the short side to the long side of each rectangular cylinder, b/d, was 0.1 and 0.6, where b is the short side of the rectangular cross-section. All the cylinders were mounted in the test section so that their axes were horizontal and perpendicular to the flow direction. The rectangular cylinders were mounted with the long side of the rectangular cross-section facing the on-coming flow. To ensure the two-dimensionality of the flow, a 30 × 50 cm plate was attached to each end of the cylinders (models), and the plates were placed 5 cm away from the side walls (**Figure 1**).

The velocity fluctuations in the wake of a cylinder were measured by an I-type hot-wire anemometer at the outer edge of the wake behind the cylinder. Spectral analyses were performed on these data with an FFT analyzer. Finally, spanwise correlations of velocity fluctuations in the wake of the cylinder were evaluated using multiple I-type hot-wires deployed simultaneously in the spanwise direction (y-direction).

2.4. Flow Field Visualization and Flow Settings

The smoke-wire technique was adopted to visualize the flow field around a cylinder (**Figure 1**). A mixture of liquid paraffin and iron powder was used to prolong the duration of the smoke release. To illuminate the visualized flows, three 1 kW projectors were installed above the test section of the wind tunnel. The visualized flow was photographed by a Nikon F3 Camera (lens: 55 mm; exposure time: 1/125 sec; f-stop: 1.2; ASA400 black and white film) and video-taped simultaneously.

In the present study, the non-dimensional parameter, k_d, is adopted to quantify the density stratification. The parameter k_d is the inverse of the Froude number and defined as $k_d = Nd/U$ where N is the Brunt-Väisälä frequency and $N^2 = -(g/\rho_0)/(d\rho dz)$. The variables g and ρ_0 are the acceleration of gravity and the reference density, respectively. The reference density, ρ_0, was set to the density value measured at the installation position of the circular cylinder, z = 30 cm above the floor of the wind tunnel. The Reynolds number is defined as $Re = \rho_0 Ud/\mu$. The coefficient of viscosity, μ, for a mixture of air and SF_6 with a volume ratio as high as 1:1, is 0.181×10^{-4} Pa·s (20°C, 1 atm), approximately the same as that for air. In the present study, experiments with density-stratified flow were conducted mostly at U = 0.7 m/s, in some cases, U = 0.3, 0.5, and 0.76 m/s. For homogeneous flow, the wind velocity was set to U = 0.83 m/s, yielding Reynolds numbers over the range of $Re = 3.5 \times 10^3 - 1.2 \times 10^4$. The range of stratification parameter k_d, varied from 0 to 2.0.

3. Experimental Results

3.1. Characteristics of Stratified Flow

Airflow with various vertical gradients of density was generated in the wind tunnel (**Figure 2**). Over this range of the vertical gradient of density, the generated density profiles were approximately linear. **Figure 2** also suggests little change in the vertical gradient over the course of the flow visualization experiments, which lasted a few minutes. Finally, the vertical profile of the mean approach velocity, U, was nearly uniform in the all the stratified flows (not shown).

3.2. Flow Field Visualization

Flow around a circular cylinder (d = 10 cm) was observed in a stratified wind field that had a nearly uniform vertical profile of wind velocity and a linear vertical profile of density as in **Figure 2**. With the flow visualization technique, change in the near-wake flow pattern was examined with respect to the stratification parameter, k_d. **Figure 3** illustrates instantaneous flow patterns that are representative for various values of k_d. The experiment was repeated a number of times for each condition, and the observed flow pattern was consistent and stable in the series of experiments. In **Figure 3**, vortex formation is observed behind the cylinder due to the roll up of the separated shear layer of the cylinder. Furthermore, with increasing values of k_d, the vortex formation occurs closer to the rear side of the cylinder, and the strength of the vortex formation and shedding increases. In particular, the vortex formation and shedding are significantly enhanced at $k_d = 0.513$ (**Figure 3(c)**) with respect to the homogeneous case. The flow pattern around the circular cylinder suddenly changes in the proximity of $k_d = 0.52$ (**Figure 3(d)**). At this stratification, strong roll-up of the shear layers that have separated from the cylinder is no longer observed, and vortex formation in the wake is highly suppressed. With a further increase in the value of

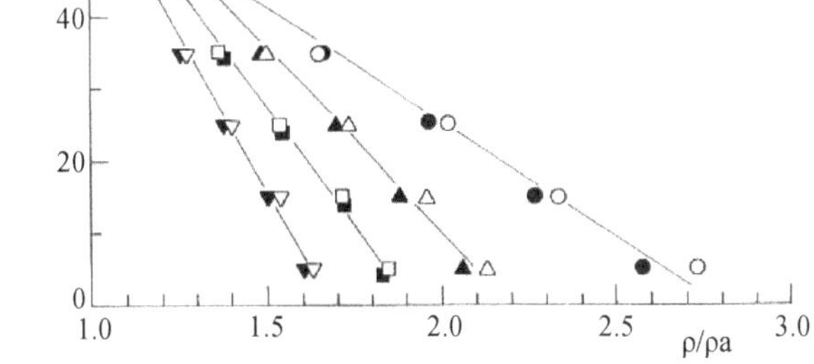

Figure 2. Example vertical profiles of the specific density, ρ/ρ_a at U = 0.7 m/s before (open symbols) and after (closed symbols) a flow visualization experiment.

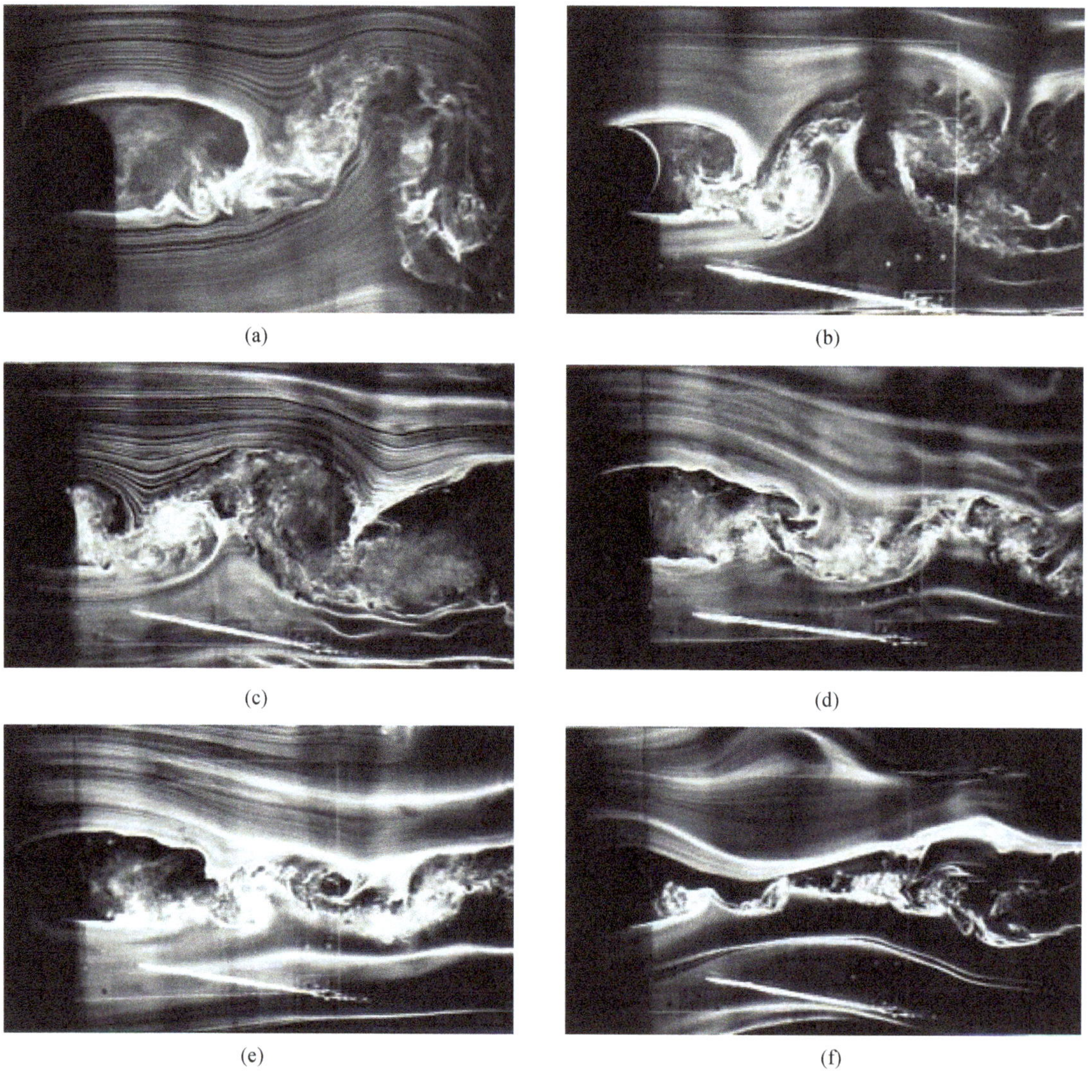

Figure 3. Instantaneous flow patterns of the near-wake of a circular cylinder (d = 10 cm) in density-stratified flows. (a) $k_d = 0$ (homogeneous flow), U = 0.83 m/s, Re = 5800, K = 0; (b) $k_d = 0.402$, U = 0.7 m/s, Re = 7700, K = 0.77; (c) $k_d = 0.513$, U = 0.7 m/s, Re = 8100, K = 0.98; (d) $k_d = 0.521$, U = 0.7 m/s, Re = 8600, K = 1.0; (e) $k_d = 0.570$, U = 0.5 m/s, Re = 5600, K = 1.09; (f) $k_d = 1.45$, U = 0.3 m/s, Re = 3500, K = 2.77.

k_d, the development of stationary lee-waves becomes evident downstream of the circular cylinder (**Figure 3(f)**; U = 0.3 m/s); that is, the pattern of the near-wake flow has shifted from a vortex flow pattern to an internal-wave flow pattern.

3.3. Measurement of the Strouhal Number

Concurrently with the flow visualization experiments, fluctuating wind velocity measurements were made at the outer edge of the wake of the circular cylinder. From the peak frequency of the wake velocity fluctuations, f_v, the Strouhal number, St(=f_v d/U) was evaluated for homogeneous and density-stratified flows (**Figure 4**). Because of the blockage effect (17%), the St value for homogeneous flow ($k_d = 0$) is slightly higher than the typical value reported in the literature (St ≒ 0.21). With increasing k_d, the value of St decreases gradually and becomes approximately 73% of that in homogeneous flow at k_d = 0.513. With further increases in k_d, the value of St increases steeply to a large value in response to the change in the flow pattern. This sudden change of the Strouhal number can be illustrated by the following example. A density-stratified flow with a large density stratification (large k_d) was initially generated, and the density stratification was progressively reduced (smaller k_d). Specifically, during an experiment with a constant wind speed, the value of k_d was nudged gradually by adjusting the volume ratio of the two gasses in the gas mixture. **Figure 5(a)** shows the temporal change of the velocity fluctuations in the wake of a circular cylinder with d = 13 cm, and is characterized by a sudden increase of the magnitude of the velocity fluctuation. The accompanying change in the predominant frequency of the vortex shedding is shown in **Figure 5(b)**, that is, the wake-flow pattern suddenly switches to that of low-frequency vortex shedding. The changes observed in this example are consistent with changes that would be expected in the transition between two flow regimes.

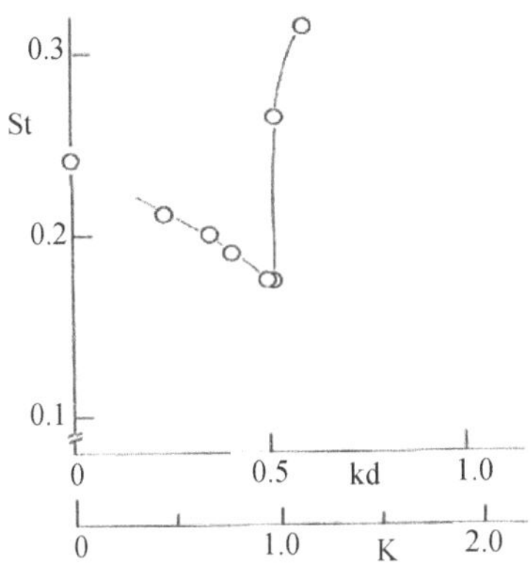

Figure 4. Relationship between the Strouhal number, St, and the stability parameters, k_d and K (refer to Section 4 for the definition of K). U = 0.7 m/s, d = 10 cm (circular cylinder), Re = 5×10^3 - 9×10^3. The value of St at $k_d = 0$ was observed in homogeneous flow (U = 0.83 m/s, Re = 5800).

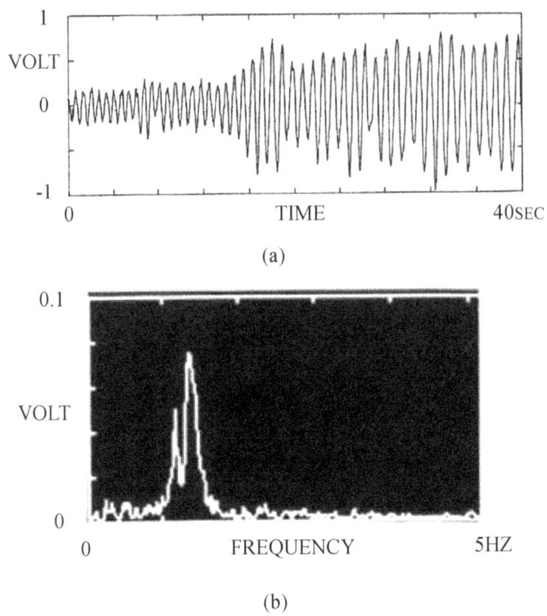

Figure 5. Sudden change in the flow around a circular cylinder at the critical stratification, $(k_d)_{cr}$; d = 13 cm (circular cylinder), U = 0.76 m/s, Re = 1.1×10^4, $k_d = 0.635$, K = 0.93. (a) Temporal fluctuations of flow velocity in the wake of the cylinder as measured in voltage; (b) Power spectrum showing the shift in the predominant frequency of vortex shedding. The predominant frequency shifted from $f_v = 1.44$ Hz (right peak) in the first part of the time series in (a) to $f_v = 1.04$ Hz (left peak) in the second part of the time series in (a).

3.4. Summary of Experimental Results and Discussion

Figure 3 shows the transition of the wake pattern with increasing stability: (b,c)→(d,e)→(f). Initially, the near-wake pattern is characterized by highly-turbulent vortices (b and c). With increasing stratification, the turbulent near-wake becomes narrow (d and e) and finally, turbulent mixing occurs in isolated regions downstream of the cylinder (f). A similar three-regime transition of the cylinder wake pattern was observed in the salt water experiment of Boyer *et al.* [7]. They refer to the isolated regions of turbulent mixing similar to those in (f) as "isolated mixed regions". In their experiment, wakes were generated behind cylinders in stably stratified water-tank flow at a Reynolds number up to 4000, which was smaller than the range of Reynolds numbers from our wind tunnel experiment (3.5×10^3 - 1.2×10^4). Because only limited experimental results are reported in Boyer *et al.* for the range of Reynolds number between 2000 and 4000, the wake flow transition is not clear in this range. However, the results from Boyer *et al.* and our experiment suggest that the transition of the cylinder wake pattern occurs in stratified flows with increasing stratification with high Reynolds numbers above approximately 400.

A further examination of the changes in the wake pattern in the present experiment (Section 3.3) reveals two flow regimes. First, with increasing k_d (increasing stability) up to a critical value, the vortex formation behind the circular cylinder is enhanced and the Strouhal number, St, decreases gradually. The corresponding critical value of k_d will be referred to as $(k_d)_{cr}$ hereafter. Second, the strength of vortex formation behind the circular cylinder reaches its maximum at $k_d \fallingdotseq (k_d)_{cr}$. With further increases in k_d, vortex formation is sharply suppressed, and vortex formation becomes weak. Furthermore, this transition is accompanied by a sharp increase in St. The rapid change in the flow pattern at $k_d \fallingdotseq (k_d)_{cr}$ will be referred as the critical phenomenon below. In the following sections, the generation mechanisms of the phenomena in these two flow regimes will be discussed. In the present experimental set-up, stably stratified flows were produced in the flow channel bounded by solid horizontal top and bottom walls. Thus, characteristics of stratified flows of finite depth are first summarized in the following section.

4. Internal Gravity Waves in Stratified Flows of Finite Depth

In this section, stratified flow of finite depth, H, is considered in a flow channel bounded by solid horizontal top and bottom walls. No wave energy penetration is allowed into the walls, and nonviscous flow with no density dispersion is assumed. For this flow, two non-dimensional numbers are relevant: k_d (=Nd/U) (or its inverse, the Froude number, F_d(=U/Nd) and K(=NH/πU) (or the blockage ratio, d/H). k_d is defined with the characteristic length of an object while F_d is defined in terms of the finite depth, H. According to linear theory, discrete vertical modes exist for the internal gravity wave that arises in stratified flow of finite depth, and the vertical modes depend on the strength of the density stratification [8,17]. The dispersion relation for the nth vertical mode (wavelength in the vertical direction = 2 H/n) can be expressed as:

$$\omega^2\left\{k_x^2+\left(n^2p^2\right)/H^2\right\}-N^2k_x^2=0 \tag{1}$$

where ω is the angular frequency and k_x is the wavenumber in the x-direction. The phase velocity, C_p, and group velocity, C_g, of the internal gravity wave can be expressed as:

$$Cp=?\frac{NH}{\left(k_x^2H^2+n^2\pi^2\right)} \tag{2}$$

and

$$C_g=\pm\frac{NHn^2\pi^2}{\left(k_x^2H^2+n^2\pi^2\right)^{3/2}} \tag{3}$$

respectively.

For $k_x \to 0$, the values of C_p and C_g attain their maxima and $C_n = C_p = C_g = \pm NH/n\pi$ where C_n is the phase velocity of the nth mode. For flow with approach velocity, U and $C_n > U$, that is, in terms of K and k_d:

$$K=\frac{NH}{\pi U}>n \text{ or } k_d>n\pi\frac{d}{H} \tag{4}$$

a columnar disturbance at the nth mode propagates upstream of the object at

$$Cn-U=\frac{NH}{n\pi}-U=\left(\frac{K}{n}-1\right)U \tag{5}$$

For $k_x \neq 0$, stationary waves develop only when $U = C_p$. Because Equations (2) and (3) yield $C_p/C_g > 1$, the group velocity, C_g, of the stationary wave that develops at $U = C_p$ is characterized by $C_g < U$ and the stationary wave can be present only downstream of the obstacle. The relationship $U = C_p$ leads to,

$$K=\left(n^2+k_x^2\frac{H^2}{\pi}\right)^{1/2} \tag{6}$$

A comparison of Equations (4) and (6) suggests that columnar disturbances can develop whenever stationary lee waves are present. Furthermore, Equation (6) implies that both stationary lee waves and columnar disturbances develop with $k_x \to 0$, *i.e.* K ≒ n, therefore, starting at K ≒ 1. The wavelength of the stationary lee waves, λ, can be expressed with K as

$$\lambda=\frac{2H}{\left(K^2-n^2\right)^{1/2}} \tag{7}$$

5. Abrupt Change in the Flow Pattern (Critical Phenomenon)

5.1. Critical Phenomenon around a Non-Circular Cylindrical Object

To investigate if the critical phenomenon at the critical stratification, k_d ≒ $(k_d)_{cr}$, is unique to flow around a circular cylinder, additional experiments were performed with two rectangular cylinders (Section 2.3). The long side of the rectangular cross-section of both cylinders (d = 10 cm) was facing the on-coming flow, and this length was identical to the diameter of the circular cylinder (d = 10 cm) for which the near wake flow was investigated in Section 3. As with the experiment with the circular cylinder, the flow around the square cylinders was investigated for various degrees of stratification, k_d, and flow visualization experiments were conducted. In addition, the fluctuating velocity in the near-wake was measured to evaluate the peak frequency, f_v. As in the flow around the circular cylinder of d = 10 cm (**Figure 4**), a sudden change in the flow pattern and a sharp increase in the value of St are observed in the proximity of $k_d = 0.52$ or K = 1 (**Figures 6** and **7**). When k_d exceeds $(k_d)_{cr}$, the pattern of the near-wake flow behind the rectangular cylinder changes and vortex formation is suddenly suppressed (change from **Figures 6(a)** to **(b)**). The present investigation suggests that the critical phenomenon in the proximity of $(k_d)_{cr}$ is not unique to the near-wake flow behind a circular cylinder. The critical phenomenon occurs at a similar value of $(k_d)_{cr}$ behind a rectangular cylinder that has the same vertical dimension as the circular cylinder.

5.2. Cases with Circular Cylinders of Various Diameters, d, and Test Sections with Various Heights

The critical stratification, $(k_d)_{cr}$, for near-wake flows is investigated behind circular cylinders of three additional diameters, d = 5, 7.5, 13 cm, in the test section of height H = 60 cm. The results are summarized together with those from the experiment with the two rectangular cylinders (Section 5.1) in **Figure 8**. For an object with a certain diameter d (or a height d facing the on-coming flow), the critical phenomenon occurs at a similar value of k_d, independent of the Reynolds number. Furthermore, the emergence of the critical phenomenon coincides with the lower limit of k_d at which stationary lee waves are predicted to occur by the linear theory (**Figure 9**).

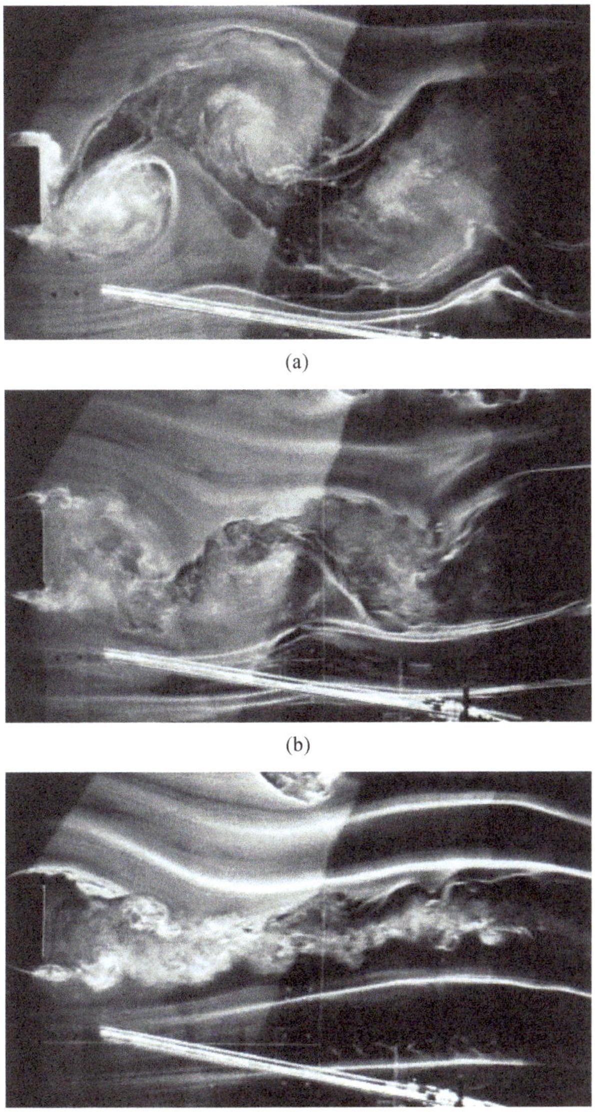

(a)

(b)

(c)

Figure 6. Instantaneous flow patterns around a rectangular cylinder of size b/d = 0.6 (ratio of short side to long side; long side was deployed vertically). D = 10 cm. (a) k_d = 0.488, U = 0.7 m/s, Re = 8400, K = 0.93; (b) k_d = 0.568, U = 0.7 m/s, Re = 8400, K = 1.08; (c) k_d = 0.860, U = 0.5 m/s, Re = 5900, K = 1.64.

To further investigate the dependency of $(k_d)_{cr}$ on d/H (or K), the height of the text section, H, was modified to H = 46.3 cm, and near-wakes behind a circular cylinder of d = 7.5 cm (d/H = 0.162) were studied. The value of $(k_d)_{cr}$ observed for this case is approximately 0.52, which falls on the predicted lower limit for the occurrence of stationary lee waves (**Figure 9**). The present result suggests 1) the value of $(k_d)_{cr}$ depends on d/H and not on the value of H itself, and 2) a close link exists between the critical phenomenon for near-wakes of horizontal objects in stratified flows and the occurrence of stationary lee waves.

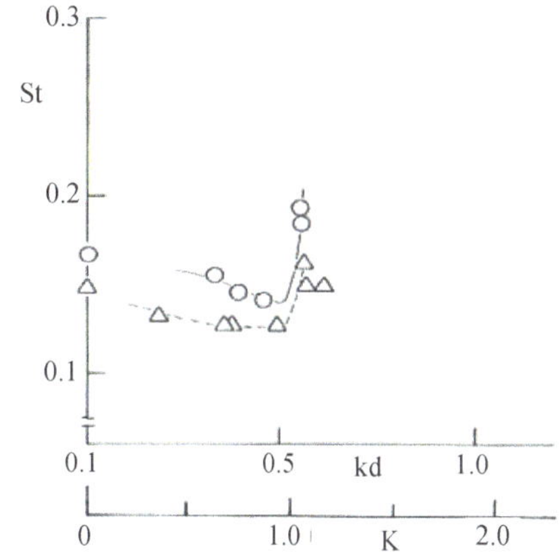

Figure 7. Relationship between the Strouhal number, St, and the stability parameters, k_d and K. d = 10 cm (d: side of the rectangular cylinder perpendicular to the streamwise direction). ○: b/d = 0.1, △: b/d = 0.6, Re = 6× 10^3 - 10^4.

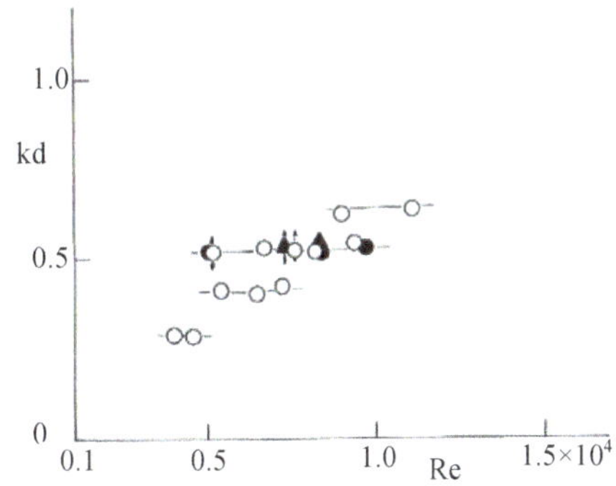

Figure 8. Relationship between the critical stratification, $(k_d)_{cr}$, and the Reynolds number, Re, for near-wake flows behind circular cylinders of 4 different diameters (○: from top to bottom, d = 13, 10, 7.5, 5 cm) and two rectangular cylinders (●: b/d = 0.1, d = 10 cm; ▲: b/d = 0.6, d = 10 cm). The Reynolds number variations are due to variations in the wind speed, U, in the range of 0.5 - 0.8 m/s.

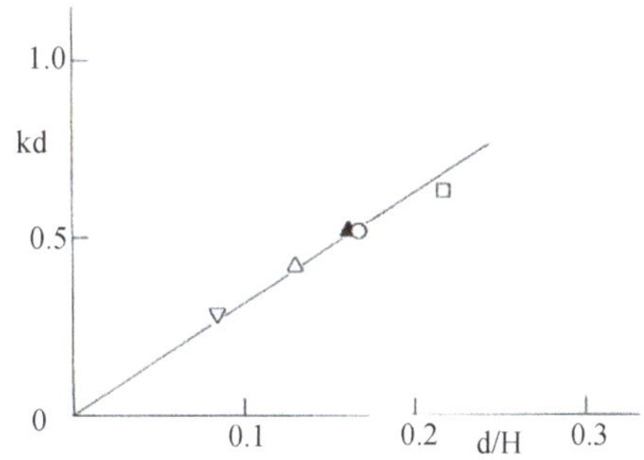

Figure 9. Relationship between the critical stratification, $(k_d)_{cr}$, and the blockage ratio, d/H for circular cylinders of 4 different diameters: ▽: d = 5 cm, △, ▲: d = 7.5 cm, O: d = 10 cm, □: d = 13 cm, H = 60 cm except for ▲ (H = 46.3 cm). The straight line, *i.e.* $k_d = \pi(d/H)$, indicates the lower limit of k_d at which first-order mode stationary lee waves can develop (development of stationary lee waves is possible at $k_d > \pi(d/H)$ or K > 1). The predictive relationship, $k_d = \pi(d/H)$, is based on the linear theory (n = 1 in Equation (4)).

5.3. Development of Stationary Lee Waves and Critical Phenomenon

As elucidated in Section 4, the linear theory predicts that stationary lee waves develop when the non-dimensional parameter K exceeds 1. K is defined in terms of the finite depth of the stratified flow, H. The wavelength of the lee wave at K = 2.77 (k_d = 1.45) in **Figure 3(f)**, when normalized by H, becomes $\lambda^*(=\lambda/H) \fallingdotseq 0.87$. This value coincides roughly with $\lambda^* = 0.78$ as predicted by the linear theory for the first-order mode stationary lee wave. At a large value of K as in this case, the presence of stationary lee waves becomes evident, and the wavelength of the lee wave can be approximated by the linear theory. At K = 1.64, an undulating lee wave with a long wavelength is observed across the entire near-wake behind the rectangular cylinder with dimensions b/d = 0.6 (**Figure 6(c)**). At K = 1.09, a lee wave is present in the upper portion of the near-wake behind the circular cylinder (**Figure 3(e)**). However, with the value of K approaching 1, the wavelength of the lee wave deviates away from that predicted by the linear theory (**Figure 10**). In the proximity of K = 1, the lee waves are not easily distinguishable (**Figure 3(d)**). For this case, the wavelength of the lee wave is estimated to be $\lambda^* \fallingdotseq 1.83$ from the wave undulation in the upper portion of the near-wake behind the circular cylinder.

The wavelength of lee waves observed in the present study deviated significantly from the predicted values at low values of K. However, a stationary lee wave initially developed in the proximity of K = 1 as predicted by the linear theory. When the stationary lee wave is present, vertical motions of fluid particles are highly suppressed downstream of the object creating the lee wave. Hence, vertical motions are highly suppressed in the vortex formation region behind the object, thus large and strong vortices cannot roll up and widen the wake. Therefore, in the proximity of K = 1, the wake pattern changes sharply to that characterized by weak vortex formation concurrent with the development of stationary lee waves.

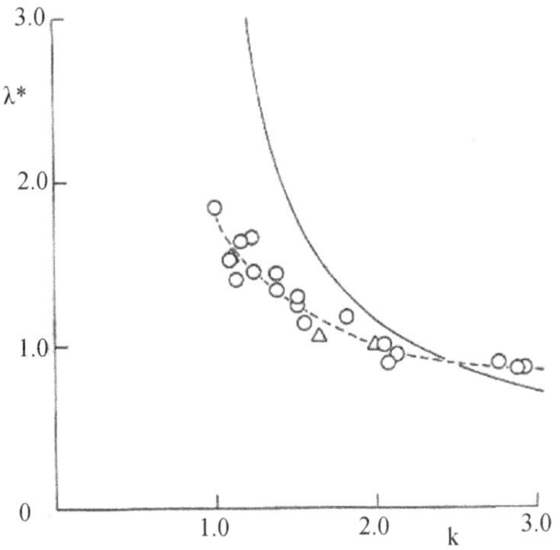

Figure 10. Relationship between the wavelength of lee waves, $\lambda^*(=\lambda/H)$, and the stratification parameter K. The solid line indicates the relationship based on the linear theory ($\lambda^* = 2/(K^2 - 1)^{1/2}$). O: circular cylinders, Δ: rectangular cylinders.

6. Intensification of Vortex Formation

6.1. Non-Circular Cylindrical Bluff Bodies

As discussed in Section 3, the vortex formation behind a circular cylinder is intensified with increasing stratification between $k_d = 0$ (homogeneous flow) and $(k_d)_{cr}$. The intensification of vortex formation with increasing stability was also examined for the near-wake flows of non-circular cylindrical bluff bodies. Generally, the Strouhal number, St, is small for the wake behind a bluff body that is conducive to vigorous vortex formation. Therefore, the value of St is negatively correlated to the formation of strong vortices. The decreasing trends of St between k_d = 0 and $(k_d)_{cr}$ for the wakes behind the rectangular cylinders of b/d = 0.1, 0.6 (**Figure 7**) are smaller than those for the wakes behind the circular cylinders (**Figure 4**). In particular, the corresponding slope for the wake behind the rectangular cylinder of b/d = 0.6 is small. The flow visualization experiments also confirmed that enhancement of vortex formation behind the rectangular cylinder of b/d = 0.6 was not as evident as that behind the circular cylinders.

6.2. Spanwise Correlation of Fluctuating u-Velocities in Near-Wake Flows

Behaviors of the spanwise correlation of the fluctuating u-velocities were investigated in the near-wakes of circular (d = 5, 10 cm) and rectangular (b/d = 0.6) cylinders. The investigations were made in homogeneous flow and stratified flow at k_d slightly less than $(k_d)_{cr}$. For the near-wakes behind the circular cylinders, the correlation of the fluctuating u-velocities is significantly higher in stratified flow at $k_d \fallingdotseq (k_d)_{cr}$ than that in homogeneous flow (**Figure 11**). In contrast, with k_d approaching $(k_d)_{cr}$, the correlation for the wake of the rectangular cylinder is nearly the same as that for homogenous flow. This result is attributable to the highly two-dimensional nature of the near-wake flow around the rectangular cylinder as will be discussed in the next subsection.

6.3. Dependency of Vortex Formation Enhancement on the Shape of the Bluff Body

As discussed earlier, enhancement of vortex formation with increasing k_d from $k_d = 0$ to $k_d = (k_d)_{cr}$ is prominent particularly in the near-wake of a circular cylinder. Recent studies on the three-dimensional structure of the

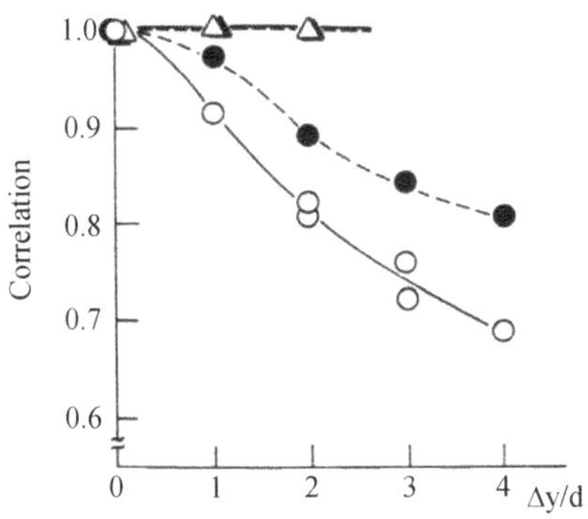

Figure 11. Relationship between the spanwise correlation of the fluctuating velocity, u, and the normalized spanwise separation distance, Δy/d. ○: circular cylinder (d = 5 cm) in homogeneous flow, ●: circular cylinder (d = 5 cm) in stratified flow of k_d = 0.27 (≒ $(k_d)_{cr}$), △: rectangular cylinder (b/d = 0.6, d = 10 cm) in homogeneous flow, ▲: rectangular cylinder (b/d = 0.6, d = 10 cm) in stratified flow of k_d = 0.49 (≒ $(k_d)_{cr}$).

near-wake flow behind circular cylinders (e.g., Yokoi and Kamemoto [18]) have revealed that the boundary layer separation from a circular cylinder is a three-dimensional phenomenon and that the separated shear layer rolls up in a non-two-dimensional manner. However, it is speculated that stable stratification suppresses the vertical velocity disturbance in the wake of a circular cylinder and causes the roll-up of the separated sheer layer to be more two-dimensional than it would be in homogeneous flow. Consistent with this speculation, **Figure 11** shows that the value of spanwise u-velocity correlation in the wake of a circular cylinder increases due to the effect of stable stratification. The increasing two-dimensionality of the near-wake flow likely enhances vortex formation similar to the enhancement observed in the near-wake flows of a vibrating circular cylinder and a circular cylinder in a periodically fluctuating flow. The increased spanwise alignment is likely the cause of the enhanced vortex formation in a two-dimensional flow. In contrast to a circular cylinder, the geometry of a rectangular cylinder naturally forces the separation of the shear layer to be aligned in the span-wise direction resulting in a wake flow that is highly two-dimensional. As a result, the effect of stratification on the strength of vortex formation is small in the near-wake of a rectangular cylinder.

7. Conclusions

Near wakes behind horizontal circular and square cylinders were investigated in linearly density-stratified flows of varying stabilities. Experiments were performed with a nearly uniform approach velocity at relatively high Reynolds numbers, Re = 3.5×10^3 - 1.2×10^4. Flow visualization experiments and measurements of the velocity fluctuation frequencies in the near-wake led to the following conclusions:

1) With increasing values of k_d(=Nd/U), *i.e.*, increasing stratification, vortex formation and shedding behind a circular cylinder are gradually enhanced with respect to those in homogeneous flow, and the Strouhal number decreases. These observations can likely be explained by the effect of increasing stratification which enhances the two-dimensionality of the near wake flow behind a circular cylinder. The enhanced two-dimensionality of the flow strengthens the roll-up of the separated shear layer.

2) The strength of vortex formation and shedding reaches its maximum at k_d ≒ $(k_d)_{cr}$, and is suddenly suppressed when k_d exceeds $(k_d)_{cr}$. This critical phenomenon is accompanied by a sudden change in the flow pattern, that is, only weak vortex formation occurs downstream of the obstacle and the Strouhal number increases sharply. In terms of the stratification parameter K(=NH/πU) which includes the depth of stratified flow, H, the critical phenomenon occurs close to K = 1 for all sizes of circular and rectangular cylinders investigated. The linear theory predicts that lee waves will develop for K > 1. Thus, in this critical phenomenon, the emergence of a stationary lee wave suppresses the vertical fluid motions of the vortices formed behind an object and causes a sudden change in the flow pattern.

3) With further increase of k_d, the wake flow pattern transitions to one dominated by internal waves.

8. Acknowledgements

The author wishes to acknowledge Kenichiro Sugitani for his extensive assistance with the present experiments.

REFERENCES

[1] W. Blumen, "Atmospheric Processes over Complex Terrain (Meteorological Monographs)," *American Meteor Society*, Vol. 23, No. 45, 1990, pp. 1-323.

[2] W. H. Snyder, "Fluid Modeling of Pollutant Transport and Diffusion in Stably Stratified Flows over Complex Terrain," *Annual Review of Fluid Mechanics*, Vol. 17, 1985, pp. 239-266.

[3] H. Honji, "Stratified Flow Experiments," *Nagare*, Vol. 6, 1987, pp. 276-284.

[4] H. Honji, "Vortex Street in the Stratified Fluids," Reports of Research Institute for Applied Mechanics, No. 60, 1984, pp. 393-401.

[5] H. Honji, "Vortex Motions in the Stratified Wake Flows," *Fluid Dynamics Research*, Vol. 3, No. 1-4, 1988, pp. 425-430.

[6] J. T. Lin and Y. H. Pao, "Wakes in Stratified Fluids," *Annual Review of Fluid Mechanics*, Vol. 11, 1979, pp. 317-338.

[7] D. L. Boyer, P. A. Davies, H. J. S. Fernando and X.

Zhang, "Linearly Stratified Flow Past a Horizontal Circular Cylinder," *Philosophical Transactions of the Royal Society of London. Series A*, Vol. 328, 1989, pp. 501-528.

[8] S. N. Wei, T. W. Kao and H. P. Pao, "Experimental Study of Upstream Influence in the Two-Dimensional Flow of a Stratified Fluid over an Obstacle," *Geophysical fluid Dynamics*, Vol. 6, No. 4, 1975, pp. 315-336.

[9] H. Hanazaki, "Upstream and Advancing Columnar Disturbances in Two-Dimensional Stratified Flow of Finite Depth," *Physics of Fluids A*, Vol. 1, No. 12, 1989, pp. 1976-1987.

[10] R. R. Hwang and S. H. Lin, "On Laminar Wakes behind a Circular Cylinder in Stratified Fluids," *Transactions of the ASME, Journal of Fluids Engineering*, Vol. 114, No. 1, 1992, pp. 20-28.

[11] N. Boulanger, P. Meunier and S. Le Dizes, "Structure of a Stratified Tilted Vortex," *Journal of Fluid Mechanics*, Vol. 583, 2007, pp. 443-458.

[12] N. Boulanger, P. Meunier and S. Le Dizes, "Tilt-Induced Instability of a Stratified Vortex," *Journal of Fluid Mechanics*, Vol. 596, 2008, pp. 1-20.

[13] P. Meunier, "Stratified Wake of a Tilted Cylinder. Part 1. Suppression of a von Karman Vortex Street," *Journal of Fluid Mechanics*, Vol. 699, 2012, pp. 174-197.

[14] P. Meunier, "Stratified Wake of a Tilted Cylinder. Part 2. Lee Internal Waves," *Journal of Fluid Mechanics*, Vol. 699, 2012, pp. 198-215.

[15] Y. Ohya and Y. Nakamura, "Near Wakes of a Circular Cylinder in Stratified Flows," *Physics of Fluids A*, Vol. 2-4, No. 4, 1990, pp. 481-483.

[16] Y. Ohya, S. Ozono and Y. Nakamura, "A Wind Tunnel for Studying Density-Stratified Flows," *Atmospheric Environment*, Vol. 28, No. 11, 1994, pp. 1895-1900.

[17] M. Funakoshi and M. Oikawa, "Non-Linear Waves in Stratified Flow," *Nagare*, Vol. 8, 1989, pp. 311-335.

[18] Y. Yokoi and K. Kamemoto, "Initial Stage of a Three-Dimensional Vortex Structure Existing in a Two-Dimensional Boundary Layer Separation Flow: Observation of Laminar Boundary Layer Separation over a Circular Cylinder by Flow Visualization," *JSME International Journal. Ser. 2, Fluids Engineering, Heat Transfer, Power, Combustion, Thermophysical Properties*, Vol. 35-II, No. 2, 1992, pp. 189-195.

Permissions

The contributors of this book come from diverse backgrounds, making this book a truly international effort. This book will bring forth new frontiers with its revolutionizing research information and detailed analysis of the nascent developments around the world.

We would like to thank all the contributing authors for lending their expertise to make the book truly unique. They have played a crucial role in the development of this book. Without their invaluable contributions this book wouldn't have been possible. They have made vital efforts to compile up to date information on the varied aspects of this subject to make this book a valuable addition to the collection of many professionals and students.

This book was conceptualized with the vision of imparting up-to-date information and advanced data in this field. To ensure the same, a matchless editorial board was set up. Every individual on the board went through rigorous rounds of assessment to prove their worth. After which they invested a large part of their time researching and compiling the most relevant data for our readers. Conferences and sessions were held from time to time between the editorial board and the contributing authors to present the data in the most comprehensible form. The editorial team has worked tirelessly to provide valuable and valid information to help people across the globe.

Every chapter published in this book has been scrutinized by our experts. Their significance has been extensively debated. The topics covered herein carry significant findings which will fuel the growth of the discipline. They may even be implemented as practical applications or may be referred to as a beginning point for another development. Chapters in this book were first published by Scientific Research Publishing Inc.; hereby published with permission under the Creative Commons Attribution License or equivalent.

The editorial board has been involved in producing this book since its inception. They have spent rigorous hours researching and exploring the diverse topics which have resulted in the successful publishing of this book. They have passed on their knowledge of decades through this book. To expedite this challenging task, the publisher supported the team at every step. A small team of assistant editors was also appointed to further simplify the editing procedure and attain best results for the readers.

Our editorial team has been hand-picked from every corner of the world. Their multi-ethnicity adds dynamic inputs to the discussions which result in innovative outcomes. These outcomes are then further discussed with the researchers and contributors who give their valuable feedback and opinion regarding the same. The feedback is then collaborated with the researches and they are edited in a comprehensive manner to aid the understanding of the subject.

Apart from the editorial board, the designing team has also invested a significant amount of their time in understanding the subject and creating the most relevant covers. They scrutinized every image to scout for the most suitable representation of the subject and create an appropriate cover for the book.

The publishing team has been involved in this book since its early stages. They were actively engaged in every process, be it collecting the data, connecting with the contributors or procuring relevant information. The team has been an ardent support to the editorial, designing and production team. Their endless efforts to recruit the best for this project, has resulted in the accomplishment of this book. They are a veteran in the field of academics and their pool of knowledge is as vast as their experience in printing. Their expertise and guidance has proved useful at every step. Their uncompromising quality standards have made this book an exceptional effort. Their encouragement from time to time has been an inspiration for everyone.

The publisher and the editorial board hope that this book will prove to be a valuable piece of knowledge for researchers, students, practitioners and scholars across the globe.

List of Contributors

Lili Liu, Chao Yu and Lu Liu
Department of Petroleum Engineering, Northeast Petroleum University, Daqing, China

Lihui Wang
Daqing Petroleum Equipment Group, Daqing, China

S. Han Aydin
Department of Mathematics, Karadeniz Technical University, Trabzon, Turkey

Mohamed Habib Sellami
Laboratory of Thermal Radiation, Department of Physics, Faculty of Science Tunis, University Tunis El Manar, Tunis, Tunisia
High School for Rural Engineering, University of Jendouba, Jendouba, Tunisia
Tunisian Association of Science, Technology and Development, Tunisia

Chengchuyue Fu
Department of Petroleum Engineering, Northeast Petroleum University, Daqing, China

Peixiang Li
Technical Development Department, The 2nd Oil Production Factory of Daqing Oilfield Company Limited, Daqing, China

Changhee Kim
Hyundai Engineering & Construction Company, Seoul, South Korea

Cheolheui Han
Department of Aeronautical and Mechanical Design, Korea National University of Transportation, Chungju, South Korea

Takafumi Morisaki and Yasuyuki Ikegami
Institute of Ocean Energy, Saga University, Saga, Japan

Basharat Salim
King Saud University, Riyadh, KSA

Mohammad Abu Taher and Heuy Dong Kim
School of Mechanical Engineering, Andong National University, Andong, South Korea

Yeon Won Lee
School of Mechanical Engineering, Pukyong National University, Busan, South Korea

Sławomir Dykas, Włodzimierz Wróblewski and Dawid Machalica
Institute of Power Engineering and Turbomachinery, Silesian University of Technology, Gliwice, Poland

Nazibuddin Ahmed and Kishor Kumar Das
Department of Mathematics, Gauhati University, Guwahati, India

A. I. Abd Al Aziz and H. I. Mohamed
Public Authority of Applied Education & Training, Kuwait, Kuwait

Hiromasa Suzuki
Graduate School of Advanced Science and Technology, Tokyo Denki University, Saitama, Japan

Masaki Endo and Yoko Sakakibara
Division of Electronic and Mechanical Engineering, Tokyo Denki University, Saitama, Japan

Xiuhua Si
Department of Engineering, Calvin College, Grand Rapids, USA

Jinxiang Xi and Jongwon Kim
Department of Mechanical and Biomedical Engineering, Central Michigan University, Mount Pleasant, USA

Dhruba Prasad Barua
Department of Mathematics, Gauhati University, Guwahati, India

Jiwan Krishna Goswami
Department of Mathematics, ADP College, Nagaon, India

Jung-Yup Kim and Ji-Seok Kim
Fire Research Center, Korea Institute of Construction Technology, Goyang-Si, South Korea

I. V. Lebed
Zhukovsky Central Institute of Aero hydrodynamics, Moscow, Russia

Donghong Wang
College of Aerospace Engineering, Nanjing University of Aeronautics and Astronautics, Nanjing, China
College of Science, Nanjing University of Aeronautics and Astronautics, Nanjing, China

Ning Zhao
College of Aerospace Engineering, Nanjing University of Aeronautics and Astronautics, Nanjing, China

Yanqing Wang
College of Science, PLA University of Science and Technology, Nanjing, China

Morteza Ghassabzadeh and Hassan Ghassemi
Department of Ocean Engineering, Amirkabir University of Technology, Tehran, Iran

Hisashi Kihara, Naoya Hirata and Ken-ichi Abe
Department of Aeronautics and Astronautics, Kyushu University, Fukuoka, Japan

Sujan Sinha
Department of Mathematics, Gauhati University, Guwahati, India

Vincent Aeschlimann
LAMH, Université Laval, Quebec, Canada

Stéphane Barre
LEGI, UMR 5519, Grenoble, France

Henda Djeridi
LPO, UMR 6523, Brest, France

Yuji Ohya, Takanori Uchida and Tomoyuki Nagai
Research Institute for Applied Mechanics, Kyushu University, Kasuga-shi, Japan